TRACE ELEMENTS AND DENTAL DISEASE

Postgraduate Dental Handbook Series, Volume 9

Volume Editors

M.E.J. Curzon, BDS, MS, PhD, MRCD(C)
Department of Caries Research
Eastman Dental Center
Rochester, New York

T.W. Cutress, BDS, MS, PhD, FRACDS
New Zealand Medical Research Council
Wellington, New Zealand

Series Editor

Alvin F. Gardner, DDS, MS, PhD, FICD
Washington, DC

TRACE ELEMENTS AND DENTAL DISEASE

M.E.J. Curzon
T.W. Cutress

John Wright • PSG Inc
Boston Bristol London
1983

Library of Congress Cataloging in Publication Data
Main entry under title:

Trace elements and dental disease.

(Postgraduate dental handbook series; v. 9)
Bibliography: p.
Includes index.
1. Teeth—Diseases. 2. Dental caries. 3. Trace elements. I. Curzon, M.E.J. (Martin E.J.) II. Cutress, T.W. III. Series. [DNLM: 1. Dental caries—Etiology. 2. Trace elements. WU 270 T759]
RK305.T62 1982 617.6′3071 82-11115
ISBN 0-7236-7035-8

Published simultaneously by:
John Wright • PSG Inc, 545 Great Road, Littleton, Massachusetts 01460, U.S.A.
John Wright & Sons Ltd, 823–825 Bath Road, Bristol BS4 5NU, England

Printed in Great Britain
Bound in the United States of America

International Standard Book Number: 0-7236-7035-8

Library of Congress Catalog Card Number: 82-11115

CONTRIBUTORS

Rashid Al-Hayali, BDS, PhD
Professor
College of Dentistry
University of Baghdad
Baghdad
Iraq

D. Beighton, PhD
Research Fellow
Research Establishment
Royal College of Surgeons
Downe, near Orpington, Kent
England

M.E.J. Curzon, BDS, MS, PhD, MRCD(C)
Chairman, Department of Caries Research
Eastman Dental Center
Associate Professor of Dental Research
The University of Rochester School of Medicine and Dentistry
Rochester, New York

T.W. Cutress, BDS, MS, PhD, FRACDS
Director, Dental Unit
New Zealand Medical Research Council
Wellington
New Zealand

Arthur D. Eisenberg, BS, PhD
Research Associate
Department of Caries Research
Eastman Dental Center
Rochester, New York

J.D.B. Featherstone, BSc, MSc, PhD, FNZIC
Senior Research Associate
Department of Caries Research
Eastman Dental Center
Rochester, New York

J.L. Hardwick, MDS, MSc, PhD, FDSRCS
Emeritus Professor of Dentistry
University of Manchester
Manchester
England

H. Steve Hsieh, PhD
Investigator and Assistant Professor
Institute of Dental Research and Department of Nutrition Science
Schools of Dentistry and Medicine,
University of Alabama in Birmingham,
Birmingham, Alabama

G. Neil Jenkins, MSc, PhD, FDSRCS, DOdont
Emeritus Professor
Department of Oral Physiology
The Dental School
Framlington Place
Newcastle Upon Tyne
England

B.E. Johnson, BS
Division of Nutrition
Department of Biochemistry
School of Dentistry
Oregon Health Science University
Portland, Oregon

Juan M. Navia, PhD
Senior Scientist and Professor
Institute of Dental Research and Department of Nutrition Science
Schools of Dentistry and Medicine
University of Alabama in Birmingham,
Birmingham, Alabama

T.R. Shearer, PhD
Professor and Director
Division of Nutrition
Department of Biochemistry and Ophthalmology
School of Dentistry
Oregon Health Science University
Portland, Oregon

Buddhi M. Shrestha, BDS, MS, PhD, Director
Division of Nutrition and Cariology
Oral Health Research Center
Fairleigh Dickinson University
Hackensack, New Jersey

M.V. Stack, MSc, PhD
Senior Scientist
Medical Research Council Dental Unit
The Dental School
Bristol
England

CONTENTS

FOREWORD

During the past decade there has been a gratifying increase in demand for new knowledge in the art and science of dentistry on the part of practitioners of dentistry and dental specialists. The biological and dental sciences are currently in a period of rather explosive growth. Stimulated by the diversification of experimental methods and dental procedures, the basic dental sciences with clinical applications and clinical dentistry are growing particularly fast.

Dental sciences are currently considered as multidimensional, and individual differences are rather great. Dental scientists and practitioners know that the adaptation of techniques from other fields of science is not simple, usually indirect, and unusable unless radically revised, extended or reformulated for use in dentistry. The basic experimental method originating in the clinical dental sciences may now be joined by the methods of the biological and physical sciences and made technically feasible by the introduction of modern electronics and computers.

It is the goal of the Postgraduate Dental Handbook Series to present critical analyses based on impressive clinical and research experience. This dental series provides basic dental sciences with clinical applications and clinical dental sciences in a correlated fashion. It likewise describes current concepts in clinical dentistry that are of direct value to dental practitioners, dental specialists, dental students and dental hygienists throughout the world. By bringing together all our knowledge, the volumes in the Postgraduate Dental Handbook Series will fill a critical need felt by dental clinicians and dental investigators alike.

Dr. M.E.J. Curzon, Chairman of the Department of Caries Research, Eastman Dental Center, Rochester, New York and Dr. T.W. Cutress, Director of the Dental Unit, New Zealand Medical Research Council, Wellington, New Zealand, present a unique work on *Trace Elements and Dental Disease* for the Postgraduate Dental Handbook Series. This book is the first review and evaluation to describe all of the available information known on trace elements and dental disease. This work, therefore, is extremely useful for all dentists, dental researchers, and allied health professionals (animal husbandry, agriculture, and the broad field of medical research).

The purpose of this book is to help professionals involved in treating dental disease and, in the interest of optimal dental care, to gain a more global and comprehensive view of the subject of dental caries. Recognizing the present need for better understanding among dental practitioners, the authors and their contributors are internationally recognized authorities who present this subject in a comprehensive manner. The result is an amalgamation of thought from all areas of the world. These contributing authorities offer guidance in a practical and stimulating manner. Treatment recommendations have been developed by a systematic process of considering every possibility and weighing the advantages and disadvantages of each to arrive at the best treatment plan for the dental patient.

This work differs from all prior undertakings concerned with dental disease. It does not present a series of minireports of research, nor does it overemphasize the biochemical, endocrinologic or technical matters. Rather, this book presents multiple points of view, disciplines, and a body of clinical and laboratory experience which constititute a course in dental caries that encompasses the relevant clinical and basic sciences. *Trace Elements and Dental Disease* provides the dental clinician with most of today's proven answers. A student of dental disease should find an up-to-date exposition, organized to provide a knowledge base to build upon for future work, or a yardstick with which to fit and assess contemporary or past works.

I wish to thank the coauthors and each contributor for assembling his entry in such an excellent manner. Its contents should still be useful a decade from today, which characterizes the contributors and authors and sets this book apart from its predecessors. The reader will find this work a useful blend of techniques and approaches to what has been an often difficult problem for the dental practitioner.

Alvin F. Gardner, Series Editor

PREFACE

In 1950, Fred Losee carried out a dental survey of caries prevalence in American Samoa. His findings led him to the conclusion that among other factors affecting caries was the influence of trace elements, originating in seawater, sea salt, or both. This work started Fred in the field of dental research, which was to occupy him for the rest of his career. By the time of his retirement in 1973, Fred Losee had accumulated a great deal of knowledge about the effects of trace elements on dental caries, and in 1974 he began writing a book on the subject. Unfortunately, with his untimely death in November 1975, the book remained only as a series of notes and outlines, together with a commitment for collaboration by the present editors.

Because of the importance of the subject, and the lack of a review on trace elements, we have produced the present book. In addition to providing for the first time under one cover the present knowledge on trace elements and dental disease, it is also a tribute to Fred Losee to commemorate his own research and his influence on many scientists.

PART I
Background and Epidemiologic Effects of Trace Elements on Dental Caries

1 Introduction

M. E. J. Curzon

From the discovery of iron and then iodine, as elements essential to life, the study of trace elements and disease has developed into an extensive field of research. It is a measure of the interest in the subject that from 1820 to 1940 the trace elements iron, iodine, copper, manganese, zinc, and cobalt were proven to be essential to life, but between 1940 and 1974 the elements molybdenum, selenium, chromium, tin, vanadium, fluorine, silicon, nickel, and others were added to the list (Schwarz, 1974). This rapid increase in our knowledge has been largely a result of much better analytic techniques with low limits of detection that enable detailed work to be done on the role or potential role of trace elements on human health.

Trace elements are variously defined depending upon the field of chemical, physical, or biologic sciences being discussed. In the field of biology, elements that are present in only minute quantities in animal tissues are called trace elements, regardless of their abundance in nature. Thus, to biologists silicon and fluorine, among the most plentiful of the elements in the earth's crust and, therefore, major elements to geologists, are "trace elements." Those trace elements that are essential for life are further characterized as microminerals or nutrients and include such elements as copper, cobalt, iron, fluorine, manganese, molybdenum, selenium, silicon, tin, vanadium, and zinc (Schwarz, 1974). There are many remaining elements for which no specific function has as yet been identified and which are also known as trace elements. Some of these may yet be shown to be essential to life.

An essential element has been defined as one required by an animal for normal development and growth. Such elements are essential nutrients because they must be ingested. Macro (major) and micro (trace) nutrients have been identified. The macro elements essential for life have long been known and are required as constituents of proteins, cell walls, and structural tissues, and they also play a part in complex biochemical reactions. Within this category are phosphorus, nitrogen, and calcium—elements required in large quantities.

However, micro elements exert their influence at very low concentrations, their role being primarily catalytic, rather than being building blocks. The elements present in animals, often in minute amounts, perform functions essential for the maintenance of healthy life, growth, and reproduction. Inadequate intake of these nutrients may impair health at the physiologic or cellular level.

The actual concentrations of elements found in different tissues are, for the "inactive" elements, probably dependent on the environmental availability of the respective elements. However, the concentrations of at least some of the essential elements are dependent on specific homeostatic mechanisms; a few elements, the heavy metals in particular (lead, mercury, arsenic, cadmium), tend to accumulate within the body and produce toxic effects. It is interesting to consider the suggestion (Schwarz, 1974) that the nutritional requirements for the essential elements correspond more closely to the concentration of the elements in seawater than in the earth's crust. Some reference is made to this in the chapter on enamel, and a list of the abundance of various elements in seawater and the earth's crust is presented.

In this review we have elected to classify the 90 naturally occurring elements, first, according to their known essentiality in human and animal life and, second, to their role as a major or minor element, rare inactive elements, very rare elements, or toxic minor elements. This arbitrary classification is seen in Figure 1-1.

For the purposes of this text trace elements are defined as all the naturally occurring elements of the periodic table excluding the major components of hydroxyapatite—calcium, phosphorus, oxygen, and hydrogen.

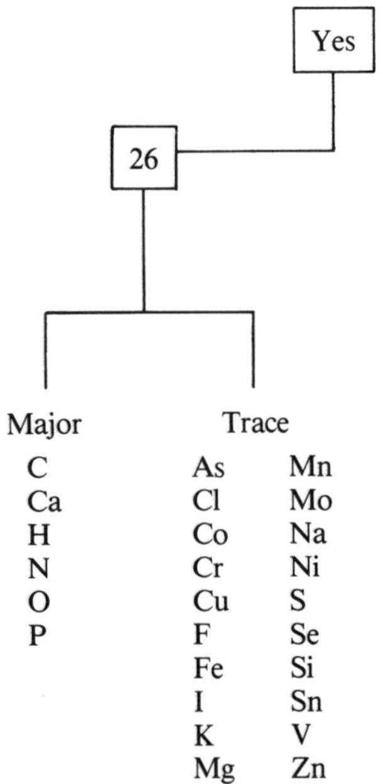

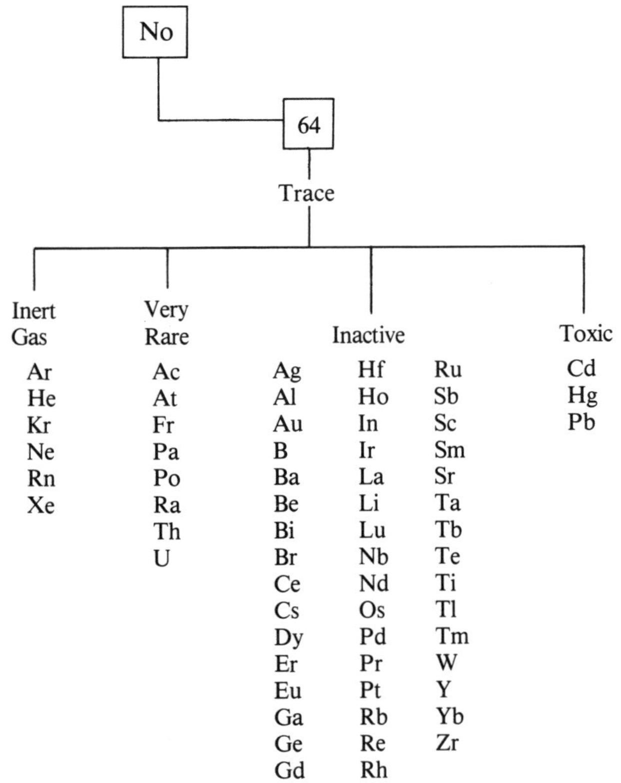

Figure 1-1 Classification of 90 naturally occurring elements essential to animal life.

TRACE ELEMENTS IN DENTISTRY

The dental profession was actively involved with the field of trace elements and dental disease at an early stage. In 1908 a mine chemist, F. S. McKay, suggested at a meeting of the El Paso (Texas) Dental Society that mottled teeth and a concomitant resistance to caries were related to a factor in drinking water. This factor was subsequently identified as fluoride, a trace element now known to be essential for calcification, growth, and fertility (Schwarz and Milne, 1972, Messer et al, 1972). Thus, the influence of trace elements on a dental disease was established early in the history of dental research.

Although there was an early start with fluoride, dental research has not seriously considered other trace elements until relatively recently. Indeed, dental research has been almost mesmerized by the element fluorine, to the virtual exclusion of consideration of other trace elements. In a recent review Hurny (1978) stated that within the past 30 years about 30,000 papers have been published on fluoride alone. It is unfortunate if fluoride alone of all the elements should be assumed to have the unique property of influencing susceptibility to caries. Such uniqueness has not yet been demonstrated. Fluoride is known to become incorporated into the enamel apatite, influencing its chemical, crystallographic, and biologic characteristics; this is known to be true of at least several other elements such as strontium, magnesium, and carbon. Therefore, it is understandable that in recent years, particularly as techniques suitable for analyzing a wide range of elements have been developed, interest in elements other than fluoride has increased.

It is the objective of this book to bring together the present knowledge and thoughts on trace elements in dental disease. The literature about fluoride as a trace element is extensive, and the reader is referred to those textbooks covering this element exclusively. In this volume, fluoride will be considered only where interactions with other trace elements occur.

Part of the problem in deciding which are the trace elements for purposes of discussion is analytical. Indeed, the original term *trace element* arose because analytical techniques were initially too crude to determine accurately the concentrations of an element present in a piece of tissue. Analytical chemists, therefore, reported the presence of an element in trace amounts, hence trace elements. The problem today is not that, but rather the opposite one of not being able to determine whether a trace element was really present in a tissue, or whether it was acquired during the analytical process as contamination.

It has generally been accepted that the term *trace* refers to a relative content of a constituent of not more than 100 ppm (100 parts/10^6 or 100 mg/l). However, for our purposes we have decided to review the literature on all elements we have defined as trace elements, although for many elements very little information is available.

Of all the dental tissues, more analytical work has been carried out on enamel than on any other because its chemical composition may materially affect the occurrence of the major dental disease, caries. Research has sought to define those elements found in enamel, and which elements affect the caries process.

Losee et al (1974) reported that at least 41 elements of the periodic table are incorporated into the dental enamel during development, Table 1-1. It is interesting that, with the exception of lead, no elements with atomic numbers above 60 are regularly present in young enamel.

One factor influencing the selection of the elements incorporated into enamel, and also related to essentiality for life, is the relationship of an atom's size and charge density. Only 3 of the 24 elements known to be essential for life in mammals have an atomic number above 34 (Losee et al, 1974). This selective process was suggested by Losee as also being applicable to enamel because of the seven elements reported in their study in the greatest concentrations (> 10 g/g dry wt), only strontium (No. 38) had an atomic number above 34.

There appears to be a selective process during enamel formation and development that restricts the incorporation of elements to those below 60, if not below 38. Of course, other elements may be acquired onto or into the surface enamel after tooth eruption. Such posteruptive uptake of some trace elements can take place in sound enamel, whereas many others will accumulate only if the surface or subsurface of the enamel is hypomineralized by developmental factors or by subsequent demineralization (Little and Steadman, 1966). Such posteruptive uptake of trace elements may also be affected by the presence of dental restorations,

Table 1-1
Summary of Trace Elements in Dental Enamel of Permanent Teeth

Concentration Range ppm	Elements
> 1000	Na, Cl, Mg
100–1000	K, S, Zn, Si, Sr, F
10–100	Fe, Al, Pb, B, Ba
1–10	Cu, Rb, Br, Mo, Cd, I, Ti, Mn, Cr, Sn
0.1–0.9	Ni, Li, Ag, Nb, Se, Be, Zr, Co, W, Sb, Hg
< 0.1	As, Cs, V, Au, La, Ce, Pr, Nd, Sm, Tb, Y
Not detected	Sc, Ga, Ge, Ru, Pb, In, Te, Eu, Gd, Dy, Ho, Er, Tm, Lu, Hf, Ta, Re, Os, Ir, Pt, Tl, Bi, Rh

prosthetic appliances, orthodontic bands, or the use of dentifrices or mouthwashes. Food composition, often reflecting variations in the geochemical environment, may also affect posteruptive accumulation of trace elements.

The papers of Losee et al (1974) and Curzon and Crocker (1978) reported on the analysis of large numbers of whole human enamel samples. We have concentrated on those trace elements listed in these two studies as the basis for our selection of elements. From the list of elements reported as detected in enamel, there is no evidence available to show that the elements cesium, cerium, praseodymium, or neodymium have any effects on dental disease. These are, therefore, arbitrarily eliminated from our consideration. On the other hand, there are two elements, cobalt and arsenic, recorded as below detection by Losee et al (1974) for which there is slight evidence of a relation to dental disease. They, therefore, are included.

The remaining trace elements will be discussed separately when there is evidence of a relationship to, or influence on, dental disease. Others will be discussed as groups, essential elements, nonessential elements, or toxic elements. Hereafter, all trace elements will be referred to by their chemical symbol.

TRACE ELEMENTS AND DENTAL CARIES

Evidence of a relationship of trace elements to caries prevalence has accumulated to a degree that indicates that further research be undertaken on this subject. Although the inverse relationship between fluoride availability and dental caries is beyond dispute, the data given in some of the early reports, such as those of Arnold and co-workers in 1948, give an indication of effects on caries by factors other than fluoride. From Arnold's and other reports, it can be seen that caries prevalence for a number of communities was substantially different than that in comparable communities despite a similarity in fluoride levels. As an example, in Clarksville (Tennessee) with 0.2 ppm F in the drinking water, the mean number of decayed, missing, and filled teeth (DMFT) in 12 to 14-year-old children was 4.58, whereas in Key West (Florida) with 0.1 ppm F in the drinking water the mean DMFT index for comparable children was 10.70 (Arnold, 1948). No cultural or racial difference appeared to explain the twofold difference in DMFT between Clarksville and Key West, and F differed by only 0.1 ppm. Other elements are suggested as possible factors.

Several trace elements in the water supplies have been claimed to be either cariostatic or cariogenic. Spectographic analyses of water from the towns with low F of Oak Park, Waukegan, and Quincy (Illinois), and the towns with high F of Aurora, Joliet, and Galesburg (Illinois) showed that

DMFT figures in the United States Public Health Reports (Arnold, 1948) are actually as closely related to levels of B and Sr as to F. From published trace element information for water supplies in the United States (Dufor and Becker, 1962), it was reported (Losee and Bibby, 1970) that high levels of Ba, Li, Mo, Sr, and V were significantly correlated with lower caries prevalence. Conversely, the same authors showed a positive correlation between Cu and Pb levels and high caries. Using caries prevalence data and analyses of water supplies in a study of statistical correlations of caries and trace elements, Adkins and Losee (1970) observed that these same elements were significantly and positively related to dental caries.

Biologic apatite, such as in enamel and dentin, differs from a pure hydroxyapatite by its inclusion of very small amounts of proteinaceous material and trace elements. While the organic fraction reflects the residue of odontogenesis, the trace elements reflect the composition of the tissue environment during the period of tooth formation and the oral environment after tooth eruption. These environments differ in many ways, including the trace element composition of ingested water and foodstuffs which in turn are dependent on geographic, dietary, and cultural factors. These same factors are often observed to be closely associated with the prevalence of caries.

Analysis of dental enamel has been carried out many times to determine its trace element composition. In 1937 Drea attempted to identify those trace elements that naturally occur in enamel. Analytical techniques at that time were not good enough to measure accurately low concentrations of elements, and in earlier studies, just as in the paper by Drea, elements were listed as present in trace quantities. With the advent of analytical techniques such as atomic absorption spectrophotometry, lower and lower detection limits have been possible, such that quantities as low as $< 0.01\ \mu g/g$ can now be easily detected. Consequently, studies of the associations between trace elements and dental caries have become possible (Curzon and Crocker, 1978; Schamschula et al, 1979); these will be discussed in detail later. It is apparent that the trace element composition of teeth may be associated with their susceptibility to caries.

Since 1953 several epidemiologic studies have examined, or reported, relationships between dental caries and trace elements such as Mo, Se, Sr, and V. Subsequent animal and in vitro studies have either supported or contradicted these findings. Overall a body of literature has thus developed within the last 20 years which we felt was now at a stage to merit review.

TRACE ELEMENTS AND PERIODONTAL DISEASE

Little attention has been given to the possibility that excesses of trace elements may affect periodontal health. It could be postulated that an element such as Zn, which influences inflammation and collagen production, could affect periodontal tissues and disease susceptibility. Likewise, if Sr affects calcification and bone metabolism, then is it possible for this element to modify the effects of bone resorption? In recent years the electrical properties of alveolar bone have been considered as possibly contributing to periodontal health. Although trace elements have not been identified as altering electrical properties of bone or tooth tissues, it is well established that the same elements become incorporated into apatite and thereby alter the properties of the crystal lattice. These questions remain unanswered, but it is our hope that this book may promote research in these fields. However, where there has been any consideration of possible effects of trace elements on dental diseases other than caries, then they will be discussed with the appropriate element.

TRACE ELEMENTS IN BONE

Interest in the trace element content of bone increases for two reasons. Both matrix and apatite phases can apparently be affected by trace element incorporation into bone with effects on the electrical as well as the physicochemical properties. Both effects may be associated with the equilibrium of the periodontium. This similarity of cementum to bone implies that the potential role of trace nutrients is equally true for cementum as for bone. Attempts have been made to separate trace elements consistently incorporated into bone from those present as contaminants resulting from dietary and environmental factors; much more research is required to gain a proper perspective of the role of trace elements in bone.

ORGANIZATION OF THE TEXT

This book has been divided into several parts to deal with the subject. In the opening section, the influence of trace elements and dental disease is reviewed, with a resume of past work and implications for clinical dentistry. A second section deals with trace elements in enamel, dentin, plaque, and saliva, while in the third section trace elements are discussed individually in greater detail, with the emphasis placed upon mechanisms of action and current research.

REFERENCES

Adkins BL, Losee FL: A study of the covariation of dental caries prevalence and multiple trace element content of water supplies. *NY State Dent J* 1970;36:618–622.

Arnold FA: Fluorine in drinking water: Its effect on dental caries. *JADA* 1948;36:28–36.

Curzon MEJ, Crocker DC: Relationships of trace elements in human tooth enamel to dental caries. *Arch Oral Biol* 1978;23:647–653.

Drea WF: Spectrum analyses of dental tissues for "trace" elements. *J Dent Res* 1937;15:403–406.

Dufor CN, Becker E: Public water supplies of the 100 largest cities in the U.S.A. US Geol. Surv. Water Supply Paper No. 1812, 1962.

Hurny TA: Fluoride and teeth, in Courvoisier B, Donath A, Baud CA (eds): *Fluoride and Bone.* Bern, Hans Hubert, 1978, pp 119–124.

Little MF, Steadman LT: Chemical and physical properties of altered and sound enamel. IV. Trace elements. *Arch Oral Biol* 1966;11:273–278.

Losee FL, Bibby BG: Geographic variations in the prevalence of dental caries in the U.S.A. *Caries Res* 1969;3:32–43.

Losee FL, Cutress TW, Brown R: Natural elements of the periodic table in human dental enamel. *Caries Res* 1974;8:123–134.

Messer H, Armstrong WD, Singer L: Fertility impairment in mice on a low fluoride intake. *Science* 1972;177:893–897.

Schamschula R, Agus H, Bunzel M, et al: The concentration of selected major and trace minerals in human dental plaque. *Arch Oral Biol* 1977;22:321–325.

Schwarz K: New essential trace elements (Sn, V, F, Si): Progress report and outlook, in Hoekstra WG (ed): *Trace Element Metabolism in Animals—II.* Baltimore, University Park Press, 1974.

Schwarz K, Milne DB: Fluorine requirement for growth in the rat. *Bioinorg Chem* 1972;1:331–335.

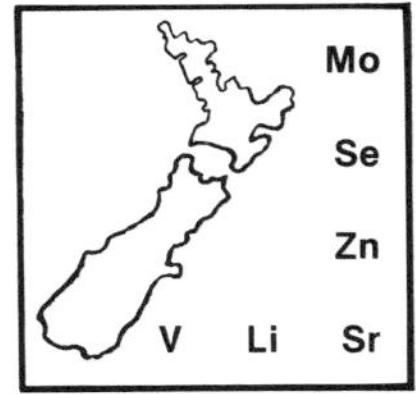

2 Epidemiology of Trace Elements and Dental Caries

M. E. J. Curzon

SOURCES OF TRACE ELEMENT INTAKE

Dental caries, like other diseases such as cancer and endemic goiter, shows marked geographic variation in prevalence. Much of this variation is probably associated with availability of dietary carbohydrates, notably sugar. However, in some geographic areas neither dietary nor fluoride variables appear to explain satisfactorily unusually high or low caries prevalence. It is in areas where known influences on caries prevalence have been examined and found not to explain the disease level that epidemiologic research has considerable potential in furthering knowledge of the etiology of caries, its initiation, and its control.

The trace element intake into the body is via a food chain originating in soils. Thus plants will extract trace elements, such as Cu, Mn, Mo, Sn, and Fe, for their own growth. The plants in the form of vegetables may be eaten directly by man or by animals as fodder. In either way the trace element intake of man is via a soil-water-plants-animal-man food chain. The sources of these trace elements are, therefore, important in any discussion of trace elements and dental disease.

Soils

Associations between caries and soils were first suggested by Sir Charles Hercus (1925) in relation to a study on endemic goiter in New Zealand. He noted that widespread variations in the incidence of dental caries and goiter could be shown to have a geographic distribution related to soil types. Olesen (1929) also showed this in his United States report concerning endemic goiter, where a high prevalence of goiter was found in the same area as high caries prevalence.

Nizel and Bibby (1944) demonstrated that the highest prevalence of caries in the United States occurred in New England, with the lowest caries prevalence in south central areas. In New England soil types are of the family of acid podzols, whereas the low caries area of the southwestern states are areas of semi-acid soils, high in mineral nutrients. It has been suggested that the carious condition of the teeth follows a pattern of soil fertility (Albrecht 1947), poorer teeth coming from areas where rainfall is high enough to encourage lush vegetation but with low mineral content. Gley soils, which are acid, are associated with high caries (Saunders, 1945). Later Zwaardemaker and Tromp (1957) found a high caries frequency with areas of "peat soils" in Holland which were acid, and in New Zealand alkaline soils were related to low

Table 2-1
Results of Dental Caries Examinations of Children Living on Various Soil Types in Eastern United States

		Mean DMFT by Age Group			
Soil Type	Area	*n*	*12*	*13*	*14*
Podzol (Spodozol)[a]	New England	580	5.72	6.49	8.03
Gray-Brown Podzol (Hapadalf)	Central Atlantic	512	4.83	5.57	7.76
Red-Yellow Podzol (Alfazols)	Southern Atlantic (Interior)	492	4.06	5.14	6.43
Humic Gleys (Aqualt)	Southern Atlantic (Coastal)	292	2.99	3.90	6.19

[a]Indicates newer geologic term not used by original authors.

DMFT: Podzol/G.B. Podzol/R.Y. Podzol/Gley: $F = 37.9$, $p = < 0.1$. Adapted from Ludwig and Bibby, 1969.

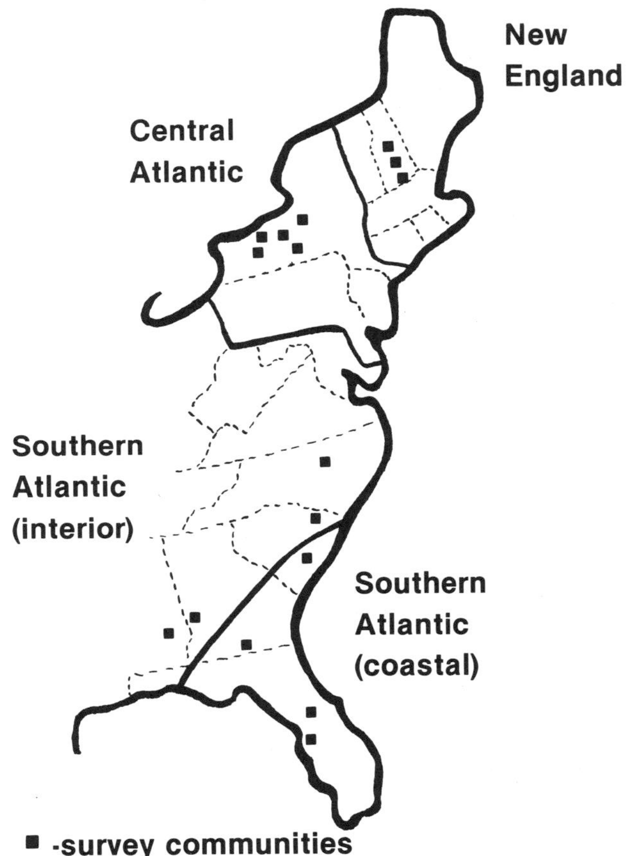

Figure 2-1 Geographic distribution of survey towns for caries-soil relationships for soil types by area (see Table 2-1).

caries (Hewat and Eastcott, 1955). Ludwig et al (1960), also in New Zealand, suggested that the uptake of a possible cariostatic element, Mo, was enhanced from an alkaline soil. Recently, the low caries areas of Florida and South Carolina in the United States were identified with yellow-brown podzolic soils as opposed to the high caries areas of New England situated on humic gleys (Ludwig and Bibby, 1969) (Figure 2-1 and Table 2-1).

However, the significance of the suggested associations between caries prevalence and soils has been questioned in a report by Pearce, Ludwig, and Darwin (1974). Following an extensive study of 3555 school children on seven soil types in New Zealand, these authors showed that low caries occurred in yellow-brown pumice soils and dry yellow-brown sands. In contrast, high caries occurred on podzolized yellow-brown earths, gleys, and wet yellow-brown sands. These findings, of course, agree closely with those of previous studies. Pearce and coauthors disagree with the significance of the findings, showing no relationship between dental caries and the consumption of locally grown vegetables and water supplies. Their conclusion was that *local* soil factors were probably not involved in producing geographic variations in caries. Other factors such as socioeconomic, fluoride tablet consumption, or attitudes to dental health might play greater roles. It must be remembered, however, that there are considerable differences between the sizes of New Zealand and the United States. The soil-caries patterns in the United States cover very large areas, and only one of the soil areas of Ludwig and Bibby's study (1969) would encompass the whole of New Zealand.

Mineral-poor topsoil (low trace element content) was shown in Sweden to be related to high caries prevalence (Åslander, 1958), in areas where there was also a humid climate, high rainfall, and severe leaching of the underlying geologic strata. The chemistry of the rock underneath the topsoil may also affect caries prevalence. This was illustrated by the finding of an area of particularly low caries in northwest Ohio, USA that was situated on a type of geologic strata known as strontianite ($SrCO_3$) and celestite ($SrSO_4$), and therefore high in Sr. Recently Pärko (1975) showed that different caries levels may be related to underlying differences in rock strata in addition to soils. In her studies, caries differences were related to types of rock, and only partially explained by F and K concentrations in such rocks. Water running through the strata forms the next link in the chain carrying trace elements to man.

Water

Drinking water is either described as ground or surface water, depending on whether it is derived from wells or rivers and streams. Water from wells reflects the geology through which it has percolated and collected and may vary in hardness. Hewat and Eastcott (1955) in New Zealand, while studying the geographic distribution of dental decay, found that there was a significant decrease in caries for those children using water of moderate hardness when compared with those waters with less than 50 ppm hardness as $CaCO_3$. The hardness of the water, although largely a reflection of Ca and its salts, is also partially due to other trace elements such as the alkaline earths. Ockerse (1949) in an area of South Africa showed that a relationship existed between caries prevalence and the concentration of fluoride (F') in water, but in addition there was an equally large difference between localities where low F′ drinking waters were used. Ockerse suggested furthermore that other trace elements in water might also act in this geographic area.

While concentrations of trace elements in water are relatively low compared with those in foods or soils, wide variations can occur and levels of individual elements can be very high. In the major drinking water supplies of the largest United States cities, for example, the levels of Sr can vary from 0.002 to 1.20 ppm (Durfor and Becker, 1962). Individual water supplies in smaller communities such as Menomonee Falls, Wisconsin, may reach as high as 34 ppm of Sr (Nichols and McNall, 1957). Consumption of a liter of such water per day would add 34 mg Sr to the diet, which would be far greater than the 2 mg normally acquired from food (Schroeder, 1965). Exceptionally high concentrations of individual trace elements in drinking water may well be related to possible caries-reducing or caries-promoting effects.

Food

Schroeder (1965) has demonstrated that foods normally make the greatest contribution of trace element intake in man. The composition of foods, and the soils they are grown in are interrelated, and it is impossible to consider them separately. The findings of Romney (1953) and Wells (1958) illustrate the importance of soil and climate on plant composition. Romney grew a strain of soybean on 20 different soil types under varying conditions, and analysis of Mn in these plants showed concentrations varying between 20 and 265 ppm. Wells, in a similar study, grew a strain of sweet vernal on a number of different New Zealand soils and showed considerable variations in trace elements such as V between 50 and 350 ppm and Fe between 30 and 160 ppm.

Analysis of plants from Vermont, Ohio, and South Carolina (Table 2-2) showed that there are considerable differences in accumulation of trace elements in relation to elements in water or soil. In addition, however, when certain vegetables were cooked in water some elements such as F, Li, Mo, and Sr were taken up by the vegetables, while others—Al, Ba, B, Cu, Mn, Ni, and Rb—were lost from the vegetables into the water (Losee and Adkins, 1969) (Table 2-3). These studies indicated that the composition of vegetables can reflect the presence of trace elements in soils and water and that the trace element content may be accentuated by cooking.

Particular types of plants (vegetables) will also show wide variations in trace elements. Thus, legumes and collards generally represent rich sources of trace elements. Differences in the availability of such foods, and food choice often based on cultural practices, may be important

Table 2-2
Trace Element Analysis of Green Beans from Various Towns in the United States

		Mean Conc. in ppm Dry Weight								
Town	State	*Mn*	*Fe*	*B*	*Cu*	*Zn*	*Al*	*Sr*	*Mo*	*Ba*
Dillon	SC	59	90	10	11	45	68	39	2.1	5
Montpelier	VT	44	88	15	12	57	51	18	0.3	5
Ft. Recovery	OH	30	79	17	6	54	68	136	2.3	9
Mean for US		36	92	13	11	42	46	26	1.9	7

Data from Losee, 1968 (unpublished)

Table 2-3
Concentrations of Trace Elements in Various Samples of Water Used to Cook Green Beans, to Show Loss of Elements from the Beans During Cooking (μg/l)

	Before Cooking	After Cooking		
Element	*Northwest Ohio Water*	*Concentration in Distilled Water*	*Distilled Water plus 1.5 ppm F*	*Northwest Ohio Water*
Al[a]	70	170	230	210
Ba[a]	15	140	120	130
B	450	380	470	980
Cu[a]	4	230	210	140
F	1510	0	910	1100
Li	25	14	14	11
Mn[a]	100	1100	900	760
Mo	95	30	32	55
Ni[a]	25	400	210	180
Rb[a]	5	390	430	380
Sr	16,000	65	170	5700
Fe[a]	950	2200	1800	930

[a]Trace elements showing a loss from the beans into the cooking waters. Adapted from Losee and Adkins, 1969.

factors in determining trace element intakes in man. Regional differences in food choice exist and have been identified in studies undertaken in the United States by the various Agricultural Research Service projects. Their findings reveal that diets in the southern states are high in green leaf vegetables, mustard greens, beet greens, and turnip greens, which are high in trace elements such as B, Mo, Sr, Al and Fe; the prevalence of caries is low there, even in areas of low fluoride in the drinking waters.

The trace element composition of water is influential in the composition of food as it is finally consumed. During cooking, processing, and reconstituting of foods there is some change in trace element content as elements are lost or absorbed. The increasing use of concentrates or freeze-dried foods, which have to be reconstituted with local water, enhances the importance of water consumption on the ultimate trace element content of foods.

STUDIES ON THE GEOGRAPHIC PREVALENCE OF CARIES

The earliest instance of a dental survey related to geography is that of Magitot, published in 1878. This work reported on the prevalence of dental decay in conscripts to the French Army during the period 1831–1847 and, as such, is probably one of the first epidemiologic

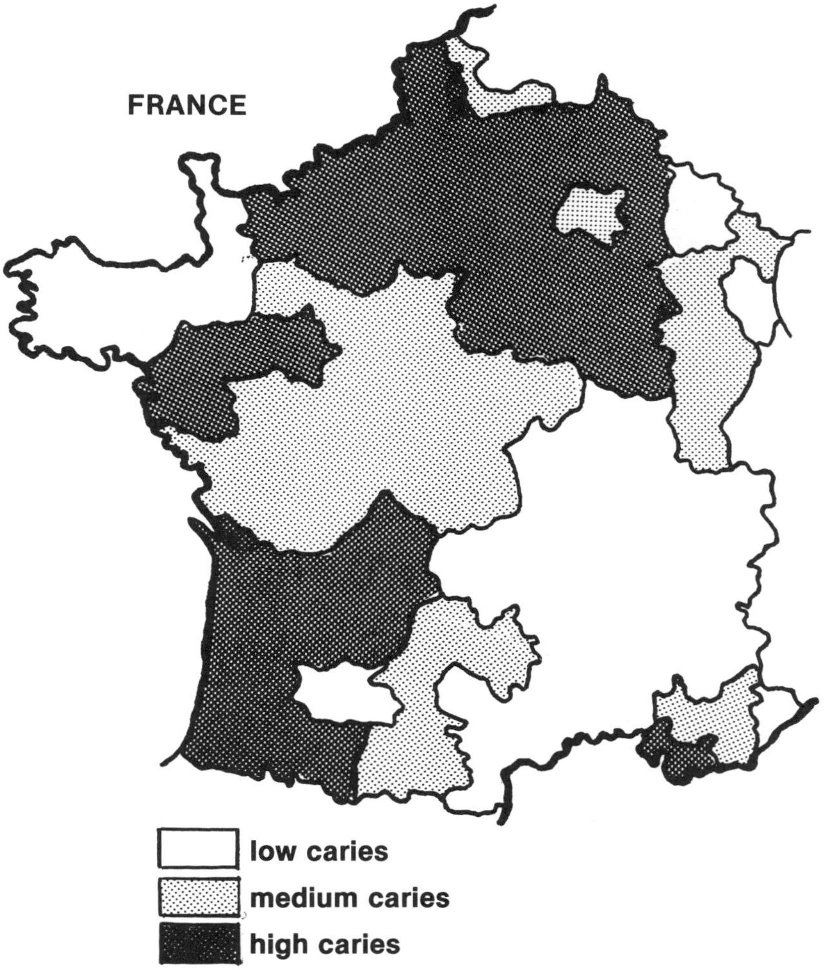

Figure 2-2 Geographic distribution of high, medium, and low caries in Army recruits in France, 1831–1847 (after Magitot, 1878).

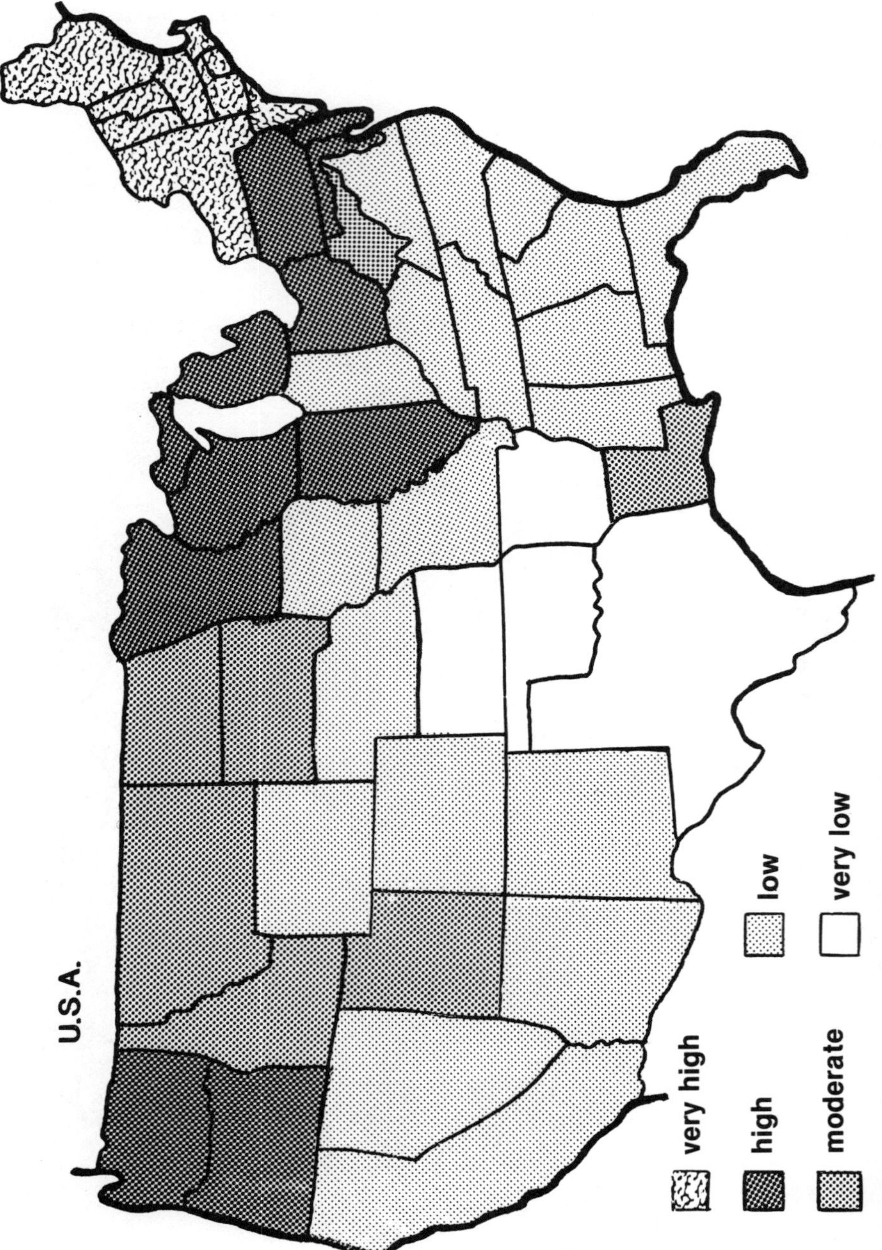

Figure 2-3 Map of United States showing caries prevalence as ranked by state (after Dunning, 1953).

studies. It predates the classic epidemiologic study of Snow on cholera, usually cited as the first such study on the distribution of a disease in a population.

There was an interesting distribution of high and low caries incidence throughout France (Figure 2-2). Magitot attributed these differences to racial background with round-headed Celts, as in Brittany, having low decay, and long-headed peoples such as in Normandy (of Viking origin) having high levels of decay. An equally good case, however, could be made to relate caries prevalence to geologic strata or soil types. The hilly areas of Brittany and the Massif Central, with ground water supplies, were areas of low decay, while the low plains along the rivers (surface water) had high decay.

It is of interest that the study of Magitot was made on military recruits, and in the United States a number of military surveys have also shown interesting geographic distribution of caries. These military studies have been reviewed by Dunning (1953), who was able to rank the states by caries prevalence and so produce a map of caries distribution, Figure 2-3. A comparison of this caries map with that of Figure 2-1, showing the results of the Ludwig and Bibby (1969) soils-caries studies, emphasizes the association of soils to caries.

Within discrete geographic areas, communities, despite common cultural practices, have been found to have significant differences in caries. The studies of Dean and co-workers (1942) on F, illustrate this point, but of greater interest is the study of Losee (1951) relating caries prevalence to trace element intake other than F. In the American dependency of Samoa there was a period following the 1939–1945 war when the Samoan culture throughout the island had only minimal

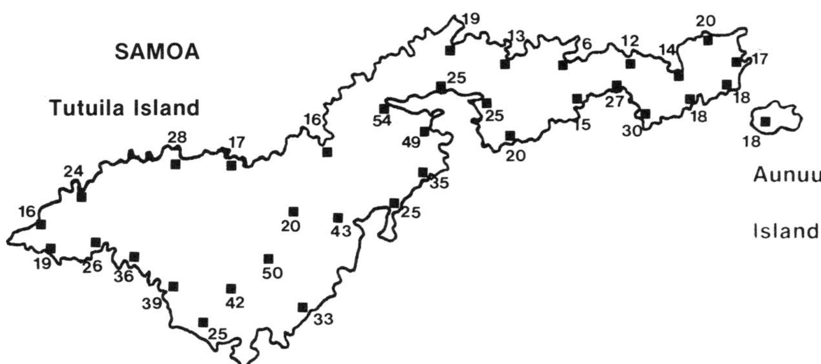

■ - 19 **site of village, mean DMFT/100 individuals.**

Figure 2-4 Caries prevalence data, as DMFT for villages in American Samoa, 1950 (after Losee, 1951).

American influences. In 1950 United States influences were largely confined to the area surrounding the main town of Pago Pago. A survey by Losee of the caries prevalence of Samoans by village showed an interesting distribution (Figure 2-4).

It was apparent that caries was lower on the north side of the island than on the south. Investigations by Losee showed that northern villages used greater quantities of sea salt in their cooking. Analysis of sea salt showed it to contain, besides F, a number of other trace elements, such as Sr, B, Mo, and Li. Losee's conclusion was that the use of sea salt markedly increased the amounts of trace elements in the diet and so affected caries incidence. The Samoan study gave an indication of a possible relationship of trace elements to dental disease and, in particular, dental caries. Subsequently, other research has indicated associations of one or more trace elements to caries.

TRACE ELEMENTS AND DENTAL CARIES

As noted earlier, data given in some of the early dental surveys concerning F can also be shown to give an indication of effects on caries by other factors. Thus, caries prevalence figures from a number of reports for several communities in the United States were substantially lower than in comparable communities where F levels were similar (Table 2-4), suggesting factors other than F played a role in lowering caries prevalence. Likewise, studies in several parts of the world have also revealed unexplained differences in caries prevalence between adjacent localities where low (< 0.3 ppm) and high (> 1.0 ppm) F drinking

Table 2-4
Comparison of Caries Prevalence in Children Aged 12–14 in United States Communities with Similar Fluoride Concentrations in Their Water Supplies

Communities	Water F conc. ppm	Mean DMFT	Percent Difference
Nashville, TN	0.0	4.61	
Key West, FL	0.1	10.70	56.9
Manning, SC	0.1	3.00	
Franklin, NH	0.1	7.20	58.3
Two Rivers, WI	0.3	6.46	
Milwaukee, WI	0.3	9.17	29.5
Joliet, IL	1.3	3.23	
Maywood, IL	1.3	2.58	20.1

Data from McClure, 1948; Ludwig and Bibby, 1969 (unpublished).

waters are used. Several of these studies have produced circumstantial evidence implicating elements other than F so the variation in caries and the epidemiologic background to these studies warrants discussion.

Individual Elements

In areas where there are trace elements at high levels, such as in soils, water, and foods, there may be an increased intake of such elements by man. Tank and Storvick (1960) showed that urinary levels of Se in children living in seleniferous areas of the northwest United States averaged 0.21 ppm. Correspondingly, children in a nearby low Se area averaged 0.05 ppm. Variations in Se content of human enamel have also been shown to vary, and are related to caries prevalence (Hadjimarkos and Bonhurst, 1961). Levels of Se intake by children in these areas reflected local soil Se originating, in all probability, from ingestion of locally produced foods such as dairy products. Thus, foods of animal origin may also show variations in trace element composition. Differences in trace element concentrations in plant foods fed to agricultural animals are reduced in extent in the animal tissues, but organ meats do reflect higher variations in trace elements than do muscle meats and fats (Söremark, 1964).

In 1953 Adler and Straub described a variation in caries experience between comparable groups of children living in adjacent villages in Hungary. Based upon analysis of drinking water and teeth, this variation was subsequently attributed to Mo. Animal experiments, in which Mo was fed as a dietary supplement to rats, appeared to confirm these findings. However, it is by no means certain that Mo was the sole agent responsible for the caries reduction because other trace elements in the diet supplement were not looked for.

A possible caries-reducing effect of Mo was described in New Zealand (Ludwig et al, 1960), during a comprehensive investigation of the caries difference between children residing in two adjacent towns of Napier and Hastings (Figure 2-5). It was planned to use these two communities as a fluoridation demonstration project, but baseline studies showed that children in Napier already had significantly lower caries prevalence scores than those in Hastings. Naturally, investigations were undertaken to try and explain these differences, and these included chemical analyses of soils, water, plants, human enamel, and urine. Vegetables grown on Napier soils contained higher concentrations of Mo, Al, and Ti and lesser amounts of Mn, Cu, Ba, and Sr than comparable crops grown in Hastings. No real differences in the Mo content of teeth from children in the two cities could be demonstrated, but analysis of 24-hour urine samples from 5- and 6-year-old boys indicated some difference in the intake of the element. Further investigation into the origin of the trace elements possibly influencing caries in Napier showed that the city was situated on an area of land which before 1931 had been a seabed. An earthquake in that year raised up the seabed, creating an area of marine sedementary soils that were alkaline and high in minerals. It was suggested that this unique situation was the underlying reason for the high intake of certain trace elements by the children of Napier. This epidemiologic study remains intriguing 20 years later. A recent survey (Cutress, 1980, unpublished data) of children in Napier and Hastings revealed that after 26 years of fluoridation (at 1 ppm) in Hastings, the caries prevalence of 13-year-old children was similar to that in Napier with Napier remaining at a low F level (< 0.3 ppm).

An investigation of a pronounced difference in caries prevalence between the areas of Langkloof and Nuwerus in South Africa was made by Pienaar and Bartels (1968). Nuwerus, where caries prevalence was relatively low, was situated on an alkaline soil where levels of available Mo were high. Analyses of selected vegetables grown in these areas

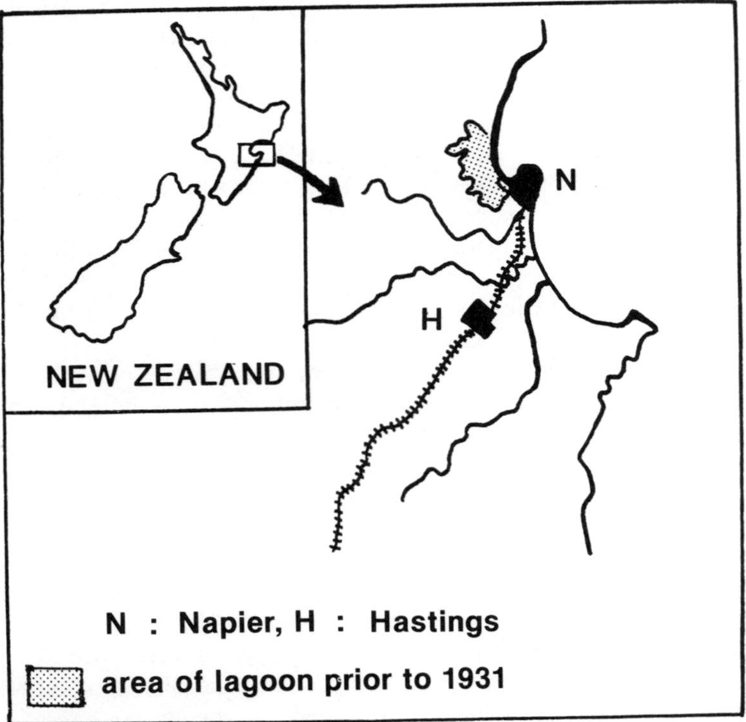

Figure 2-5 Location of survey towns of Napier and Hastings (New Zealand) to show their relationship to the lagoon raised by an earthquake in 1931.

showed that the Mo content of samples from the low caries area of Nuwerus averaged 77 $\mu g/100$ g.

Selecting the molybdenosis pasture areas of the San Joaquin Valley in the United States as a study area, Curzon et al (1971) could find no differences in caries prevalence between children living on these high Mo soils and children from a control area where molybdenosis has never been reported.

The demonstration of some association, still not completely proven, between Mo and caries serves to illustrate several points of interest in trace element dental research. First, there is the evidence of worldwide interest in possible relationships of trace elements to caries. Epidemiologic studies on Mo have now been reported from New Zealand, Hungary, the United States, England, and South Africa, mostly supporting the hypothesis that Mo reduces susceptibility to caries. Second, there is the difficulty of trace element analysis. For a good trace element investigation, samples must be collected from individuals with a known caries history and then individually analyzed. It is valid to pool samples from an individual, but not samples from many individuals. The analytical techniques chosen can mean the difference between a successful study and one showing interesting trends but no clear results. It is unfortunate that too many analytical results have been reported as "< 10 ppm." At high detection limits such as this, significant effects can be overlooked. After all, if one group averages 9 ppm and another group 0.5 ppm, these differences are substantial. For example, for F in the range 0.5 to 10 ppm, substantial differences in caries reduction and fluorosis can be shown. But, with a detection limit of < 10 ppm, no differences would be seen. Finally, apart from the New Zealand study, the Mo caries reports seemed to concentrate exclusively on the one element. Again, lack of availability of good analytical techniques may have prevented study of a large number of trace elements. In a Mo study in Somerset, England (Anderson, 1969), Mo was the main effector of low caries—or was there another element, not sought or reported on? For trace element studies, multiple elements must be considered.

Multiple Trace Element Studies

Many of the investigations reviewed above plus many described in later chapters, with the exception of the Napier-Hastings study, concentrated their analytical findings to one trace element. Most of the epidemiologic studies have been noted for work on single trace elements, such as the Se studies of Hadjimarkos. Two studies looked at as many trace elements as possible, and as far as many different human tissues or environmental variables as possible. These two studies, in Papua-New

Guinea (Schamschula et al, 1978) and in Colombia, South America (Glass, 1973), were initiated by the findings of unexpected differences in caries prevalence levels between two or more groups of people living within a fairly circumscribed geographic area.

Papua–New Guinea Early observations and dental epidemiologic studies in Papua–New Guinea showed evidence that caries prevalence was generally low, but that distinct differences, and even sharp contrasts, existed between population groups living in close geographic proximity (Barmes, 1969). The majority of New Guineans reside in village communities scattered along rivers, often isolated because of the mountainous terrain and with communication being via the waterways. There is a history of inbreeding, and there are many areas still without much exposure to European civilization. The diet of these peoples consists en-

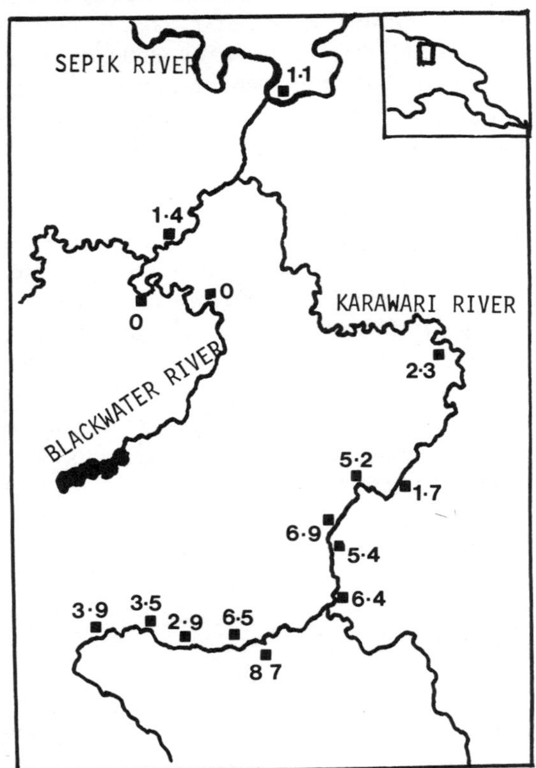

Figure 2-6 Geographic location of Papua-New Guinea study area, showing location of villages and caries prevalence.

tirely of locally grown or locally produced foods from land surrounding each village. The influence of the local geochemical environment on caries is, therefore, probably at its maximum with these people.

During 1968–1969, dental examinations were carried out under the auspices of the United States Public Health Service to test the hypothesis that the variations in caries experience were related to concentrations of major trace elements in soil, food, and water. The area concerned included a number of villages along the Sepik and Fly Rivers and some of its tributaries. These were located in part of Papua–New Guinea (Figure 2-6).

Results showed a strong inverse association between caries and concentrations of alkali and alkaline earth elements, especially Sr, Ba, K, Mg, Ca, and Li in garden soils (Barmes, 1969). There was also a direct association between caries prevalence and concentrations of Cu and Pb in village garden soils. Finally, there was evidence of an inverse association between caries and V, Mo, Mn, Al, Ti, and P, and a direct association between Pb, Cu, Cr, Zn, Se and the composition of staple foods, such as sago, used in the area.

These initial findings prompted continuation of the work with a much more extensive study (Schamschula et al, 1978). This work concentrated on a smaller area of the Sepik River, and certain hypotheses were formulated which were to be tested. The collection of large numbers of samples for trace element analysis and subsequent statistical evaluation was felt necessary to answer a number of questions such as, "Did trace elements in tooth enamel, plaque, or saliva influence dental caries?" In this second study, sampling and dental examinations were restricted to just 15 villages. The area of study together with the mean caries scores, by village, are shown in Table 2-5.

Table 2-5
Caries Distribution by Village Along the Sepik River Tributaries in Papua-New Guinea

Village	n	Mean DMFT	Village	n	Mean DMFT
1. Latoma	15	3.9	9. Tangambot	17	5.2
2. Danyig	9	3.5	10. Amongabi	33	6.9
3. Yentabak	14	2.3	11. Kundiman	41	1.7
4. Sikaum	10	8.7	12. Manjamai	15	2.3
5. Barapidgin	25	6.5	13. Sangriman	30	0.0
6. (not used)			14. Kabriman	20	0.0
7. Chimbot	9	6.4	15. Mumeri	13	1.4
8. Maramba	16	5.4	16. Mindimsit	32	1.1

Adapted from Schamschula et al, 1978.

The prevalence of dental caries was fairly low near the Timbunke Mission and in a village near the second tributary river system south of the Sepik River. But there was no caries at all in villages 13 and 14 beyond the confluence of the Karawani and Blackwater Rivers. As the Karawani was followed toward its source, the level of caries increased until village 4 was reached. There was then a drop in caries prevalence in the last three villages of the study, along the Wogupmeni River and in the mountains.

This second Papua study found a number of interesting findings and implications related to many of the looked-for variables. For caries there was a clustering of villages to give four distinct groups, with a general trend for caries to increase with proximity of the mountains. One group of villages nearest to the source of the river (villages 1, 2, and 3) had relatively low caries and had a comparatively high exposure to environmental F and Li as reflected in their concentrations in surface enamel, plaque, and saliva.

Plaque quantity was consistently and significantly associated with individual caries experience. As this was a positive association, this finding supported the concept that plaque is definitely related to caries. There were beneficial associations for the trace elements F and Sr together with the major minerals Ca and Ca/P ratio in plaque related to caries. For saliva it was found that a high mineral content appeared to have a beneficial effect on caries with the strongest and most consistent associations being for Li. Other trace elements occurring in saliva and related to caries were Mn, Mg, and F.

Obviously, considerable interest was centered on the results of the enamel analyses on samples collected in vivo. Whereas high concentrations of minerals in plaque and saliva were generally beneficial in their associations with caries, the opposite appeared to be true for enamel. The outstanding exception was Li, which showed a consistent and significant inverse association with caries experience. The most significant variables related to dental caries identified in the statistical analysis are indicated in Table 2-6.

Colombia, South America Dental caries prevalence is relatively high in Colombia, South America, and overall levels correspond to those found in nonfluoridated communities of New England within the United States. However, Glass et al (1973) published a report that identified a community with substantially lower caries prevalence than in surrounding villages. Dental examinations showed mean DMFT counts for children in this village, Heliconia, to be 5.5 compared with a nearby village, Don Matias, with 13.9 mean DMFT (Figure 2-7).

Subsequently, multiple trace element analyses were completed on soils and water, and showed that there were significant differences between the two towns (Table 2-6). It appeared that the low caries village of

Table 2-6
Statistically Significant Differences for Dental Caries and Trace Element Concentrations Between Two Colombian Villages

Element (ppm)	Heliconia	Don Matias	p = <
Mean D.F.S. score	3.84	15.20	0.01
Teeth			
Mn	3.82	10.00	0.001
Plaque			
Cr	0.01	0.002	0.01
Mn	0.004	0.006	0.01
Saliva			
Al	2100	2900	0.01
Fe	1100	2300	0.05
Mn	19	27	0.05
Ni	45	5.5	0.05
Water			
Ca	20,000	4000	0.001
Mg	60,000	800	0.001
Fe	200	400	0.001
V	10	5	0.001
Mo	6	1	0.001
Cr	4.2	39	0.001
Soils			
Cr	100	20	0.001
Cu	20	0	0.001
Ni	70	0	0.001
Sr	109	3.10	0.01
Mn	46	61	0.01
Ag	22	12	0.01
Ca	15,900	4380	0.01
Mg	12,000	4830	0.01

Adapted from Glass et al, 1973.

Heliconia had harder water, reflected in the significantly higher concentrations of Ca and Mg; water supplies were also higher in Mo and V. Conversely, the drinking water of the high caries village of Don Matias was significantly higher in Cu, Fe, Mn, and Ti. For soils, there was again an inverse association of caries with Ca and Mg, but also with Ag, Cr, Cu, Ni, and Sr. Positive associations were identified with Al and Mn.

The major finding from the Colombian study was, once again, an interaction of soil and water apparently related to caries. It is not clear precisely which of the trace elements identified was exerting the main effect upon caries. The content of the F′ in the water supplies in both villages was less than 0.1 ppm, and, therefore, would not affect caries. The alkaline earth elements, Ca and Mg, and to a lesser extent Sr, appeared to be major factors in the geochemical environment. It was not

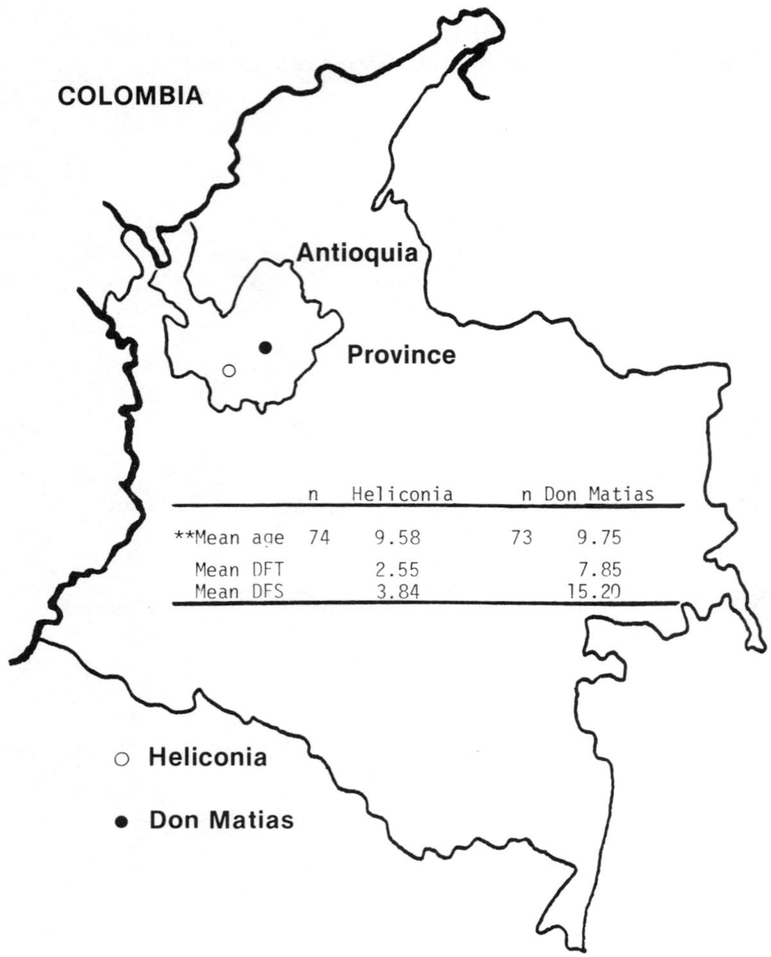

	n	Heliconia	n	Don Matias
**Mean age	74	9.58	73	9.75
Mean DFT		2.55		7.85
Mean DFS		3.84		15.20

Figure 2-7 Map of Colombia (South America) showing location of survey towns and mean caries prevalence (after Glass et al, 1973).

possible to separate the effects of these elements, and it may be that the primary effect was simple water hardness. High levels of Ca and Mg in drinking water are the major elements affecting water hardness, and these levels may also be related to remineralization. Nevertheless, the Colombian study points up the difficulties of trace element research where a number of factors may be effecting a change in caries pattern but where it is difficult to separate individual trace element actions.

CONCLUSION

For the past 30 years interest in the role of trace elements on dental disease, mostly caries, has centered on the demonstration of relationships of individual trace elements to caries. Overall positive and negative relationships have been reported for a number of elements which include Mo, Se, Sr, V, and Zn. Interactions of trace elements undoubtedly occur, and these have been considered to a limited extent, such as Mo-F, Mo-Cu and Sr-F.

The evidence from the epidemiology studies has not been conclusive for any particular element to show a causative effect, although it can be considered substantial for the three trace elements Mo, Se, and Sr. The mechanism of action of these elements is still little known, and much further experimental work is required to determine the role of trace elements in dental disease. Much of the work to date has concentrated on the identification of trace elements in the dental tissues of enamel, dentin, and cementum and to a lesser extent of plaque and saliva. Many of the trace elements that have been described as related to dental caries in epidemiologic studies are known to occur in human teeth. It is therefore pertinent now to discuss the identification of trace elements in enamel.

REFERENCES

Adler P, Straub J: Waterborne caries protective agent other than fluoride. *Acta Med Acad Sci Hung* 1953;4:221–225.

Albrecht WA: Our teeth and our soils. *Ann Dentistry* 1947;6:199–213.

Anderson RJ: The relationship between dental conditions and the trace element molybdenum. *Caries Res* 1969;3:75–87.

Åslander A: Tooth formation in the light of plant nutrition. *Roy Inst Tech Div Agric Res* (Stockholm) 1958.

Barmes DE: Caries aetiology in Sepik Villages—Trace element, micronutrient and macronutrient content of soil and food. *Caries Res* 1969;3:44–49.

Curzon MEJ, Kubota J, Bibby BG: Environmental effects of molybdenum on dental caries. *J Dent Res* 1971;50:74–77.

Durfor CN, Becker E: Public Water Supplies of the 100 Largest Cities of the U.S.A. Water Supply Paper 1812, Washington, DC, Government Printing Office, 1964.

Dean HT, Arnold FA, Elvove E: Domestic water and dental caries. V. Additional studies of the relation of fluoride domestic waters to dental caries experience in 4,425 white children aged 12-14 years of 13 cities in 4 states. *Pub Health Repts* 1942;57:1155–1179.

Dunning JM: The influence of latitude and distance from the sea coast on dental disease. *J Dent Res* 1953;32:811–829.

Glass RL et al: The prevalence of human dental caries and waterborne trace elements. *Arch Oral Biol* 1973;18:1099–1104.

Hercus CE et al: Epidemic goitre in New Zealand and its relation to the soil iodine. *J Hygiene* 1925;24:321–402.

Hewat REJ, Eastcott DF: Dental Caries in New Zealand. Monograph. Medical Research Council of New Zealand, 1955.

Hadjimarkos DM, Bonhurst CW: The selenium content of eggs, milk, and water in relation to dental caries in children. *J Pediatr* 1961;59:256–259.

Losee FL: Results of a year's research on dental caries in American Samoa—1950. *J Am Acad Appl Nutrition* 1951;5:258–265.

Losee FL, Adkins BL: A study of the mineral environment of caries resistant Navy recruits. *Caries Res* 1969;3:23–31.

Ludwig TG, Healy WB, Losee FL: An association between dental caries and certain soil conditions in New Zealand. *Nature* 1960;186:695–696.

Ludwig TG, Bibby BG: Geographic variation in the prevalence of dental caries in the U.S.A. *Caries Res* 1969;3:32–43.

Magitot E. (Chandler's trans.) Treatise on Dental Caries, Paris, 1878.

Nichols MS, McNall DR: Strontium content of Wisconsin Municipal Waters. *J Am Water Works Assoc* 1957;49:1493–1501.

Nizel AE, Bibby BG: Geographic variations in caries prevalence in soldiers. *JADA* 1944;31:1619–1626.

Ockerse T: Dental Caries: Clinical and Experimental Investigations. Dept. of Health, Pretoria, S Africa, 1949.

Olesen R: Epidemic goiter in Tennessee. US Pub. Health Reports 1929;44: 865–897.

Pärko A: Dental Caries Prevalence in the Rapakivi Granite and Olivine Diabase areas of Laitila, Finland. Academic Dissertation. University of Turku, Finland, 1975.

Pearce EP, Ludwig TG, Darwin J: Dental caries prevalence in seven soil areas of New Zealand. *Archs Oral Biol* 1974;19:157–163.

Pienaar WJ, Bartels EE: Molybdenum content of vegetables and soils in the Vredendal and Langkloof areas. *J Dent Res S Afr* 1968;23:242–244.

Romney EM: *Manganese Content of Soybeans Grown on Twenty New Jersey Soils.* Ph.D. Thesis, Rutgers University, 1953.

Saunders JD: Dental caries: A study of regional variations in New Zealand. *Trans Roy Soc NZ* 1945;75:57–64.

Schamschula RG, et al: *WHO Study of Dental Caries Aetiology in Papua-New Guinea.* WHO Publication No 40, Geneva, Switzerland, 1978.

Schroeder HA: The biological trace elements. *J Chron Dis* 1965;18:217–224.

Söremark R: Studies on the concentration of vanadium in some biological specimens. Proc. 3rd. Int. Colloq on Biology. Paris, France. 1964, pp 223–237.

Tank G, Storvick CA: Effect of naturally occurring selenium and vanadium on dental caries. *J Dent Res* 1960;39:473–488.

Wells N: *Mineral Content of a Standard Strain of Sweet Vernal Grown on Different N.Z. Soils.* Soil Bureau Bulletin. N.Z. Dept. Sci. and Indust. Res. Wellington, NZ, 1958.

Zwaardemaker JB, Tromp SW: The geographic distribution of dental caries in the Netherlands. *J Dent Res* 1957;36:795–799.

PART II
Trace Elements in Teeth (Enamel and Dentin), Bone, Calculus, Saliva, and Plaque

The mineralized dental tissues are enamel, dentin, cementum, and alveolar bone. Most studies carried out on the trace element content of these tissues are concerned with dental enamel and to a much lesser degree dentin. There is a little information on the trace element content of bone and virtually none on cementum.

The trace element composition of the mineralized tissues is related to those elements incorporated into the apatite crystal lattice during the mineralizing period and to those which diffuse into the tissue after the completion of mineralization. It follows that the trace element content of teeth reflects both the trace element biological environment during the time of tooth development and the oral environment (for enamel) or the vascular environment (for dentin) associated with the erupted tooth. For example, Pb and Sr concentrations in surface enamel are age dependent, the levels being associated with the availability of these elements in the environment. Absorption occurs from dietary food and water or oral habits such as smoking or chewing. On the other hand, Mg is preferentially lost with age apparently because of the selective reactivity of the apatite with the oral fluids. The pupal layer of dentin has been noted to reflect the uptake of trace elements into the vascular system.

Many and varied studies have described the influence of age and environmental factors on the trace element content of tooth tissue. Although some studies have been concerned with identifying the range and types of trace elements in dental tissues, the majority had an ultimate objective, directly or indirectly, of determining the influence, if any, of one or more trace elements on the susceptibility of the teeth to dental caries. Some trace elements influence the chemical properties of mineralized tissues, others are important cofactors in enzymatic transformations and play a role in the nucleation of mineralized tissues.

Hydroxyapatite is not a single entity but rather a continuous series of apatites each differing by atomic substitutions for some of the basic stoichiometric elements. For instance, Sr, Sn, Mo, Cd, Pb, rare earths, Na, and Mg may substitute for Ca, and V, As, and S for phosphorus. Evidence gradually accumulated to confirm the variable nature of biologic apatite which may result from environmental factors present at the time of apatite formation. Different patterns of crystallinity have been observed in hydroxyapatite synthesized by methods which use drinking waters obtained from different geographic areas in New Zealand. X-ray diffraction evaluation of the apatites showed variation in the quality of crystal structure associated with the element content of the water phase. It has also been proposed that poor crystal structure associated with Ca-deficient enamel may be improved with mineralizing solutions containing Zn and Sr.

The large number of naturally occurring trace elements; a variety of analytical techniques; and the distribution of elements between people, teeth, depth of tissue, and other variables have, together, resulted in considerable accumulation of data. A few elements such as Sr and Pb have been extensively studied from many aspects whereas most others have unfortunately attracted little interest. Even where epidemiologic evidence has implicated several elements (Mo, Se, V, Li) with susceptibility to caries the occurrence of these elements in dental tissues has received only cursory study.

In Part II of this book, current knowledge of trace element distribution in mineralized dental tissues is reviewed. The available data are extensive and have been gathered for reasons including basic qualitative and quantitative evaluation of enamel for a few or many elements in any one study, also for epidemiologic purposes, in some cases as a measure of environmental pollution (Pb, Cd, Hg), and for forensic purposes.

The information reviewed here is from teeth which differed in age and tooth type as well as in the form of the enamel samples—whole enamel, surface enamel, or individual tooth samples or pooled samples. Nevertheless, the total data represent a considerable body of literature concerned with the trace element composition of dental tissues.

3 Teeth, Calculus, and Bone

T. W. Cutress

TRACE ELEMENTS IN ENAMEL OF PERMANENT TEETH

As the most highly mineralized of biological tissues dental enamel shows certain characteristics not found elsewhere in the body. Dental enamel and dentin, unlike other biological tissues, once formed do not metabolize or have a turnover rate. The originally formed apatite mineral remains basically unchanged throughout life apart from changes occurring at its surface, in contact with oral tissues, where diffusion processes operate. Enamel reflects the trace element environment present in tissue fluids at the time of tooth development.

Most enamel studies on trace element composition have been concerned with analysis of whole enamel, that is, the whole thickness of enamel separated from other tooth tissues by physical methods. The analysis of this whole enamel would be representative of the mean concentration of element(s). However, a change in the element composition of enamel with depth from the surface (Brudevold et al, 1960) and also from various areas of the same enamel (Weatherell et al, 1966) has been reported. Only in recent years has a methodological approach become possible for investigating the full range of elements in mineralized tissues.

Before the use of techniques suitable for multiple element analyses, such as neutron activation and spark source mass spectrometry, data on the elemental composition of enamel accumulated from studies concerned with analyses of elements of specific interest (for example, a potential association with caries and Mo, V, Sr, Se, F) or a wider range of elements for general screening (for example, using spectrographic techniques). The methods of analysis used vary widely and include those capable of analyzing for only single elements, or a small range of elements, or almost the complete range of elements in the periodic table. Methods used include proton induced X-ray emission, X-ray fluorescence, spark source mass spectrometry, charged particle activation, anodic stripping voltametry, and high resolution spectrometry. The sensitivity of these methods varies widely. For single element analysis the analytical conditions can be chosen to maximize the accuracy and sensitivity of estimation. As these conditions usually vary from element to element, a compromise of analytical conditions is often necessary for multiple element analysis which results in less than optimal sensitivity and accuracy for individual elements. Because of the widely varying objectives and analytical procedures used in acquiring trace element data a comprehensive review unfolds as a patchwork of studies which can nevertheless be given form under the following headings of trace elements in enamel, dentin, bone, and calculus.

Where applicable each mineral will be considered for trace element variation by age, depth, geographic source, oral environment, and dental caries. At the risk of repetition and for the sake of completeness some of the trace elements in enamel will also be discussed in the individual chapters concerned with specific elements.

In the past ten years mass spectrometry and atomic absorption technology have provided an opportunity for extensive screening studies. Although a survey of the available literature reveals a wealth of data, comparisons can be made only with difficulty because of differences in the source of teeth, the wide variation in sample preparation, sample size, analytic methodology, and the manner of presentation of the findings.

For enamel the available information will be considered with respect to whole enamel, variation with depth, age, geographic source (of teeth), oral environment, and dental caries.

Whole Enamel

A general perspective of the trace element composition of whole enamel is provided in Table 3-1, summarizing the elements which have

Table 3-1
Summary of Trace Elements in Dental Enamel of Permanent Teeth

Concentration Range ppm	Elements
> 1000	Na, Cl, Mg
100–1000	K, S, Zn, Si, Sr, F
10–100	Fe, Al, Pb, B, Ba
1–10	Cu, Rb, Br, Mo, Cd, I, Ti, Mn, Cr, Sn
0.1–0.9	Ni, Li, Ag, Nb, Se, Be, Zr, Co, W, Sb, Hg
< 0.1	As, Cs, V, Au, La, Ce, Pr, Nd, Sm, Tb, Y
Not detected	Sc, Ga, Ge, Ru, Pd, In, Te, Eu, Gd, Dy, Ho, Er, Tm, Lu, Hf, Ta, Re, Os, Ir, Pt, Tl, Bi, Rh

been detected and not detected in whole enamel following quantitative analyses. The median concentrations of the various elements detected together with their equivalent micromole concentrations are also shown (Table 3-2) and are based on the studies listed (Table 3-3). In the approximately 30 studies carried out in the past 25 years, from which this summary has been compiled, about 69 elements have been studied but only a few elements have received considerable attention.

The source of information is geographically wide, with enamel samples derived from teeth from the United States, Norway, Sweden, New Zealand, South Africa, Egypt, Scotland, England, Finland, New Guinea, Yugoslavia, and several other countries. The chosen analytical methods have also varied to include volumetric analysis, neutron activation and X-ray spectrometry, mass spectrometry, voltametry, flame photometry, atomic absorption, and ion probe techniques. Other variables which need to be considered in this summarized information are the variety and age of teeth from which the enamel was obtained, the method of enamel separation, the number of samples analyzed and whether the samples related to enamel of single teeth or composite enamel samples.

Presentation of data varies between studies, and has been shown in terms of concentration range, means with variability, or median values, all of which tend to add confusion in comparative studies. In particular, perusal of studies using mean and variability statistics indicates that the distribution of element concentrations is not normal. The element concentrations (Table 3-2) have been presented as median values based on an approximation of the mid-range concentration of elements quoted in quantitative studies (Table 3-3).

Table 3-2
Concentration of the More Commonly Observed Trace Elements in the Enamel of Permanent Teeth

Element	Median Concentration ppm[a]	Micromoles	Element	Median Concentration ppm[a]	Micromoles
Na	7200	313.0	Mn	1	0.02
Cl	3200	90.0	Cr	1	0.02
Mg	2800	115.0	Sn	1	0.008
K	270	7.7	Cd	1	0.009
S	214	6.7	Li	0.9	0.1
Zn	210	3.2	Ag	0.8	0.006
Si	190	6.8	Ni	0.6	0.02
Sr	115	1.3	Nb	0.6	0.006
Fe	28	0.5	Se	0.4	0.005
Al	23	0.9	Be	0.3	0.03
B	15	1.4	Zr	0.3	0.003
Ba	15	0.1	Co	0.2	0.003
Cu	7	0.1	W	0.2	0.001
Pb	5.5	0.03	Sb	0.1	0.001
Rb	5	0.06	Hg	0.1	0.0005
Br	4	0.05	As	0.06	0.0008
Mo	2	0.03	Cs	0.04	0.0003
I	2	0.02	V	0.02	0.0003
Ti	1.5	0.03	Au	0.02	0.0001

[a]By dry weight

The variation in concentration of different elements reported for specific investigations is discussed in more detail later under the respective element headings. However, despite lack of agreement on the "normal" range of element occurrence in enamel, it is evident that the elements reported to occur in enamel can be grouped according to their frequency of occurrence by weight. This grouping (Table 3-1) shows clearly that Na, Cl, and Mg contribute more substantially to enamel composition than do all other elements with the exception, of course, of the major components. Each of these three elements alone contributes more to the composition of enamel than the other minor elements combined. Despite their relatively high concentration no evidence has been presented to show that Na or Cl and possibly Mg influence the integrity of enamel or its susceptibility to caries, erosion, or other defects.

The next most commonly occurring elements are grouped within the range 100 to 1000 ppm. This arbitrary grouping should not obscure the fact that K, with the highest median concentration at 300 ppm, is substantially lower in concentration than Mg (2800 ppm). Also in this group is Sr which has been an element of interest both because of a postulated association with dental caries susceptibility and also as an indicator of radioactive contamination through its radioactive isotope, Sr^{90}.

Summary of All Reports on Trace Elements in Enamel of Human Permanent Teeth

			Enamel Samples		Method of		Concentration		
Element and Author(s)	Year	n	*Age*	*Origin*	Analysis		*Mean*	± *SD*	*Range*
Na									
Söremark and Samsahl	1961	15	14–16	Sweden	NA		11,600	4000	
Lundberg et al	1965a	10	14–25	Sweden	NA		9700	1200	
Retief et al	1971	8	NG	South Africa	NA		7000	100	
Derise et al	1974	175	10–90	United States	NA		7200	200	
Lakomaa and Rytömaa	1977	25	14	Finland	NA		6900	800	
Frostell et al	1977	1	NG	Sweden	IP		(0.2–0.3%)		
Naujoks	1967	NG	NG	Germany	FP		8380		
						Median	7200		4500–23,600
Cl									
Söremark and Samsahl	1961	15	14–16	Sweden	NA		6500	3000	
Retief et al	1971	8	NG	South Africa	NA		3200	100	
Losee et al	1974b	28	< 20	United States	MS		6022	723	
Derise et al	1974	175	10–90	United States	NA		2800	100	
Lakomaa and Rytömaa	1977	25	< 14	Finland	NA		3200	220	
Frostell et al	1977	1	NG	Sweden	IP		1200 (est)		
						Median	3200		2500–15,500
Mg									
Burnett and Zenewitz	1958	96	NG	United States	wet		2100	100	
Retief et al	1971	8	NG	South Africa	NA		2800	100	
Derise et al	1974	175	10–90	United States	NA		3900	600	
Losee et al	1974b	28	< 20	United States	MS		1670	120	
Lakomaa and Rytömaa	1977	25	< 14	Finland	NA		2800	700	
Frostell et al	1977	1	NG	Sweden	IP		1300 (est)		
						Median	2800		800–4500
K									
Losee et al	1974a	56	< 20	United States	MS		458	23	180–1000
Derise et al	1974	175	10–90	United States	NA		270	10	
Lakomaa and Rytömaa	1977	25	< 14	Finland	AA		210		
Curzon and Crocker	1978	337	10–20	US/NZ	MS		961		60–4056
Frostell et al	1977	1	NG	Sweden	IP		250 (est)		
						Median	270		

Table 3-3 (continued)

Element and Author(s)	Year	n	Enamel Samples		Method of Analysis		Concentration		
			Age	*Origin*			*Mean*	± *SD*	*Range*
S									
Losee et al	1974a	56	< 20	United States	MS		294	15	110–56
Curzon et al	1975	36	11	New Zealand	MS		214	14.8	
Curzon and Crocker	1978	336	10–20	US/NZ	MS		24		0–200
						Median	214		
Zn									
Söremark and Samsahl	1961	15	14–16	Sweden	NA		276	106	
Lundberg et al	1965a	10	14–25	Sweden	NA		215	41	
Retief et al	1971	8	NG	South Africa	NA		263.4	14.8	
Losee et al	1974c	93	< 20	United States	AA		241.4	12.9	129–1197(461)
Derise and Ritchey	1974	173	> 10	United States	AA		185.7	3.8	
Losee et al	1974a	56	< 20	United States	MS		203	12	42–510
Lakomaa and Rytömaa	1977	25	< 14	Finland	AA		126	21	
Oehme and Lund	1978	8	< 20	Norway	V				76–542
Attramadal and Jonsen	1978	26	Ancient	Norway	V		740		300–1500
Curzon and Crocker	1978	336	10–20	US/NZ	MS		153		9.9–806
Curzon et al	1975	36	11	New Zealand	MS		151.4	14.2	
						Median	210		
Si									
Losee et al	1974a	56	< 20	United States	MS		243	18	100–450
Losee et al	1974c	93	< 20	United States	ES		136	20.4	26–1155(642)
Steadman et al	1959	12	Ancient	United States	ES		33		
						Median	190		
Sr									
Söremark and Samsahl	1961	15	14–16	Sweden	NA		93.5	21.9	
Lundberg et al	1965a	10	14–25	Sweden	NA		83	0.5	
Retief et al	1971	7	NG	South Africa	NA		111.2	9.9	

Table 3-3 (continued)

Element and Author(s)	Year	n	Enamel Samples: *Age*	Enamel Samples: *Origin*	Method of Analysis		Concentration: *Mean*	± *SD*	*Range*
Losee et al	1974a	56	< 20	United States	MS		121		21–280
Losee et al	1974c	93	< 20	United States	ES		76.2	7.9	14–450(280)
Derise and Ritchey	1974	173	> 10	United States	NA		285.6	25	
Curzon et al	1975	36	11	New Zealand	MS		92.6	4.9	
Nixon and Helsby	1976	120	NG	England	AA		112	8	80–181
Curzon and Losee	1977b	147	11–19	United States	MS		183	15.4	21–1200(700)
Curzon and Losee	1977a	108	< 20	United States	MS		128.9	8.4	
Curzon and Losee	1977a	100	< 20	United States	MS		156.7	10.9	
Curzon and Crocker	1978	337	10–20	US/NZ	MS		157.1		13–1400
Frostell et al	1977	1	NG	Sweden	IP		200 (est)		
						Median	115		
Fe									
Söremark and Lundberg	1964	15	14–16	Sweden	NA		338	109	
Retief et al	1971	9	NG	South Africa	NA		118.3	71.7	
Losee et al	1974a	56	< 20	United States	MS		9.5	0.96	0.8–26
Losee et al	1974c	93	< 20	United States	ES		163.8	9.8	11–759(165)
Curzon et al	1975	36	11	New Zealand	MS		18.3	1.9	
Lakomaa and Rytömaa	1977	25	< 14	Finland	AA		2.77		
Curzon and Crocker	1978	337	10–20	US/NZ	MS		27.95		0–157
						Median	28		
Al									
Retief et al	1971	8	NG	South Africa	NA		86.13	4.54	
Derise and Ritchey	1974	173	10–90	United States	AA		89.8	2.4	
Losee et al	1974a	56	< 20	United States	MS		14.0	1.5	1.5–54
Losee et al	1974c	93	< 20	United States	ES		74.7	10.0	8–325
Curzon et al	1975	36	11	New Zealand	MS		17.6	3.9	
Lakomaa and Rytömaa	1977	25	< 14	Finland	NA		240	80	

Table 3-3 (continued)

Element and Author(s)	Year	n	Enamel Samples		Method of Analysis	Concentration		Range
			Age	*Origin*		*Mean*	± *SD*	
Curzon and Losee	1977a	38	< 20	United States	MS	19.1	3.5	
Curzon and Losee	1977a	26	< 20	United States	MS	8.3	1.3	
Curzon and Crocker	1978	335	10–20	United States	MS	22.9		0–510
					Median	23		
Ba								
Retief et al	1971	7	NG	South Africa	NA	125.1	23.7	
Losee et al	1974a	56	< 20	United States	ES	15.3	0.75	4.2–44(32)
Curzon et al	1975	36	11	New Zealand	MS	17.6	3.9	
Curzon and Losee	1977a	108	< 20	United States	MS	6.8	2.26	
Curzon and Losee	1977a	100	< 20	United States	MS	2.1	0.57	
Curzon and Crocker	1978	334	10–20	US/NZ	MS	18.8		0–510
					Median	15		
B								
Losee et al	1974a	56	< 20	United States	MS	18.2	2.65	0.5–69
Losee et al	1974c	93	< 20	United States	ES	20.3	2.59	< 2–141(93)
Curzon et al	1975	36	11	New Zealand	MS	11.8	2.26	
Curzon and Crocker	1978	337	10–20	US/NZ	MS	8.4		0–190
					Median	15		
Cu								
Söremark and Samsahl	1961	15	14–16	Sweden	NA	0.26	0.11	
Nixon and Smith	1962	100	6–65	Scotland	NA	9.5	7.8 (0)	
						11.3	9.0 (6)	
Losee et al	1974a	56	< 20	United States	MS	6.8	4.0	0.07–208

Table 3-3 (continued)

Element and Author(s)	Year	n	Enamel Samples		Method of Analysis		Concentration		
			Age	*Origin*			*Mean*	± *SD*	*Range*
Losee et al	1974c	93	< 20	United States	ES		12.2	1.46	2–126
Derise and Ritchey	1974	173	> 10	United States	AA		10.1	1.9	1.5–53.9
Curzon and Losee	1977a	35	< 20	United States	MS		1.64	0.44	
Curzon and Losee	1977a	26	< 20	United States	MS		0.51	0.06	
Lakomaa and Rytömaa	1977	25	< 14	Finland	AA		1.38		
Attramadal and Jonsen	1978	26	Ancient	Norway	V		< 5 to 115		
Oehme and Lund	1978	8	< 20	Norway	V		1.3	16.8	
Curzon and Crocker	1978	336	10–20	US/NZ	MS		1.5		0–30.0
Curzon et al	1975	36	11	New Zealand	MS		0.17	0.02	
						Median	7.64		
Pb									
Losee et al	1974a	56	< 20	United States	MS		3.1	0.16	1–6.5
Derise and Ritchey	1974	173	10–90	United States	NA		180.7	0.9	
Losee et al	1974c	93	< 20	United States	ES		19.2	10.7	< 2–1000(56)
Curzon et al	1975	36	11	US/NZ	MS		5.5	0.9	
Attramadal and Jonsen	1978	26	Ancient	Norway	V		5		1–140
Oehme and Lund	1978	8	< 20	Norway	V		3		1.8–4.9
Curzon and Crocker	1978	335	10–20	US/NZ	MS		19.6		0–156
Malik and Fremlin	1974	7		England			34.2		24.5–41.7
						Median	5.5		
Rb									
Söremark and Lundberg	1964b	15	14–16	Sweden	NA		4.9	2.2	
Lundberg et al	1965a	10	14–25	Sweden	NA		73	56	
Losee et al	1974a	56	< 20	United States	MS		0.41	0.02	
Curzon et al	1975	36	11	US/NZ	MS		0.44	0.03	
Curzon and Crocker	1978	256	10–20	US/NZ	MS		4.61		0–30
						Median	4.6		

Table 3-3 (continued)

Element and Author(s)	Year	n	Enamel Samples		Method of Analysis	Concentration		Range
			Age	*Origin*		*Mean*	± *SD*	*Range*
Br								
Söremark and Samsahl	1961	15	14–16	Sweden	NA	4.6	1.1	
Retief et al	1971	8	NG	South Africa	NA	33.8	5.7	
Losee et al	1974a	56	< 20	United States	MS	1.1	0.09	< 0.1–2.9
Rasmussen	1974	43	All ages	Scandinavia	NA			0.87–7.3
Curzon et al	1975	36	11	New Zealand	MS	3.4	0.72	
Curzon and Crocker	1978	287	10–20	US/NZ	MS	4.5		0–33.2
					Median	4		
Mo								
Losee et al	1974c	93	< 20	United States	EM	1.1	0.3	0.6–22
Losee et al	1974a	56	< 20	United States	MS	5.5	0.7	0.7–39
Curzon et al	1975	36	11	New Zealand	MS	1.0	0.1	
Curzon and Crocker	1978	334	10–20	US/NZ	MS	2.37		0–32.0
					Median	2.0		
Cd								
Lundberg et al	1965a	15	14–25	United States	NA	< 1 × 10		
Losee et al	1974a	56	< 20	United States	MS	0.99	0.15	0.03–6.7
Losee et al	1974c	93	< 20	United States	AA	14.9	3.05	0.4–268
Curzon et al	1975	36	11	New Zealand	MS	3.3	0.59	
Attramadal and Jonsen	1977	26	Ancient	Norway	V	0.2	0.4	
Curzon and Spector	1977	335	9–20	United States	MS	1.86		< 0.03–27
Oehme and Lund	1978	8	< 20	Norway	V	0.04	16.2	
Shearer et al	1980	10	11–22	United States	AA	1.09	1.87	
					Median	1		
I								
Losee et al	1974a	28	< 20	United States	MS	0.04		0.01–0.07
Derise and Ritchey	1974	173	10–90	United States	AA	5.6	0.58	
Curzon and Crocker	1978	92	10–20	United States	MS	2.0		0–9.9
					Median	2		

Table 3-3 (continued)

Element and Author(s)	Year	n	Enamel Samples		Method of Analysis		Concentration		Range
			Age	*Origin*			*Mean* ± *SD*		
Ti									
Losee et al	1974a	56	< 20	United States	MS		0.46	0.13	0.1–4.8
Losee et al	1974c	93	< 20	United States	ES		4.0	0.99	0.6–66
Curzon et al	1975	36	11	New Zealand	MS		0.89	0.33	
Curzon and Crocker	1978	243	10–20	US/NZ	MS		1.93		0–31.4
						Median	1.5		
Mn									
Steadman et al	1959	12	Ancient	United States	ES		27		
Söremark and Samsahl	1961	15	14–16	Sweden	NA		0.54	0.11	
Lundberg et al	1965a	10	14–25	Sweden	NA		1.12	0.48	
Nixon et al	1966	13	11–53	United States	NA		1.1		0.30–2.01
Battisone et al	1967	106	8–30	United States	NA		0.60	0.04	0.15–1.16
Retief et al	1971	4	NG	South Africa	NA		0.59	0.04	
Losee et al	1974a	56	< 20	United States	MS		0.32	0	0.08–1.1
Losee et al	1974c	93	< 20	United States	ES		3.6	0.93	< 0.6–63
Derise and Ritchey	1974	173	10–90	United States	AA		7.0	0.12	
Curzon et al	1975	36	11	New Zealand	MS		0.83	0.2	
Lakomaa and Rytömaa	1977	25	< 14	Finland	NA		1.42	0.17	
Curzon and Crocker	1978	336	10–20	US/NZ	MS		0.60		0–6.7
						Median	1		
Cr									
Söremark and Lundberg	1964a	15	14–16	Sweden	NA		0.004	0.002	
Retief et al	1971	6	NG	South Africa	NA		1.02	0.51	
Losee et al	1974a	56	< 20	United States	MS		2.3	0.69	0.1–31
Losee et al	1974c	93	< 20	United States	ES		6.4	0.71	< 0.6–33
Curzon and Crocker	1978	236	10–20	US/NZ	MS		0.45		0–18
						Median	1.0		

Table 3-3 (continued)

Element and Author(s)	Year	n	Enamel Samples		Method of Analysis	Concentration		Range
			Age	*Origin*		*Mean*	± *SD*	*Range*
Sn								
Losee et al	1974a	56	< 20	United States	MS	0.53	0.14	0.03–8.1
Losee et al	1974c	93	< 20	United States	ES	2.04	1.06	< 2–93
Curzon et al	1975	36	11	New Zealand	MS	0.55	0.09	
Curzon and Losee	1977a	21	< 20	United States	MS	0.24	0.08	
Curzon and Losee	1977a	8	< 20	United States	MS	1.35	0.37	
Curzon and Crocker	1978	245	10–20	US/NZ	MS	1.6		0–44
					Median	1		
Ni								
Losee et al	1974a	56	< 20	United States	MS	0.32	0	0.08–1.1
Losee et al	1974c	93	< 20	United States	ES	1.57	0.27	< 0.06–13
Curzon et al	1975	36	11	New Zealand	MS	0.21	0.05	
Curzon and Crocker	1978	249	10–20	US/NZ	MS	0.90		0–9.0
					Median	0.6		
Li								
Losee et al	1974a	56	< 20	United States	MS	1.6	0.12	1.4–4.2
Curzon et al	1975	36	11	New Zealand	MS	1.3	1.1	
Curzon and Losee	1977a	13	< 20	United States	MS	0.35	0.02	
Curzon and Losee	1977a	7	< 20	United States	MS	0.07	0.02	
Curzon and Crocker	1978	246	10–20	US/NZ	MS	0.92		0–13.2
					Median	0.9		
Ag								
Söremark and Lundberg	1964a	15	14–16	Sweden	NA	0.005	0.001	
Retief et al	1971	9	NG	South Africa	NA	0.56	0.29	
Losee et al	1974a	56	< 20	United States	MS	0.14		0.01–0.77
Losee et al	1974c	93	< 20	United States	ES	0.70	0.16	< 0.1–9
Curzon et al	1975	36	11	New Zealand	MS	0.79	0.22	
Curzon and Losee	1977a	3	< 20	United States	MS	1.11	1.04	

Table 3-3 (continued)

Element and Author(s)	Year	n	Enamel Samples: *Age*	Enamel Samples: *Origin*	Method of Analysis		Concentration: *Mean*	± *SD*	*Range*
Curzon and Losee	1977a	21	< 20	United States	MS		7.43	2.7	
Curzon and Crocker	1978	244	10–20	US/NZ	MS		3.44		0–37
						Median	0.8		
Nb									
Losee et al	1974a	56	20	United States	MS		0.31	0.02	0.1–0.76
Curzon et al	1975	36	11	New Zealand	MS		0.60	0.05	
Curzon and Crocker	1978	245	10–20	US/NZ	MS		0.17		0–1.0
						Median	0.6		
Se									
Hadjimarkos and Bonhorst	1959	85	< 20–> 50	United States	V		0.43	1.6	
Nixon and Myers	1970	6	NG	United Kingdom	NA		0.87		0.21–2.08
Losee et al	1974a	56	< 20	United States	MS		0.43	0.03	0.12–0.90
Derise and Ritchey	1974	173	10–90	United States	AA		0.38	0.60	
Retief et al	1974	10	NG	South Africa	NA		0.08	0.03	
Curzon et al	1975	36	11	New Zealand	MS		0.40	0.03	
Retief et al	1976	10	NG	South Africa (B)	NA		0.08	0.03	
				South Africa (W)			0.01	0.004	
Curzon and Crocker	1978	326	10–20	US/NZ	MS		1.47		0–18.1
						Median	0.40		
Be									
Losee et al	1974a	10	< 20	United States	MS		0.09		< 0.01–0.97
Losee et al	1974c	93	< 20	United States	ES		0.31	0.06	< 0.02–1.9
Curzon and Crocker	1978	241	10–20	US/NZ	MS		1.36		0–15.9
						Median	0.3		
Zr									
Losee et al	1974a	56	< 20	United States			0.53	0.08	< 0.02–2.6
Losee et al	1974c	93	< 20	United States	ES		1.8	0.3	< 0.60–12

Table 3-3 (continued)

Element and Author(s)	Year	n	Enamel Samples		Method of Analysis		Concentration		Range
			Age	*Origin*			*Mean*	± *SD*	
Curzon and Losee	1977a	24	< 20	United States	MS		0.27	0.1	
Curzon and Losee	1977a	29	< 20	United States	MS		0.16	0.09	
Curzon and Crocker	1978	244	10–20	US/NZ	MS		0.08		0–0.8
						Median	0.3		
Co									
Söremark and Lundberg	1964a	15	14–16	Sweden	NA		0.0002	0.0001	
Retief et al	1971	10	NG	South Africa	NA		0.13	0.13	
Derise and Ritchey	1974	173	> 10	United States	AA		34.3	0.8	
Losee et al	1974a	56	< 20	United States	MS		126		< 0.01–0.26
Losee et al	1974c	93	< 20	United States	ES		2.31	0.57	< 0.6–30(16)
Curzon and Crocker	1978	246	10–20	US/NZ	MS		0.27		0–1.5
Little and Steadman	1966	330	> & < 30	United States	ES		ND		
						Median	0.2		
Sb									
Nixon et al	1967	29	14–61	Scotland/Egypt	NA		0.04		0.005–0.67
Retief et al	1971	9	NG	South Africa	NA		0.96	0.69	
Losee et al	1974a	28	< 20	United States	MS		0.13	0.01	0.02–0.34
Rasmussen	1974	49	ALL	Scandinavia	NA		0.001	1.59	
Curzon et al	1975	36	11	New Zealand	MS		0.09	0	
Curzon and Crocker	1978	225	10–20	US/NZ	MS		0.20		0–3.0
						Median	0.1		
Hg									
Nixon et al	1965	40	9–78	United Kingdom	NA		2.6		0.14–16
Losee et al	1974a	56	< 20	United States	MS		< 0.11		
Rasmussen	1974	52	ALL	Scandinavia	NA		0.14		< 0.001–1.9
						Median	0.1		

Table 3-3 (continued)

Element and Author(s)	Year	n	Enamel Samples		Method of Analysis		Concentration		
			Age	*Origin*			*Mean* ± *SD*		*Range*
W									
Söremark and Samsahl	1961	15	14–16	Sweden	NA		0.24	0.12	
Losee et al	1974a	56	< 20	United States	MS		< 0.08		
						Median	0.16		
As									
Nixon and Smith	1960	25	NG	Scotland	NA		0.06		0.031–0.145
Rasmussen	1974	46	ALL	Scandinavia	NA		0.07		0.001–0.406
Losee et al	1974a	56	< 20	United States	MS		< 0.02		
						Median	0.06		
Cs									
Losee et al	1974a	56	20	United States	MS		0.04		0.02–0.10
V									
Losee et al	1974a	56	< 20	United States	MS		0.03		0.01–0.14
Losee et al	1974c	93	< 20	United States	ES		< 0.06		< 0.06
Curzon et al	1975	36	11	New Zealand	MS		0.02		
Curzon and Crocker	1978	239	10–20	US/NZ	MS		0.02		0–0.2
						Median	0.02		
Au									
Söremark and Samsahl	1961	15	14–16	Sweden	NA		0.02	0.01	
Lundberg et al	1965a	10	14–25	Sweden	NA		1×10^{-4}		
Retief et al	1971	7		South Africa	NA		0.11	0.07	
Losee et al	1974a	56	20	United States	MS		0.02		
						Median	0.02		
Y									
Losee et al	1974a	2	< 20	United States	MS		0.007		0.01–0.17
Losee et al	1974b	2	< 20	United States	MS		0.007		< 0.01–0.17

Table 3-3 (continued)

Element and Author(s)	Year	n	Enamel Samples: *Age*	Enamel Samples: *Origin*	Method of Analysis	Concentration: *Mean* ± *SD*	*Range*
Pr							
Losee et al	1974a	24	< 20	United States	MS	0.03	0.01–0.07
Losee et al	1974b	24	< 20	United States	MS	0.027	< 0.01–0.07
Nd							
Losee et al	1974a	24	< 20	United States	MS	0.05	0.02–0.09
Losee et al	1974b	24	< 20	United States	MS	0.045	< 0.01–0.09
La							
Losee et al	1974a	1	< 20	United States	MS	0.004	
Losee et al	1974b	1	< 20	United States	MS	0.004	< 0.02–0.12
Steinnes et al	1974	5	21–72	Norway	NA	0.005	0.004–0.008
Sc							
Steinnes et al	1974	5	21–72	Norway	NA	0.0007	0.0003–0.001
Tb							
Steinnes et al	1974	5	21–72	Norway	NA	< 0.005	
Yb							
Steinnes et al	1974	5	21–72	Norway	NA	< 0.01	

NA = neutron activation
IP = ion probe
FP = flame photometry
MS = spark source mass spectrometry
wet = wet chemistry
AA = atomic absorption spectrophotometry
V = voltametry
ES = emission spectrometry

Five elements more, Fe, Al, Pb, B, and Ba group together in the next arbitrary range of the scale. Of these Fe has the highest median value at 28 ppm and is substantially lower than Sr (115 ppm). Two of these elements have aroused interest: Al, because of its ability to interact with F to form a series of complex ions, and Pb, a toxic element cumulative in animals and humans and studied in enamel as a means of predicting or diagnosing body exposure to Pb.

The arbitrary nature of this scale for grading elements becomes more evident for the remainder of the elements detected in enamel. No obvious groupings are apparent, and the element contribution to the composition of enamel decreases from Cu at 8 ppm to Au at 0.02 ppm through 24 intermediary elements. However, four of these elements, Mo, Mn, Li, and Se, have been elements of specific interest because of epidemiologic observations linking them, separately, as factors influencing tooth susceptibility to caries. Reference to the periodic table shows no interesting distribution linking the more commonly occurring 13 trace elements (10 ppm). Nine are metals and 4 nonmetals; 11 elements have atomic numbers below 40. Seven of the elements are considered essential to animal life, but six have no known physiologic function in biologic systems.

The frequency of abundance of elements in the earth's crust is compared (Table 3-4). The 13 elements are spread throughout the list of the 38 most common elements in the earth's crust and 10 within the 20 most common elements. Elements occurring in concentrations in enamel between 0.03 and 10 ppm, with just a few exceptions, are found to occur most commonly in the less abundant elements in the earth's crust. The exceptions are Ti, Mn, Zr, and V which occur among the 20 more abundant elements.

Occurrence of elements by weight and by frequency of atoms Analytically it has been most common to report the contribution of various components of enamel in proportion by weight of sample. However, the chemical and crystallographic properties of enamel will be dependent on the number of atoms of an element rather than on the weight of element present in the apatite. Heavier atoms will, according to weight, appear to contribute more to enamel composition than lighter atoms where similar numbers of atoms are present. A ranking in terms of micromoles (or "gram atoms") is easily compiled from the ratio of the median weight to the atomic mass number because the relative atomic weights of elements have the same number of atoms.

A listing of elements based on the frequency of occurrence of atoms is closely allied to the listing by concentration by weight of the elements found in enamel. There are, however, several significant changes in relative order. Thus Li is more common by frequency than by weight (14th instead of 25th) while Pb occurs a little less frequently than its weight values suggest (17th instead of 14th). The trace element Be is more common by frequency than by weight (19th instead of 29th) and Sn and Ag are marginally less common by frequency than by weight.

The points of interest here are the increased importance of Li and the decreased potential importance of Pb when considered on the relative numbers of atoms of the elements in enamel lattice. It may be that Li is more important than its weight contribution to enamel would suggest. However, taken overall, the presentation of elements by weight does not appear to obscure the relative contribution of element distribution in the

Table 3-4
Average Abundance of Elements in the Earth's Crust (ppm)

	Element	Crust		Element	Crust
1.	0	46.4×10^4	36.	Pb	12.5
2.	Si	28.2×10^4	37.	B	10
3.	Al	8.1×10^4	38.	Th	8.5
4.	Fe	5.4×10^4	39.	Pr	8
5.	Ca	4.1×10^4	40.	Sm	7
6.	Na	2.4×10^4	41.	Gd	7
7.	Mg	2.3×10^4	42.	Dy	6
8.	K	2.1×10^4	43.	Er	3.5
9.	Ti	5000	44.	Yb	3.5
10.	H	1400	45.	Be	3
11.	P	1100	46.	Cs	3
12.	Mn	1000	47.	Hf	3
13.	F	650	48.	U	2.7
14.	Ba	500	49.	Br	2.5
15.	Sr	375	50.	Sn	2.5
16.	S	300	51.	Ta	2
17.	C	220	52.	As	1.8
18.	Zr	165	53.	Ge	1.5
19.	Cl	130	54.	Mo	1.5
20.	V	110	55.	Ho	1.5
21.	Cr	100	56.	Eu	1.2
22.	Rb	90	57.	W	1.2
23.	Ni	75	58.	Tb	1
24.	Zn	70	59.	Tl	0.8
25.	Ce	70	60.	Lu	0.6
26.	Cu	50	61.	Tm	0.5
27.	Y	35	62.	Sb	0.2
28.	La	35	63.	I	0.2
29.	Nd	30	64.	Cd	0.15
30.	Co	22	65.	Bi	0.15
31.	Li	20	66.	In	0.06
32.	N	20	67.	Ag	0.07
33.	Sc	20	68.	Se	0.05
34.	Nb	20	69.	Hg	0.02
35.	Ga	18	70.	Au	0.003

enamel crystal lattice—with the exception of Li. Because the results of the great majority of studies are expressed as concentration of elements by weight of enamel the data in this review present and discuss the studies on the basis that the relative significance of elements with few exceptions is unchanged from that based on relative frequency of atoms.

Elements not detected in enamel or at very low concentrations (< 0.01 ppm) There are 23 elements that have not been reported as detected in whole enamel even at very low concentrations. Seven other elements have been quantitated in only one study and at very low levels. The elements Ce, Pr, Nd, and Y were reported by Losee, Cutress, and Brown (1974a) at median values of 0.07, 0.03, 0.05, and 0.01 ppm; Pr and Nd were found in 24 of 28 samples, and Y in only 2 of 28 samples.

In one study using highly specific analytic and separation techniques, developed for detection of rare earths, Sc, Tb, and Yb were reported (Steinnes et al 1974). Concentrations of each element were much less than 0.01 ppm. Total rare earth concentrations register low on the list of abundance of elements of the earth's crust. All but six (Ce, Y, Nd, Sc, Ga, and Pr) are listed in the 40 less abundant elements (Table 3-4). Certainly the 30 less common trace elements contribute not more than 2 ppm by weight of enamel.

Elements detected but occurring at concentrations below 1 ppm Twenty-two elements have been identified in studies with a median concentration of less than 1 ppm. Seven of these elements, only occasionally detected or detected at concentrations less than 0.1 ppm, have been referred to previously.

Lanthanum (La) Like other elements of the lanthanide series this element has only occasionally been reported in enamel and at very low concentrations. Losee et al (1974b) detected La by mass spectrometry in only 1 of 56 samples in the concentration range 0.02 to 0.1 ppm. Steinnes et al (1974) in a neutron activation study restricted to analysis of lanthanide elements in five enamel samples, reported La within the concentration 0.004 to 0.008 ppm, and it should therefore be considered a contaminant.

Gold (Au) This noble metal was detected by neutron activation techniques in several studies (Söremark and Samsahl, 1961; Lundberg, Söremark, and Thilander, 1965a; Hardwick and Martin, 1967; Retief et al, 1971) in the range 0.02 to 0.1 ppm; two studies indicated a mean value of 0.06 ppm but Losee et al (1974b) failed to detect Au by a mass spectrometry technique with a sensitivity not less than 0.02 ppm.

Vanadium (V) This element is always detected in studies using suitable analytic techniques. It has, however, the lowest concentration of the transitional group of metals. Mean concentrations are of the order 0.02 to 0.06 ppm with a range 0 to 0.2 ppm (Losee et al, 1974a; Losee, Curzon, and Little, 1974c; Curzon and Crocker, 1978).

Cesium (Cs) The heaviest alkali metal, Cs, has been reported in only one study (Losee et al, 1974a) as occurring in enamel. However, it was detected in only 27 of 56 enamel samples and then at a very low concentration of about 0.04 ppm, with a range of 0.01 to 0.07 ppm.

Arsenic (As) A semimetal and metabolic poison, As has been reported by several investigators although Losee et al (1974a) failed to detect As in 56 enamel samples within the sensitivity (0.02 ppm) of the mass spectrometry method used. However, using neutron activation methods, Nixon and Smith (1960) specifically studied As and reported a mean value of 0.06 ppm and range of 0.03 to 0.15 ppm for 25 teeth. Rasmussen (1974) analyzing for As levels in teeth from neolithic, Celtic, Roman, Viking, middle, and modern ages found a range of values consistently between 0.001 to 0.41 ppm (mean 0.07 ppm).

Mercury (Hg) Mass spectrometry analysis (Losee et al, 1974a) failed to detect Hg above levels of 0.1 ppm in 56 enamel samples but Nixon, Paxton, and Smith (1965), using a neutron activation technique, reported levels as high as 12 ppm in enamel with the mean value for 40 enamel samples from sound erupted teeth being 2.6 ppm. In unerupted teeth the Hg content was much lower (< 0.1 ppm), but enamel in contact with silver amalgam fillings showed Hg levels up to 1600 ppm, as would be expected.

In another analysis Rasmussen (1974), using neutron activation techniques on enamel from 52 ancient and contemporary teeth, found the Hg content varied between 0.001 and 1.9 ppm with a mean value of about 0.1 ppm. Losee et al (1974b) failed to detect Hg in enamel (above 0.11 ppm) using mass spectrometry analysis on 56 tooth samples obtained from various geographic locations in the United States. Riabuhik et al (1972) reported an Hg concentration of 0.03 ppm in tooth enamel collected in the Soviet Union.

Antimony (Sb) A relatively rare element in nature, Sb was reported at a median concentration of 0.1 ppm with a range between 0 and 3 ppm in six studies. In one investigation Sb was detected in half of the 56 samples. In those samples in which it was detected Sb varied between 0.02 and 0.34 ppm. Rasmussen (1974) reported a Sb content of 0.001 to 1.6 ppm in 52 enamel samples obtained from ancient and contemporary teeth.

Nixon, Livingston, and Smith (1966) compared tooth enamel from Scottish sources with that obtained from Egyptian sources being treated with Sb therapy for bilharzia infection. They found slightly higher Sb in enamel of bilharzia subjects with the overall range being 0.005 to 0.6 ppm. The highest Sb concentrations were reported by Retief et al (1971) at a mean concentration of 0.96 ppm. Curzon and Crocker (1978) considered the Sb content of 225 teeth samples from a variety of locations in the United States and also samples from New England to be of the order 0.2 ppm.

Wolfram (W) Losee et al (1974a) using mass spectrometry analysis failed to detect the element at concentrations above 0 to 0.8 ppm in 56 enamel samples. However, Söremark and Samsahl (1961) using neutron activation analysis detected a mean content of W in 15 samples of 0.24 ppm.

Cobalt (Co) A relatively common element, Co occurs about twice as frequently in the earth's crust as Pb and B. A light metallic element, Co, however, is generally found in low concentrations in enamel. Eight recent reports indicate a median value of 0.3 ppm. There are substantial differences in the Co content of enamel reported by different investigators.

Mass spectrometry analysis of 56 tooth samples by Losee et al (1974a) detected Co above 0.01 ppm in enamel of only one tooth—the Co level being 0.26 ppm. Little and Steadman (1966) detected no Co in 300 tooth samples using emission spectroscopy. Retief et al (1971) and Söremark and Lundberg (1964) likewise reported very low concentrations of Co at 0.13 and 0.0002 ppm respectively following neutron activation analysis. However, in one study much higher levels of Co in enamel were reported by Derise and Ritchey (1974). Atomic absorption analyses of 173 teeth indicated mean concentrations of Co at 34.3 ppm. A mass spectrometry analysis of 93 enamel samples (Losee et al, 1974c) showed mean Co content of 2.3 ppm with individual tooth concentrations ranging from 0.6 to 30 ppm.

Zirconium (Zr) The 18th most abundant of elements in the earth's crust, Zr, like Co, occurs at a low concentration in enamel. The median concentration of Zr is 0.27 ppm within the range 0 to 12 ppm. Individual studies report concentrations of 0.53, 1.8, 0.27, 0.16, and 0.08 ppm (Losee et al, 1974a, 1974c; Curzon and Losee, 1977a; Curzon and Crocker, 1978).

Beryllium (Be) The lightest of the alkaline earth metals, Be occurs only in traces (3 ppm) in the earth's crust. In enamel it has been found at concentrations of 0.09, 0.31, and 1.36 ppm (Losee et al, 1974a, 1974b; Curzon and Crocker, 1978). The range covers 0 to 15.9 ppm. In Losee's study (1974a) Be was detected only in 10 of 56 enamel samples.

Selenium (Se) Of increasing interest as a possible anticancer element, Se is a rare element (about the same as Au, Ag, Hg) in the earth's crust. However, it has attracted particular interest in dentistry because of reports linking it with caries prevalence. Mass spectrometry (Losee et al, 1974a; Curzon et al, 1975; Curzon and Crocker, 1978), neutron activation (Nixon and Myers, 1970; Retief et al, 1974, 1976), atomic absorption (Derise and Ritchey, 1974), and volumetric analyses (Hadjimarkos and Bonhurst, 1959) generally agree on the mean concentration of Se in enamel at about 0.40 ppm.

Retief et al (1974, 1976) have twice reported the lowest Se concentration in enamel samples in South African whites at 0.08 ppm and South

African Bantus at 0.1 ppm. The highest mean levels were reported by Curzon and Crocker (1978) for 326 teeth from the United States and New Zealand at 1.47 ppm.

Niobium (Nb) Analysis for this element has been by mass spectrometry by the Losee-Curzon group (1974, 1978) which reported 0.31 ppm for 56 tooth samples and 0.17 ppm for 245 samples respectively. The range appears quite small at 0 to 1.0 ppm.

Silver (Ag) This is a rare element in the earth's crust (0.07 ppm). Several recent studies report mean Ag values ranging from 0.005 ppm (Söremark and Lundberg, 1964) to 7.43 ppm (Curzon and Losee, 1977b). The median value is about 0.7 ppm. The highest value reported is 37 ppm. The true values in enamel may well be affected by the widespread use of silver amalgam.

Lithium (Li) The lightest of the alkali metals, Li is moderately common in the earth's crust (31st most common, about the same as N). Dental interest in Li has increased in recent years following reports linking it to caries prevalence (Schamschula et al, 1978). It is a regular but very minor component of enamel with mean values variously reported ranging from 0.07 to 1.6 ppm (Curzon and Losee, 1977a; Losee et al, 1974a). The overall individual tooth variations are between 0 to 13 ppm, with a median value of about 0.9 ppm. Because of its very light atomic weight Li occurs more frequently in enamel than its concentration by weight would indicate. It is about the 14th most frequently occurring element in enamel.

Nickel (Ni) This element is one of the more common transition metals that are usually detected in enamel. Reports place the concentration of Ni in enamel between 0.2 ppm (Curzon et al, 1975) and 1.56 ppm (Losee et al, 1974c) with a median concentration of about 0.6 ppm. Orthodontic wire and oral appliances may contribute to levels of the element in teeth, by contamination of the enamel surface.

Elements occurring at concentrations between 1 and 10 ppm

Tin (Sn) Ranked the 50th most abundant element in the earth's crust, Sn occurs at about 1 ppm in enamel. Studies using mass spectrometry analysis reported the Sn content of enamel as 0.5, 0.6, 0.2, 1.4, and 1.6 ppm (Losee et al, 1974a; Curzon et al, 1975; Curzon and Losee, 1977b; Curzon and Crocker, 1978). Brudevold and Steadman (1956a), using emission spectroscopy, reported Sn concentrations at 2.0 ppm. The variation between enamel samples is of the order 0.03 to 93 ppm, which is probably related to the use of stannous fluoride (SnF_2) in topical fluoride applications.

Chromium (Cr) Although it is a commonly occurring element of the earth's crust, Cr cannot always be identified in enamel. However, the median value found from five studies indicates mean enamel concentrations of about 1 ppm. Individual studies agree on the low concentrations at 1.0, 2.3, 6.4, and 0.5 ppm (Retief et al, 1971; Losee et al, 1974a; Losee et al, 1974b; Curzon and Crocker, 1978), but tooth samples have been reported with Cr concentrations between 0 and 33 ppm, while Söremark and Lundberg (1964a) found a particularly low level of Cr at 0.004 ppm from analysis of 15 teeth.

Manganese (Mn) One of the most studied minor elements of enamel is Mn. Apart from the relative ease of analysis, interest in Mn concentration and variation in enamel was increased by epidemiologic work which showed an association between increased Mn in enamel and decreased caries susceptibility. The role of Mn in enzymatic activity in calcification processes has heightened this interest.

In the 11 more recent enamel studies the Mn content is of the order of 1 ppm; mean values of the studies range between 0.3 (Losee et al, 1974a) to 7.0 (Derise and Ritchey, 1974) ppm. Individual tooth analysis indicated a Mn variation of 0.6 to 63 ppm. Mass spectrometry, neutron activation, atomic absorption, and emission spectroscopy have all been used for Mn analyses and, apart from atomic absorption, which has indicated a mean concentration of 7 ppm, the studies agree on the mean concentration of Mn in enamel and also indicate that Mn concentrations are influenced by the geographic source and age of the enamel.

Titanium (Ti) Apart from recent interest in titanium tetrafluoride (TiF_4) as an anticaries agent, the element Ti has not been of significance in dentistry. Data from four studies give a median concentration of 1.5 ppm. Mean concentrations of the various studies are 0.5 ppm, 4 ppm (only 29 of 56 samples detected Ti), 0.9 ppm, and 1.9 ppm. Individual enamel samples varied between 0 and 66 ppm of Ti (Losee et al, 1974a; Curzon and Crocker, 1978).

Iodine (I) An essential trace element for man and the heaviest of the stable halogens, I, like others in the group, is a regular constituent of enamel but in low concentrations. According to the small number of studies reporting the element, I occurs at about 2 ppm; Losee et al (1974a) reported 0.04 ppm; Derise and Ritchey (1974), 5.6 ppm; Curzon and Crocker (1978), 2.0 ppm.

Cadmium (Cd) An element similar to Zn in properties and structure, Cd is only about 1/1000 as abundant as Zn in nature, and its concentration in enamel reflects this. Although Cd in enamel has been well documented in various studies there is not a great deal of evidence to consider it relevant in dentistry. Curzon et al (1977) studied the association between Cd and caries and concluded that a positive association existed beween them, although it was not possible to exclude the influence of other metals.

The reported mean values for Cd vary widely between 1×10^{-4} ppm (Lundberg et al, 1965a) by neutron activation analyses and 14.9 ppm (Losee et al, 1974c) by mass spectrometry. Some of this variability may be attributed to geographic variation and environmental pollution, and some to analytic technique. Variation in Cd between teeth was wide even within studies, as, for example, the range of 0.4 to 268 ppm reported by Losee et al (1974c).

The three mass spectrometry analyses of enamel involving 427 teeth indicate a mean value of about 2 ppm Cd with a range of 0.03 to 27 ppm, whereas the other major study (Losee et al, 1974c) of 93 teeth by optical emission analysis indicated about 15 ppm. A low value of 1×10^{-4} ppm Cd was found in unerupted teeth.

Molybdenum (Mo) A transitional element of the Cr group which has been cited as an essential element for mammals and an influence on caries susceptibility in man, Mo has had considerable attention in dental research (see Chapter 6). Despite several epidemiologic studies exploring the Mo–caries relationship, data on the Mo content of enamel are limited to a few studies. The concentration range of Mo in enamel is small, according to reports, with mean values varying between 1.0 ppm and 5.5 ppm. Individual tooth samples vary between 0 and 39 ppm (Losee et al, 1974a; Curzon et al, 1975).

Bromine (Br) This element is a nonessential halogen which occurs at about 4 ppm in enamel. The natural frequency of the halogens relates directly to their relative frequency of occurrence in enamel. Only the study of Retief et al (1971) reports a level of Br in enamel as high as 34 ppm (in teeth of South Africans) which is exceptional compared with studies from Sweden, United States, and New Zealand, which report mean concentrations of 4, 3.4, and 4.5 ppm (Rasmussen, 1974; Curzon et al, 1975; Curzon and Crocker, 1978).

Rubidium (Rb) An alkali metal of moderate abundance (22nd in the earth's crust), Rb appears to be a normal constituent of enamel at 4 to 5 ppm mean value. The concentrations of Rb and other alkali metals are associated directly with their respective abundance in the earth's crust. The mean values reported from five studies vary between 0.4 and 4.9 ppm for erupted teeth and a very high 73 ppm for (ten) unerupted teeth (Söremark and Lundberg, 1964a; Lundberg et al, 1965a; Losee et al, 1974a; and Curzon et al, 1975).

Copper (Cu) It is perhaps not surprising that Cu, one of the earliest metals discovered by man, has been well studied for its occurrence in enamel. It is far more common in the earth's crust (50 ppm) than are other members of the transitional subgroups Ag and Au. A median of reported mean values indicates a concentration of about 7 ppm in enamel with surprisingly moderate variations considering that analytic methods vary between mass spectrometry, emission spectroscopy, neutron activation, atomic absorption, and voltametry. In addition, the widespread use of copper piping and utensils might be expected to influence significantly oral and systemic Cu levels. Enamel from teeth originating in Sweden, Scotland, United States, Finland, Norway, and New Zealand varied in Cu content from 0.2 ppm to 10 ppm. Individual enamel specimens have been quoted between 0 and 208 ppm (Derise and Ritchey, 1974; Losee et al, 1974c; Curzon et al, 1975). Nixon and Smith (1962) specifically analyzed for Cu (using neutron activation) and estimated an overall mean value of about 10 ppm for 100 samples. Of the samples 60% had Cu concentrations of between 0 and 10 ppm. Only 10% of samples had concentrations exceeding 20 ppm. Losee et al (1974a) reported a mean value of about 7 ppm for 56 samples although Curzon and Crocker (1978), also using mass spectrometry analysis, reported a lower value of 1.5 ppm.

Elements occurring at concentrations between 10 and 100 ppm

Lead (Pb) The heaviest element by far to occur in appreciable quantities in enamel, Pb is the highest numbered element in the periodic table to be found routinely in teeth. Its compounds are poisonous, significant not because of the likelihood of ingesting a toxic single dose but principally because Pb tends to accumulate in the body's central nervous system and to interfere with metabolic and enzymatic processes. Thus, Pb has been studied in teeth with much interest because of the concentrations in teeth as a potential indicator of environmental exposure by ingestion and absorption.

Estimates of Pb in enamel vary considerably with the median value of mean values being 14 ppm by weight. Because of its high atomic weight, ranking of the element by weight gives it more prominence than when it is ranked by the frequency of number of atoms (gm atoms). Thus, by weight, Pb is the 11th most common minor element whereas by atomic frequency it is 15th, suggesting less prominence than Li, Cu, Ba, and B.

Individual studies have reported low Pb concentrations of 3 to 6 ppm (Losee et al, 1974a; Curzon et al, 1975; Attramadal and Jonsen, 1978; Oehme and Lund, 1978); moderate values from about 20 to 40 ppm (Malik and Fremlin, 1974; Curzon and Crocker, 1978); and a high value, 180 ppm (Losee et al, 1974c). Individual teeth have been reported as containing between 0 and 1000 ppm, although most studies report much lower maximum values of from 140 ppm to 156 ppm. Several reports indicate much less variability between tooth samples, 1 to 6.5 ppm, 1.8 to 4.9 ppm (Oehme and Lund, 1978). The presence of Pb in enamel is discussed fully in Chapter 21. Surface enamel concentrations have been particularly studied and are reported on, as has Pb in dentin and primary teeth.

Boron (B) The lightest nonmetallic element detected in enamel, B has received some special attention because of potential influence on caries susceptibility. Analyses agree quite closely on the mean value of B in enamel as 18, 20, 12, and 8 ppm, the median concentration being 15 ppm. Individual tooth ranges appear considerable, 0 to 190 ppm, 22 to 141 ppm, 6 to 69 ppm (Losee et al, 1974a, 1974b; Curzon et al, 1975; Curzon and Crocker, 1978).

Barium (Ba) An alkaline earth ranked as the 15th most abundant element in the earth's crust, Ba is one of four chemically similar elements (Mg, Ca, Sr, and Ba) of a group which have significance or potential significance in dentistry. The median value of reported mean values in several countries is 15 ppm. Retief et al (1971) using neutron activation analysis found 125 ppm Ba in South African teeth whereas all other investigators reported mean values below 20 ppm. Individual tooth enamel values vary, for example, between 0 and 510 ppm (Curzon and Crocker, 1978) and from 4 to 44 ppm and 0.8 to 17 ppm by Losee et al (1974c) who also noted geographic and age influences on Ba concentration in enamel.

Aluminum (Al) The most abundant metallic element, Al is ranked the third most abundant element in the earth's crust. There is interest in Al in dentistry because of a suggested association with caries prevalence and its strong complexing properties with F.

Substantial data on Al concentration in enamel are available from several studies. The median value of mean concentration reported is 23 ppm. High concentrations of between 70 and 120 ppm (Lakomaa and Rytömaa, 1977; Retief et al, 1971; Derise and Ritchey, 1974; Losee et al, 1974c) and low concentrations between 8 and 23 ppm (Losee et al, 1974a; Curzon et al, 1975; Curzon and Losee, 1977b; Curzon and Crocker, 1978) have been reported. Individual enamel samples have on analysis been found to vary in Al content between 0 and 510 ppm, 8 and 325 ppm, and 1.5 and 54 ppm.

Iron (Fe) The fourth most abundant element in the earth's crust and second most abundant metal and an essential element for mammals, Fe has been reported in widely varying levels in enamel with a median value at 28 ppm. Mean values from separate studies vary between very high 338 ppm (Söremark and Lundberg, 1964a), high 118 ppm (Retief et al, 1971), moderate 64 ppm (Losee et al, 1974c), and low 30 ppm (Lakomaa and Rytömaa, 1977). Individual tooth enamel has been found to have an Fe content of between 0 and 157 ppm, 11 and 759 ppm, and 0.8 and 26 ppm with at least part of this variation related to geographic and age factors. The concentration of Fe in foods and water varies widely and this must be a major factor in the wide variation reported for the element in enamel.

Elements occurring at concentrations between 100 and 1000 ppm

Strontium (Sr) Another of the alkaline earth group, Sr is of increasing interest in dentistry because of strong indications that it influences susceptibility to caries. Many studies of an epidemiologic nature have in recent years investigated correlations between Sr in water and teeth, and caries prevalence. The alkaline earths Sr and Ca are closely allied chemically and in biologic activity. Their absorption and deposition in calcified tissues are well recognized, and Sr can substitute for Ca in apatite structures. This element is considered in detail in Chapter 15.

Early screening analyses using mass spectrometry analysis identified the Sr content of enamel between 10 and 1000 ppm (Calonius and Visäpäa, 1965; Hardwick and Martin, 1967). More precise analyses have reported mean Sr values of between 70 and 100 ppm (Söremark and Samsahl, 1961; Lundberg et al, 1965a; Curzon et al, 1975) and 286 ppm (Derise and Ritchey, 1974) with a median value of about 115 ppm. Most investigators report mean values between 100 and 200 ppm. Individual tooth content of Sr appears to vary widely between 13 and 1400 ppm (Curzon and Crocker, 1978), 21 and 280 ppm (Losee et al, 1974a), and 80 and 181 ppm (Nixon and Helsby, 1976).

The Sr content of enamel is dependent on geographic factors with water probably being the most important influence (Nixon and Helsby, 1976; Spector and Curzon, 1978); it also varies by age. Although the apparent mineral correlation of Sr in teeth and drinking water has been studied, it has not been substantially proven that the element is cariostatic because of the difficulty of distinguishing the effect of Sr from that of F.

Silicon (Si) Si is the second most common element in the earth's crust, yet surprisingly the significance and occurrence of Si in teeth have received little attention. Two recent studies (Losee et al, 1974a, 1974c) from one laboratory using mass spectrometry and optical emission spectroscopy reported mean values of 243 ppm (56 teeth) and 136 ppm (93 teeth) respectively. The variation between samples was high, 100 to 450 ppm, and 26 to 1155 ppm. A median value for these studies is 190 ppm.

In an earlier study, Steadman et al (1959) included analysis for Si in ancient Indian teeth. Emission spectrometry of enamel chips revealed Si concentrations of 15 to 33 ppm for Pueblo Indians dated 1000 AD and Knoll Indians dated 3000 BC. However, the number of teeth analyzed was relatively small, and variability was not given. Hardwick's (1967) semiquantitative study also placed Si in enamel in the range of 100 to 1000 ppm.

Zinc (Zn) A metal ranked as the 24th most common element in the earth's crust, Zn is considered an essential trace element in mammals. The ease of analysis and relatively high concentration in teeth have made this a well-studied element in enamel. A median value from ten recent studies indicates an enamel content of 210 ppm. Most studies report similar mean values regardless of methods used (atomic absorption, mass spectrometry, neutron activation analyses), ranging from 126 to 276 ppm. An exceptionally high value using voltametry indicated 740 ppm. In ancient teeth Zn ranged from 100 to 700 ppm. Individual enamel samples showed wide variation from 10 ppm to 1197 ppm (Curzon and Crocker, 1978) and 1500 ppm (Attramadal and Jonsen, 1978).

Sulfur (S) An essential trace element in man, S plays a varied biologic role. A nonmetallic element much less common than O, it is the only other element of importance in group VI.

Most of S in enamel appears to be present as sulfide in the organic phase. Despite its essential role in animals the S content of teeth has not been well reported. Three reports, from mass spectrometry screening analyses, found mean values of 294, 214, and 24 (Losee et al, 1974a; Curzon and Crocker, 1978; Curzon et al, 1975). Individual enamel samples range from 0 to 560 ppm. Hardwick's (1967) semiquantitative analyses placed the S content of enamel between 10 and 100 ppm. The possible role of S in dentistry is considered in more detail in Chapter 12.

Potassium (K) Like Na, K is a commonly occurring alkali metal in the earth's crust—ranked eighth of all elements. Another element found in all enamel at relatively high concentrations, K has not been subjected to extensive investigation. Brudevold and Söremark (1967) in their review suggested a range of concentration of K in enamel between 0.05 and 0.03%. Derise and Ritchey (1974) analyzed 175 samples by atomic absorption and reported a concentration of 0.03% while Lakomaa and Rytomaa (1977) reported 0.02% and Haataja et al (1972) 0.04%.

Elements occurring at concentrations exceeding 1000 ppm Three elements, Na, Cl, and Mg, form a distinct small group of enamel components which occur at relatively high concentrations. Table 3-2 clearly demonstrates that these elements stand out from others by nature of their high concentration in weight and atomic frequency in the apatite lattice, in the same way that elements occurring between 10 and 100 ppm stand apart from other elements. However, the elements are all essential for man and occur commonly in the earth's crust, Na and Mg being ranked 6th and 7th and Cl 19th.

Sodium (Na) The major contributing element to the minor composition of enamel, Na is found at a level of about 7100 ppm. It is more than twice as common, by weight, as the next common minor element, Cl. In contribution to the apatite lattice Na is even greater, about three times as much as Mg and Cl.

Mass spectrometry and atomic absorption studies show mean concentration of Na in enamel at about 7100 ppm with individual studies reporting between 6900 ppm and 11600 ppm—all by neutron activation analysis (Söremark and Samsahl, 1961; Lundberg et al, 1965a; Retief et al, 1971; Derise et al, 1974; Lakomaa and Rytömaa, 1977). Frostell et al (1977) using an ion probe profile, demonstrated a Na gradient of 2 to 3000 ppm.

Magnesium (Mg) A regular component of enamel, Mg is found at a level of about 2800 ppm. Reports vary between mean values of 1670 ppm and 3900 ppm (Losee et al, 1974b; Derise et al, 1974) from mass spectrometry analyses of 28 and 175 teeth respectively, to 2800 ppm

(Retief et al, 1971; Lakomaa and Rytömaa, 1977) for neutron activation (8 teeth), and atomic absorption analysis (25 teeth). An unusually high value of 8000 ppm (estimated) from concentration profiles was reported by Frostell et al (1977) using an ion probe technique.

Chlorine (Cl) As with other halogens, Cl is a regular component of enamel at a level of about 3200 ppm and is the most commonly occurring halogen. The reported differing mean values range from 720 ppm by Lakomaa and Rytömaa (1977) of 25 teeth to 6500 ppm by Söremark and Samsahl (1961) of 15 teeth, both studies using neutron activation analysis. Others using this technique reported 2800 ppm (Derise et al, 1974) and 3200 ppm (Retief et al, 1971), while Losee et al (1974b) reported a Cl content of 6022 ppm for mass spectrometry analysis. Frostell et al (1977) in an ion probe profile clearly showed a decreased level from the enamel surface, but an estimate from the profile indicated an average level of 2000 ppm Cl.

Rare earths and human dental enamel There is some confusion about what a rare earth is. The term *rare earth* is misleading since the rare earths are not earths, but metals, and they are not really rare. The Commission on Nomenclature of the International Union of Pure and Applied Chemistry recommends that the term *rare earth metals* may be used for Sc (No. 21), Y (No. 39), and elements 57 through 71. The Commission further recommends that elements 57 through 71 also be known as the "lanthanum series." Although the occurrence and concentration of rare earths in enamel have not been widely studied (Table 3-3), it is evident that they are present at very low levels, mostly at the limits of detection. In a semiquantitative mass spectrometric study Hardwick and Martin (1967) determined Y in enamel in the range 1 to 10 ppm and La at less than 1 ppm. A more specific study of rare earths (Steinnes et al, 1974), using neutron activation, showed that La, Sm, Tb, and Yb were present in very low concentrations, indicating a total rare earth content of the order of 0.1 ppm or less in enamel of the five teeth analyzed.

Losee et al (1974b), using a spark source mass spectrometry screening analysis for 64 elements, have provided most of the limited information available on the rare earths in dental enamel. Individual analyses of whole enamel samples obtained from people raised in widely differing geographic areas were found to contain only the rare earths Ce, Pr, and Nd in the majority of samples. The mean concentration of the three elements were of the order 0.007 to 0.07 ppm with no maximum values exceeding 0.2 ppm. The element Y was detected in only 2 of the 56 samples and all other rare earths, if present, were below their respective detection limits by this method. The detection limits vary for different elements but for the rare earths the limits were between 0.1 and 0.02 ppm.

Dental interest in the rare earths is minor, but interest has developed from observations that rare earths occur in the apatite phase of fish bone

(Hogdahl, Melsom, and Bowen, 1968) and that some rare earths compounds can reduce the solubility of tooth enamel (Manley and Bibby, 1949; Shrestha, Mundorff, and Bibby, 1977). Of the 13 elements making up the *true* rare earth series only three have been detected in human enamel. Interestingly these three, Ce (No. 58), Pr (No. 59), and Nd (No. 60) are the three lightest of the rare earth series and among the most abundant in the earth's crust. Also they are the heaviest elements found rather consistently in enamel except for Pb (No. 82).

Variation of Trace Elements Within Enamel

The stability and reactivity of the exposed surface of teeth with the oral environment are important factors in maintaining the integrity of the teeth. The "resistance" of the enamel surface may determine the susceptibility of a tooth to caries, erosion, and staining, particularly as caries is essentially a surface and immediate subsurface phenomenon. Early studies identified certain changes in properties of enamel with distance from the tooth surface; these included changes in chemical composition of the minor elements. Interest in the chemical composition of enamel has been particularly concentrated on the influence of F since it was observed that F was much higher in surface enamel than deeper enamel. In addition, the recognition of posteruptive uptake or loss introduces the prospect of enamel susceptibility changing with the influence of the environment such as food and water.

Analytic problems pose a major obstacle to determination of trace elements in surface enamel. The sampling of enamel to a specified depth is difficult; sample size restricts accurate analysis to a few elements with confidence or a semiquantitative analysis of a wider range of elements. In most studies the depth (or layer) of enamel analyzed, the method of layering, and even the method of analysis have produced data which are not easily comparable. For instance, the most extensively studied minor element, F, is much higher in the outer enamel layer than in deeper layers. The marked difference with depth of F has been shown to a lesser degree with other elements. Hence, in trying to evaluate the role and significance of trace elements in enamel, the best information is probably that which shows the concentration profile of an element through the complete depth of enamel.

In all probability the most significant layer of enamel is that from the surface to 30 to 50 μm. Within this layer have been recorded: the subsurface initial caries lesion, the changes in density, mineral content, and the larger changes in trace element concentration. Therefore, before considering the variation of trace elements concentration with depth, it is of great interest to compare the composition of the outer 50 μm with whole enamel.

The literature on this subject is not extensive. Most information is confined to analyses of small groups of elements and only one study, Cutress (1979), has attempted a full screening of the trace element content of the outer enamel layer. The results of this particular study must be considered semiquantitative rather than quantitative.

The thickness of outer enamel layers analyzed for trace elements varies: very thin, 2.7 μm to 3.6 μm (Brudevold et al, 1977), obtained using in vivo biopsy technique; moderately deep layers, 15 to 50 μm which include concentration profiles in vitro across section of teeth (Frostell et al, 1977) to an estimated depth of 15 μm; in vitro acid etch of surface enamel to a depth of 30 μm and 42 μm (Cutress, 1972, 1979) on teeth from known sources; in vivo biopsy acid etch on teeth of primitive tribes with varying prevalence of caries (Schamschula et al, 1978). Data on thicker outer layers of 90 μm to 200 μm have also been reported (Little and Steadman, 1966; Malik and Fremlin, 1974).

A comparison of concentrations of elements on whole and surface enamel has been made by Schamschula et al (1978). Tabulated reports of various elements estimated as being present in the outer layer of enamel show much variation due at least in part to differences in the thickness of the enamel layer samples. However, geographic variation in tooth source, age of tooth source, and method of analysis also contribute to the variation in element concentration. Certainly for some elements such as Sr and Zn the geographic source of the teeth samples contributed significantly to the concentration of the elements occurring in these outer enamel layers. Data on the elements in the outer 50 μm are available for 27 elements, and several other elements have been analyzed but not detected. With a few exceptions (namely, Mg, Zn, and Sr) quantitative analysis is meager for the 27 elements which, taking into consideration the techniques and other factors influencing analysis, prevents a confident statement of their concentration range.

The Sr content of surface enamel has been the most widely studied of the trace elements with mean values showing fairly good agreement 120, 204, 239, 100, 225, 200, and one low value 67, giving a median value of all studies of 200 ppm. Zn appears to occur at a median value of 900 ppm although mean values vary between 468 ppm and 2100 ppm. Mg occurs at about 1650 ppm with reports of concentrations between 145 ppm and 4000 ppm.

Substantially different concentrations of Pb were reported by Brudevold et al (1975, 1977) from thin layer biopsy (2.7 and 3.6 μm layers respectively) methods which revealed that Pb concentrations were 1790 ppm and 2480 ppm; Cutress (1979) found 24 ppm at 42 μm and Schamschula et al (1978), 18 ppm at 16 μm depth. Brudevold and coworkers demonstrated a rapid decrease in Pb even with the outer 3 μm layer, which may explain the difference in Pb concentration noted in these studies. If so, this indicates a strong tendency for environmental Pb to be taken up into the surface of enamel.

Comparison of the median values of elements found in the outer 50 μm of enamel with whole enamel values is of interest. Median values show that the ranking order of concentrations is similar except that surface concentration of Al, Cu, Mn, and Li would place them at a higher level than given by whole enamel. The concentration of Al (at median value 343 ppm, surface layer) is much higher than for whole enamel (23 ppm). Similarly Cu is reported at 37 ppm and 282 ppm in surface enamel but only 7 ppm from whole enamel; Mn is reported at 8 ppm and 50 ppm as compared with 1 ppm in whole enamel. On the other hand, the Si concentration in the outer layer appears lower (at about 40 ppm) than for whole enamel (190 ppm).

The significance of the concentration of trace elements in surface enamel stated in isolation appears of doubtful value unless compared with a standard or reference source. In vitro analysis on the profile of concentrations with depth provides a more valuable measure of element distribution; several studies have provided this information for a few elements and some have been specifically concerned with single element analyses of Zn, Sr, Pb, Cu, Mn, Sb, and Sn.

Steadman et al (1959), using a laborious and exacting grinding technique, successfully demonstrated the variation of 11 trace elements in enamel. They reported on the variation in five enamel layers approximately 50 to 100 μm thick from pooled samples of Pueblo Indian teeth (about 800 years old). The decreased concentration of most of the seven elements with distance from the enamel surface was quite evident (Table 3-5), and the gradient appeared to differ between elements. In the same study similar analyses on teeth from the 5000-year-old Indian Knoll source demonstrated similar gradients in concentration of elements in enamel. In this latter analysis Sr, Fe, and Al were included. Although concentrations of both Fe and Al dropped from a surface level of 400 to 40 ppm and 380 to 35 ppm respectively the concentration of Sr appeared relatively similar at about 200 ppm throughout the enamel.

Frostell et al (1977), using the completely different analytic technique of ion probe microanalysis, demonstrated the gradient concentrations of several trace elements across the full thickness of enamel and dentin. This method provides a continuous scan that can be standardized to give qualitative results. However, in this study the technique was demonstrated on one tooth section only which showed little or no gradient for Na and K, a substantial decrease of Mg immediately near the enamel surface, and a steady decrease of Cl from the surface to deeper enamel whereas Sr was more or less homogeneously distributed. The potential of ion probe microanalysis seems well suited for study of element concentration in enamel but as yet the technique has not been exploited.

Table 3-5
Trace Element Concentrations in Successive Layers of Enamel (ppm)

Layer	Trace Element						
	Zn	*Si*	*Mn*	*Pb*	*Ag*	*Cu*	*Sn*
1	1300	870	90	80	25	5	1
2	1300	48	35	25	20	2	0
3	1300	25	20	23	15	3	0
4	800	32	35	13	15	2	0
5	740	32	42	14	13	1	0

After Steadman et al, 1959.

Little and Steadman (1966) in a comparison of elements in altered (carious) and sound enamel produced data of element concentration with depth. The technique for sampling was similar to that used by Steadman et al (1959), but the concentration trends found for respective elements differed. Although Zn, Pb, Cu, and Sn showed a similarity to Al and decreased with depth, Si and Sr remained about the same and Fe and Cr showed no distinctive changes with depth.

Zinc (Zn) Brudevold et al (1963) showed that for pooled enamel samples from teeth of known ages, collected from a defined geographic area, Zn concentrations decreased with distance from the enamel surface, for both unerupted and erupted teeth (Figure 3-1). For unerupted teeth Zn concentrations decreased from 1300 ppm to 610 ppm; for

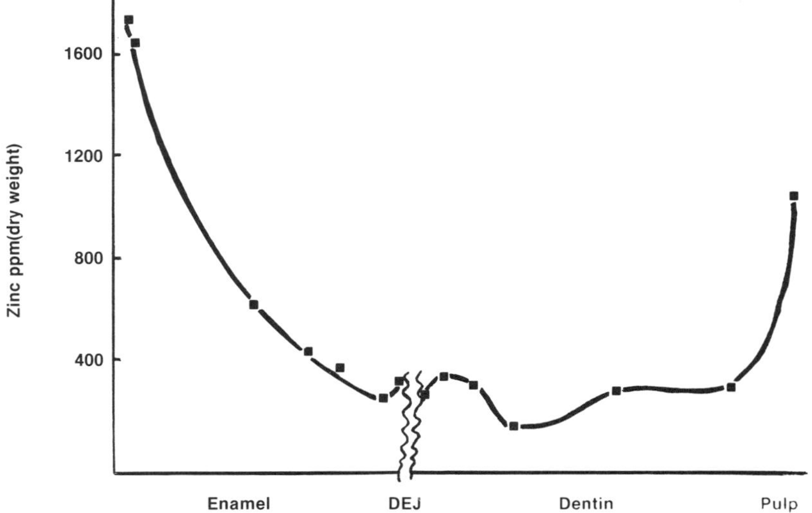

Figure 3-1 Zinc distribution in teeth from the surface enamel to the pulp in unerupted teeth (after Brudevold et al, 1963).

erupted teeth under age 20 years Zn concentrations decreased from 2100 ppm to 180 ppm and for older teeth aged 30 to 49 years the Zn change was 1500 ppm to 210 ppm. Teeth examined from other geographic areas showed similar decreasing concentration gradients.

Strontium (Sr) Using pooled enamel samples of successive layers Steadman et al (1958) found, for enamel of specific age and geographic source, no change with depth or age but a distinct difference between geographic sources.

Lead (Pb) The distribution of Pb within enamel has received more study than that of any other trace element. According to Brudevold and Steadman (1956b), Pb in both unerupted and erupted enamel shows a decrease in concentration with increasing depth of enamel following analysis of pooled ground samples of enamel (Table 3-6).

It appears that Pb is incorporated into enamel during development but that at all ages the greatest Pb concentrations are found in the outermost layer. In 1974 Malik and Fremlin, using charged particle activation analysis, produced a density profile of Pb in a tooth section. This confirmed the presence of high Pb concentration on the outer enamel layer rather than in deeper layers. The Pb concentration in the outer 200 μm layer was between 30 and 55 ppm and on the interior 20 μm layer between 22 and 52 ppm. The outer Pb levels are lower than those reported by other studies. The Pb concentration profile in individual teeth showed distinct differences.

A very steep gradient of Pb in the outer enamel layer has been reported by Brudevold et al (1977). Successive thin layers of enamel were etched for a depth of about 3 μm from teeth obtained from a community with a high level of exposure to Pb (Table 3-7). The outermost layer (0.75 μm) showed on average a surprisingly high Pb content of 4430 ppm, and at 2.8 μm this had fallen to 1250 ppm.

Table 3-6
Lead Concentrations in Successive Layers of Erupted and Unerupted Teeth

Enamel Layer	Unerupted Teeth	Erupted Teeth Age, Years			
		< 20	*20–29*	*30–39*	*> 50*
1	210	350	360	520	550
2	130	220	260	430	460
3	67	189	105	280	420
4	60	74	85	200	310
5	55	35	65	–	156
6	47	0	54	–	152

After Brudevold and Steadman, 1956b.

Table 3-7
Lead Concentrations of Enamel Biopsies Grouped by Depth

	Biopsy Depth (μm)						
	0.6 < 0.9	*< 1.2*	*< 1.5*	*< 1.8*	*< 2.1*	*< 2.4*	*2.4–3.6*
No. of samples	3	4	16	35	77	61	55
Mean depth (μm)	0.75	1.1	1.4	1.7	2.0	2.3	2.8
Mean Pb (ppm)	4430	3550	2830	2060	1810	1430	1250
SD	1440	1080	1650	1110	990	1060	810

After Brudevold et al, 1977.

Mercury (Hg) From activation analysis Nixon et al (1965) reported the Hg content of outer enamel of 40 sound erupted teeth as 2.79 ppm and ovine enamel at 2.34 ppm.

Manganese (Mn) A greater concentration of Mn in outer than in deeper enamel was reported by Nixon et al (1966), the mean value being 1.04 ppm and 0.57 ppm respectively for samples taken from the mid-region of the tooth.

Antimony (Sb) Teeth from two geographically distinct areas showed no difference in Sb content of inner or outer enamel even from teeth obtained from subjects on Sb therapy (Nixon et al, 1967).

Tin (Sn) Spectrographic analysis of successive layers of enamel ground from groups of unerupted and erupted teeth of different ages revealed little difference of Sn concentration with depth except in the outermost layer. This small change in Sn concentration was considered to be associated with uptake from the Sn component of amalgam restorations (Brudevold and Steadman, 1956a). The widespread use of SnF_2 in toothpastes and prophylaxis pastes must surely also affect the Sn content of surface enamel.

Copper (Cu) Nixon and Smith (1962) analyzed 100 enamel samples from subjects aged from 6 to 65 years and found only small differences in Cu content of outer and inner enamel layers. Outer enamel showed 9.5 ppm and inner enamel 11.3 ppm Cu. Brudevold and Steadman (1955) similarly concluded that Cu in enamel does not appear to show a significant gradient with depth of enamel. The concentrations reported for outer and inner enamel of unerupted teeth were 7 ppm and 5 ppm, and for teeth from subjects over 50 years, 20 ppm and 12 ppm respectively. This is surprising since the use of Cu piping and the availability of the Cu depending on the pH of the water supplies would be expected to bring about changes in the composition of the surface enamel.

Cadmium (Cd) Shearer, Johnson, and Desart (1980) determined the Cd gradient in enamel using layer-by-layer etching and found a fairly constant level of the element in successive layers from the surface to the dentin enamel junction, 1.87 ppm in the outer 100 μm and 1.09 ppm at 600 μm. They concluded that most of the enamel Cd was deposited during tooth development.

Variation with Age

Enamel samples from teeth of various ages show a different distribution of some elements. The most significant studies are those concerned with element concentrations at different enamel depths in which it was observed that the outermost enamel layer showed changes with age.

Because change with age could be expected to be associated mostly with environmental influence, it follows that variations which occur reflect cumulative environmental factors.

Brudevold (1957) reviewed changes in enamel with age and remarked that the clinical observations of darkening and variation in the natural color of teeth of different individuals with age may be related to the concentration of metallic ions in enamel or dentin. However, despite several studies that have considered change of elemental concentration with age, the conclusions are indeterminate with the exception of those for Pb.

The concentration of Pb in enamel has been found to increase with age based on analyses of both whole and surface enamel. Brudevold and Steadman (1956b) reported Pb present in enamel from unerupted and erupted teeth, the Pb being in higher concentration in the outer enamel layer and increasing correspondingly with age (Table 3-6).

Little and Steadman (1966), although reporting lower concentrations of Pb than Brudevold and Steadman, also found concentrations higher for people over 30 years of age than for those under 30, for both outer and inner layers of enamel. Derise and Ritchey's (1974b) study of whole enamel of individuals aged between 10 and 24 years found only a gradual increase with age—about 44 ppm compared with about 49 ppm for those over the age of 25 years. Malik and Fremlin (1974) found Pb concentration in whole enamel of 28 ppm at 19 years and 53 ppm at 57 years. Enamel biopsy data (Brudevold et al, 1977) from children in a community with high exposure to Pb reported no increase in Pb among 9- to 12-year-old children.

Variations by age and geographic origin As an individual ages, changes occur in the composition of the tooth enamel. Various factors cause this, such as wear, food composition, smoking, and cultural habits which affect food composition. The major changes occur at the enamel surface where elements, mainly trace elements, are selectively taken up by the enamel from the oral environment. At the same time the enamel will be worn away as a result of chewing and tooth brushing. In some racial groups other factors such as betel nut chewing will also play a role. Thus, changes in surface enamel composition affect total enamel composition and reflect cumulative environmental variables.

This subject has been previously reviewed by Brudevold (1957) and our knowledge has changed little since that time. The main interest was for Pb, no doubt because of the use of teeth as a record of toxicologic studies on Pb poisoning. Little and Steadman (1966) originally noted Pb enamel concentrations much higher in people over the age of 30 than in younger subjects. Malik and Fremlin (1974) found Pb concentrations in whole enamel of 28 ppm in 19-year-old persons compared with 53 ppm in the teeth of persons 57 years old.

The concentration of Sn does not appear to change with age

(Brudevold and Steadman, 1956a; Little and Steadman, 1966; Nixon and Smith, 1962), except when exposed to Ag-Sn amalgam. Other elements studied and reported not to change with age are Mg, Fe, Si and Cr, Hg, and Sb (Derise et al, 1974; Derise and Ritchey, 1974; Little and Steadman, 1966; Nixon et al, 1965, 1967).

The metal Mn was reported unchanged with age by Little and Steadman (1966) and Nixon et al (1966), but a slight decrease with age was noted by Derise and Ritchey (1974), also a significant posteruptive decrease by Battisone et al (1967). The latter study also found a decrease of Sn in enamel with age.

Change in concentration in enamel with age is not apparent for most other elements, at least for those studied. Although Zn levels in unerupted teeth are lower than those in erupted teeth, little or no change is noted after tooth eruption. This appears to be true also for Al (Derise and Ritchey, 1974; Steadman et al, 1958; Little and Steadman, 1966).

Evaluation of data for 23 elements leads to the conclusion that the concentrations of 15 elements in enamel do not change with age; four elements (Sr, Mn, Zn, Cu) decrease, and five elements (Pb, Na, K, Co, Se) increase (Table 3-8).

Table 3-8
Summary of Trace Element Changes in Enamel with Age

No Change	Decrease	Increase
Sr,[1] Al,[1,2] Si[1]	Sr[2]	Pb[1,2,13,14]
	Mn[2,10]	Na[11]
Cr,[1] Hg,[3] Sb[4]	Zn[2]	K[11]
	Cu[1]	Co[2]
Ag,[1] Mn,[1,5] Fe[1,2]		Se[2]
Sn,[1,6] I,[2] Zn[1,7,8]		
Cu,[9,2] Mg,[10] Se[11]		

1. Little and Steadman, 1966
2. Derise and Ritchey, 1974
3. Nixon et al, 1965
4. Nixon et al, 1967
5. Nixon et al, 1966
6. Brudevold and Steadman, 1956a
7. Brudevold et al, 1963
8. Steadman et al, 1958
9. Brudevold, 1957
10. Battisone et al, 1967
11. Derise et al, 1974
12. Hadjimarkos and Bonhorst, 1959
13. Brudevold and Steadman, 1956b
14. Malik and Fremlin, 1974

Trace Element Variation with Geographic Source

With the discovery that the geographic distribution of F in drinking water was associated with the prevalence of caries, it followed that other elements present in the diet and environment might influence tooth enamel composition. Many studies have directly or indirectly provided data on the trace element content of tooth enamel from various geographic sources. Although the whole range of elements has not been sufficiently studied to identify the effect of their availability on pre- or post-eruptive uptake into enamel, it is quite evident that the composition of enamel for some elements is much influenced by geographic origin of the tooth. In this category are Zn, Sr, Pb, and Mn (and probably Ba and Cd), each of which has been identified in at least two separate studies as varying in concentration of enamel according to geographic origin of the tooth.

Information from single studies has also indicated that other elements in enamel vary geographically, but substantiating studies are still required. One study (Cutress, 1972) clearly showed the relatively high accumulation of several elements (F, Si, Al, Cu, Fe, Ba, and Sr) in tooth enamel from one defined geographic source. Table 3-9 is a compilation of data about elements which have been investigated in at least two independent studies concerned with geographic variations in tooth composition.

Zinc (Zn) Brudevold et al (1963) found that large amounts of Zn are acquired pre-eruptively in tooth enamel and that the amount varied widely according to the geographic origin of the teeth. The teeth were obtained from various states within the United States as well as geographically isolated areas such as Tonga, Nauru, and Greenland. Although they presented data for successive layers of enamel, the outer surface suitably demonstrated the variation in Zn content which varied from 430 ppm (Greenland) to 2100 ppm (Maine).

Table 3-9
Summary of Reports on Trace Element Variation in Enamel of Teeth from Different Geographic Sources

No Difference	Difference		Possible Differences
Sb[1,2]	Zn[2,3-5,7]	Ba[1,5,8]	K[2,4]
Mg[3-5]	Sr[2,3,8-11]	Pb[2,3,12]	Li[2,3]
Ba[1]	Cd[2,12]	B[2]	Mo[2,5]
Se[2,6]	Br[2]	Cu[2,5]	Na[4,5]
Ni[2]	Fe[2,5]	S[2]	Al[2,5]
Rb[2]	Ag[2]	Mn[2,4,5]	Ti[2,5]
Sn[2]	Nb[2]	Cl[4]	
U[2]	Si[5]		

1. Nixon et al, 1967
2. Curzon et al, 1975
3. Schamschula et al, 1978
4. Lakomaa and Rytömaa, 1977
5. Cutress, 1972
6. Nixon and Myers, 1970
7. Brudevold et al, 1963
8. Curzon and Losee, 1977a
9. Steadman et al, 1958
10. Spector and Curzon, 1978
11. Curzon and Losee, 1977b
12. Curzon and Spector, 1977

The United States-New Zealand comparative study of Curzon et al (1975) found relatively low Zn concentrations (151 ppm compared with 203 ppm) in enamel of teeth from New Zealand. However, within and between country comparisons of Zn content in the outer 90 μm enamel layer by Cutress (1972) reported Zn between 60 and 600 ppm, with higher levels in the outer enamel but no substantial difference between teeth from different sources. Likewise Schamschula et al (1978) reported relatively small variations (720 ppm to 839 ppm) in the Zn content of the outermost enamel layer (2.7 μm) obtained by biopsy of teeth of primitive people in the Sepik area of Papua-New Guinea.

Strontium (Sr) Interest in this element has increased in recent years and of all elements its association with geographic differences has been most substantiated. Steadman et al (1958) determined the Sr in teeth from several geographic areas and observed considerable variation in the Sr content of comparable enamel layers. Lowest Sr concentrations occurred in tooth enamel from the United States (Maine and New Mexico), 25 to 100 ppm. Highest concentration was from Greenland (Gödhab) and Tonga, 300 to 320 ppm. Cutress (1972), using an acid etching technique to obtain successive layers (30 μm each), also reported wide differences in Sr content of enamel; Sr concentration varied between 20 and 600 ppm. Enamel from Texas and Western Samoa showed the higher level of 400 to 600 ppm, and the lowest Sr, 97 to 107 ppm was found in enamel from European New Zealanders.

Teeth from 13 towns and cities in England showed variations in enamel Sr between 80 and 181 ppm (Nixon et al, 1976), and the enamel Sr concentration was found to be associated with Sr in drinking water supplies. Curzon et al (1975) compared tooth enamel from New Zealand and the United States and found a small but significantly different Sr content, 92.6 ppm in New Zealand teeth and 121.0 ppm in United States teeth. Curzon and Losee (1977a) also found differences in Sr content of enamel of teeth from two regions of the United States — 129 ppm in New England and 157 ppm in South Carolina. It was considered that the teeth were representative of the areas and that the prevalence of caries was lower in South Carolina than in New England. The same investigators in a different study reported wider variations in a nine-state study with mean Sr in enamel varying from 81 to 342 ppm between states.

Spector and Curzon (1978) analyzed 233 teeth for enamel Sr using a surface-etching technique on extracted teeth from nine different communities in Ohio and Wisconsin. Portsmouth (Ohio) had the lowest mean enamel Sr — 82 ppm (water Sr was 0.2 ppm) — and Coldwater (Ohio), the highest median values — 471 ppm (water Sr 15.0 ppm). The median value was 282 ppm (Delphos, Ohio, with water Sr at 7.8 ppm). The Sr concentration in enamel, therefore, showed a clear association with Sr in water supplies.

Schamschula et al (1978) included enamel biopsy procedures in the WHO/NIH study of caries etiology in Papua-New Guinea. They found the Sr content of tooth enamel of subjects from villages geographically close to each other varied only slightly, mean values ranging from 83 ppm to 116 ppm.

Lead (Pb) In their comparison of enamel from the United States and New Zealand Curzon et al (1975) reported a small but significant difference in Pb content of enamel. Later Curzon et al (1977) identified distinctive and wide variations in the mean Pb content of whole enamel obtained from eight widely dispersed areas in the United States. The lowest mean value of 2.7 ppm was found in tooth enamel from South Carolina, the median mean values of 24.4 ppm and 31.0 ppm were from Ohio and Florida, and the highest values of 55.0 ppm were from Montana. The Papua-New Guinea study reported variable Pb content in tooth biopsy (2.7 μm) samples from the four villages; the range was between 4.6 ppm and 15.1 ppm.

Manganese (Mn) Variations in Mn content of enamel obtained from widely diverse sources were evident but not greatly different in the study reported by Cutress (1972). However, Mn values were found to be lower in tooth enamel from New Zealand and higher in teeth from Niue and Texas. Curzon et al (1975) reported very low values for enamel from both New Zealand and the United States but levels of Mn were significantly higher in United States teeth. Lakomaa and Rytömaa (1977) obtained and analyzed tooth enamel from six localities in Finland and concluded that geographic differences occurred in Mn content of enamel.

Cadmium (Cd) Geographic variations in the Cd content of enamel are apparent from two studies of Curzon et al (1975, 1977). Higher enamel Cd was found in tooth enamel from New Zealand compared with United States teeth. The eight-state study showed considerable differences in mean Cd values ranging from 0.7 ppm in California to 7.3 ppm in Montana.

Copper (Cu) and Iron (Fe) Five of six geographic areas listed in one study (Cutress, 1972) reflected similar Cu content (10 to 31 ppm) of enamel. However, teeth from Texas had considerably higher levels of this element with a range from 103 to 122 ppm. The study of Curzon et al (1975) comparing New Zealand and United States tooth enamel found a higher mean value from United States teeth.

The distribution of Fe was similar to that found for Cu. The Texas teeth had the highest Fe content (17 to 47 ppm) and other areas had between 7 and 47 ppm. However Curzon's study reported Fe to be higher in New Zealand teeth. Lakomaa and Rytömaa (1977) reported no differences in either Cu or Fe content of tooth enamel from the various localities they sampled in Finland.

Other Elements Evidence of geographic influence on the enamel content of other trace elements is available from single studies. Curzon et al (1975) in their United States-New Zealand comparative study reported significant differences for B, Br, S, Ag, and Nb between the two countries. Lakomaa and Rytömaa (1977) reported that Na varied in tooth enamel according to the origin of the tooth in Finland. Cutress found exceptionally high Si levels, 34 to 55 ppm in tooth enamel from Texas compared with 1 to 8 ppm from other areas studied.

Other elements found to occur at similar concentrations in enamel despite diverse geographic origins are Ba, Ni, Rb, Sn, V, Sb, Mg, and Se (Nixon et al, 1967; Cutress, 1972; Nixon and Myers, 1970; Curzon et al, 1975; Lakomaa and Rytömaa, 1977; Schamschula, 1978).

On the other hand, evidence is contradictory on the variation of tooth composition with the geographic source of the enamel for the elements K, Li, Mo, Ni, Al, and Ti (Cutress, 1972; Curzon et al, 1975; Lakomaa and Rytömaa, 1977; Schamschula, 1978). Overall, studies indicate that 8 elements are not influenced by the geographic source of the tooth, 15 elements are influenced, and evidence on 6 elements is equivocal (Table 3-9).

Trace Element Variation and Prevalence of Caries

A number of elements have been studied for their possible role in caries, and, although several have been implicated, their mode of action can only be speculated on, even if a relationship is confirmed. The presence or absence of the respective trace elements in enamel above or below specified concentrations has to be considered as the potential means of influencing the caries effect. The deposition of elements in enamel results from uptake during mineralization of the tooth or posteruptively.

Studies of caries prevalence and trace element content of enamel are few despite much discussion and speculation; only a small number of elements have been implicated as having a possible correlation; namely, Mn, Cu, Cd, Al, Fe, Se, Sr, Ag, Sn, Ba, Li, Zr, and Pb. Of these only Sr, Se, and Li appear to have sufficient evidence to support their potential role in influencing caries susceptibility by their presence in enamel (Table 3-10).

Curzon and Crocker (1975) analyzed 451 whole enamel samples from teeth of 10- to 20-year-old people with high or low caries prevalence from diverse geographic locations in the United States. Complex statistical evaluation of the 30 trace elements detected in whole tooth enamel demonstrated that 8 elements had significant correlations with caries prevalence of the tooth donor. Four elements, Al, Fe, Se, and Sr in addition to F, showed a negative relationship, and three elements, Mn, Cu,

Table 3-10
Summary of Reports on Proposed Associations Between Caries Prevalence and Trace Element Content of Enamel

Reference	No. Samples	Trace Element	Type of Enamel	Effect on Caries of Increased Trace Element Content
Curzon and Crocker, 1978	451	Mn, Cu, Cd	Whole	Increased
		Al, Fe, Se, Sr	Whole	Decreased
Curzon and Losee, 1977a	208	Sr, Ag, Sn	Whole	Decreased
		Al, Ba, Cu, Li, Zr	Whole	Increased
Curzon and Losee, 1977b	147	Sr	Whole	Decreased
Retief et al, 1976	20	Se	Whole	Decreased
Brudevold et al, 1977	251	Pb	Outer 2.1 μm	Increased
Schamschula et al, 1978	301	Li	Outer 2.7 μm	Decreased
		Pb		Increased
Vrbic and Stupar, 1980	16	Sr	Whole	Decreased

and Cd had a positive relationship to DMFT. Curzon and Losee (1977a) from analysis of 208 whole enamel samples of teeth from communities with high and low caries prevalence in the United States detected 30 trace elements. Apart from F, Sr in enamel was found to be strongly related to low caries prevalence. Ag and Sn showed a weak inverse relationship to caries prevalence, whereas Al, Ba, Cu, Li, and Zr were positively related to caries when the data were considered irrespective of geographic source; only 3 elements—F, Sr, and Zr—were found to be significantly correlated with caries prevalence in teeth from a low caries area. The hypothesized association between the Sr content of enamel and dental caries prevalence was also studied by Curzon and Losee (1977b) by analysis of 147 samples of whole enamel of teeth obtained from subjects 11 to 19 years old with known caries history. A strong association was found between the Sr content of enamel and caries prevalence.

Retief et al (1976) determined Se in tooth enamel of 16- to 17-year-old black and white South African students. Although black students had low prevalence of caries (2.1:2.5, female:male) and white students a high prevalence of caries (11.1:8.7, female:male) with Se concentrations of 0.08 and 0.01 ppm respectively, the investigators did not consider this supportive of the hypothesis that Se is caries-promoting in enamel. More recently the same group of workers reported on caries associations with nine trace elements (Table 3-10).

In two studies (Brudevold et al, 1977; Schamschula et al, 1978) concerned with analysis of the outer surface of enamel and caries prevalence, Pb was implicated with a contradictory role of both increasing and decreasing caries prevalence with increase in enamel Pb content. Although Brudevold et al mention the possibility of Pb as a caries-inducing agent, the enamel F levels of the high Pb group were higher than in the low Pb group.

Schamschula et al in a different geographic location, Papua-New Guinea, reported lower concentrations of Pb from analysis of enamel biopsy samples (average depth 20.3 μm) of slightly older 16- to 18-year-old subjects. In this study the variation of DMFT could more readily be explained on the basis of Pb content of surface enamel than on its F content.

The most recent study from Yugoslavia (Vrbic and Stupar, 1980) identified two towns with significant differences in caries prevalence in children. After studying a number of possible factors it was reported that enamel from children in the town with lower caries had significantly higher Sr. Analysis for Al showed higher concentrations in soil and water from the low caries town, but they were not reflected in enamel concentrations.

Studies concerned with trace elements in enamel and caries are difficult to interpret. Enamel may be acting as an environmental recorder of the presence of a trace element. Low caries prevalence may be associated with an increased enamel uptake of a particular trace element, without any causal relationship. Nevertheless, enamel composition studies serve a useful method of identifying those trace elements which merit further study. When several studies all show similar findings, such as for Pb, then there is a sounder basis to infer a cause.

TRACE ELEMENTS IN ENAMEL OF PRIMARY TEETH

Only a small number of elements have been studied for their presence in the enamel of primary teeth, and the evidence suggests that Na, Cl, Mg, K, Zn, Pb, Ba, Cu, and Fe occur at similar concentrations to those in permanent enamel (Table 3-11). An exception appears to be Al which, on the basis of two studies (Cutress, 1972; Lakomaa and Rytömaa, 1977), is present at higher levels in primary than in permanent enamel. However, in the two studies mentioned, analyses of permanent enamel also revealed unusually high Al levels. A summary of all data available on trace element concentration in primary enamel is given in Table 3-12. Perusal of the literature reveals very little of interest except that the Sr content of primary enamel is associated with the geographic source of teeth (Cutress, 1972; Nixon and Helsby, 1976).

Although many studies of Pb in primary teeth have now been completed, the majority are concerned with the Pb content of whole teeth as opposed to enamel or dentin. The data obtained are more pertinent to dentin than enamel. Because of the particular importance of Pb in teeth as an indication of exposure to this element the Pb content of teeth and its significance are referred to in detail in Chapter 20. Some analyses for Pb in whole teeth point to much lower concentrations than have been found in enamel. For example Oshio (1973) found an average tooth Pb concentration of 6.5 ppm on a remarkably large sample of 795 teeth; Needleman et al (1972) reported 11.1 ppm Pb; and Pinchin et al (1978) showed 4.4 ppm Pb for whole primary teeth. However, others (Habercam et al, 1974; Proud, 1976) considered 92.4 and 28.8 ppm Pb as average values for their samples.

Table 3-11
Summary of Trace Elements in Enamel of Primary Teeth

Concentration Range ppm	Elements
> 1000	Na, Cl, Mg
100–1000	K, Al, Zn
10–100	Sr, Pb, Ba, Cu, Ni, Ti, Fe
1–10	Mn, Se, Cd, Si, Mo

Table 3-12
Trace Elements in Enamel of Primary Teeth

Author	Year	Enamel	n	Age	Geographic Location	ppm Concentration
Na						
Naujocks et al	1967	Whole	39	NG	Germany	8250
Cutress	1972	Surface	6	9–13	New Zealand	4400
Haataja et al	1972	Crown	93	5–13	Finland	7200
Lakomaa and Rytömaa	1977	Whole	104	< 14	Finland	6300
					Median	6800
Cl						
Haataja et al	1972	Crown	93	5–13	Finland	3600
Lakomaa and Rytömaa	1977	Whole	104	< 14	Finland	2700
					Median	3200
Mg						
Haataja et al	1972	Crown	93	5–13	Finland	2900
Lakomaa and Rytömaa	1977	Whole	104	< 14	Finland	3100
Cutress	1972	Surface	18	9–13	New Zealand	1600
					Median	3000
K						
Haataja et al	1972	Crown	93	5–13	Finland	420
Lakomaa and Rytömaa	1977	Whole	104	< 14	Finland	270
					Median	345
Al						
Cutress	1972	Surface (40 μm)	18	9–13	New Zealand	200
Lakomaa and Rytömaa	1977	Whole	104	< 14	Finland	330
					Median	265

Table 3-12 (continued)

Author	Year	Enamel	n	Age	Geographic Location	ppm Concentration
Zn						
Cutress	1972	Surface (40 μm)	18	9–13	New Zealand	181
Stack et al	1976	Whole	4	Fetal	United Kingdom	79
Lakomaa and Rytömaa	1977	Whole	104	< 14	Finland	145
					Median	145
Sr						
Cutress	1972	Surface (40 μm)	18	9–13	New Zealand	81
Nixon and Helsby	1976	Whole	114	NG	United Kingdom	94
					Median	87
Pb						
Shapiro	1972	Whole	NG	NG	United States	38
		Outer				235
		Inner				25
Stack et al	1976	Whole	4	Fetal	United Kingdom	31
					Median	35
Fe						
Cutress	1972	Surface (40 μm)	18	9–13	New Zealand	22
Stack et al	1976	Whole	4	Fetal	United Kingdom	14
Lakomaa and Rytömaa	1977	Whole	104	< 14	Finland	6
					Median	14

Table 3-12 (continued)

Author	Year	Enamel	n	Age	Geographic Location	ppm Concentration
Mn						
Cutress	1972	Surface (40 μm)	18	9–13	New Zealand	4
Stack et al	1976	Whole	4	Fetal	United Kingdom	6
Lakomaa and Rytömaa	1977	Whole	104	< 14	Finland	3
					Median	4
Ba						
Cutress	1972	Surface	18	9–13	New Zealand	19
Cu						
Cutress	1972	Surface	18	9–13	New Zealand	17
Ni						
Stack et al	1976	Whole	4	Fetal	United Kingdom	16
Ti						
Cutress	1972	Surface	18	9–13	New Zealand	16
Se						
Hadjimarkos and Bonhurst	1959	Whole	6	NG	United States	5
Mo						
Cutress	1972	Surface	18	9–13	New Zealand	3
Cd						
Stack et al	1976	Whole	4	Fetal	United Kingdom	2
Si						
Cutress	1972	Surface	18	9–13	New Zealand	2

NG = not given.

TRACE ELEMENTS IN DENTIN OF PERMANENT AND PRIMARY TEETH

The distribution of trace elements in dentin has not been extensively studied; no data are available on a wide range of elements. Nevertheless, the presence of 26 trace elements has been reported in dentin from permanent teeth and 11 elements in primary dentin. The great variation in the source of the dentin, and in methods of analysis introduces difficulties in comparing studies, and, as with enamel, the concentrations of elements appear to vary with the depth of the dentin layer.

With consideration of these limitations, a summary of element distribution in dentin has been prepared (Table 3-13) which categorizes trace elements in concentration ranges similar to the summarized information on elements in enamel. Data from individual studies are shown in Table 3-14 for permanent teeth and in Table 3-15 for primary teeth. The most common elements reported in dentin of permanent teeth are Mg, N, Cl, Si, K, Zn, Sr, and Ba which resemble quite closely the profile observed in enamel. The few (11) trace elements studied in primary dentin—with the exception of Pb—are restricted to the reports of two studies (Lakomaa and Rytömaa, 1977; Haataja, 1972).

Table 3-13
Summary of Trace Elements Reported in the Dentin of Permanent and Primary Teeth

Concentration Range, ppm	Elements	
	Permanent Dentin	*Primary Dentin*
> 1000	Na, Mg	Na, Mg
100–1000	Cl, Zn, Sr, Si, K, Ba	Cl, Zn, Al, K
10–100	Pb, Br, Fe, Al, Rb	Pb
1–10	W, Cu, Co, Cr, Ag, Sn, I	Cu, Mn, Fe, Se
0.1–0.9	Mn, Sb, Se	
< 0.1	Au, Cd, Pt	

Permanent Dentin

Elements occurring at concentrations < 0.1 ppm Three metals have been identified in dentin although at very low concentrations with median values of the order 0.03 ppm for Au and below detection for Cd and Pt (Söremark and Samsahl, 1962a; Lundberg et al, 1965b; Retief et al, 1971).

Elements occurring at concentrations between 0.1 and 0.9 ppm Analyses of Mn in dentin in several studies are in agreement that

Table 3-14
Summary of Trace Element Reports on Dentin of Permanent Teeth

Element	Author	Year	n	Age	Origin		Concentration ppm
Mg	Burnett and Zenewitz	1958	144	NG	United States		8700
	Retief et al	1971	6	NG	South Africa		8700
	Derise et al	1974	175	10–90	United States		7400
	Frostell et al	1977	1	NG	Sweden		13,000 (Est)
	Lakomaa and Rytömaa	1977	104	NG	Finland		8500
						Median	8700
Na	Söremark and Samsahl	1962a	15	14–16	Sweden		7500
	Lundberg et al	1965b	10	14–25	Sweden		8500
	Retief et al	1971	6	NG	South Africa		5500
	Derise et al	1974	175	10–90	United States		5500
	Frostell et al	1977	1	NG	Sweden		1100 (Est)
	Lakomaa and Rytömaa	1977	104	NG	Finland		5400
						Median	5750
Cl	Söremark and Samsahl	1962a	15	14–16	Sweden		3900
	Retief et al	1971	6	NG	South Africa		350
	Derise et al	1974	175	10–90	United States		600
	Frostell et al	1977	1	NG	Sweden		150
	Lakomaa and Rytömaa	1977	104	NG	Finland		720
						Median	600
Si	Steadman et al	1959	12	NG	United States		860
			23	NG	United States		37
						Median	450

Table 3-14 (continued)

Element	Author	Year	n	Age	Origin	Concentration ppm
K	Derise et al	1974	175	10–90	United States	180
	Frostell et al	1977	1	NG	Sweden	200 (Est)
	Lakomaa and Rytömaa	1977	104	NG	Finland	190
					Median	190
Zn	Steadman et al	1959	12	NG	United States	2200
			23	NG	United States	570
	Söremark and Samsahl	1962a	15	14–16	Sweden	199
	Lundberg et al	1965b	10	14–25	Sweden	160
	Retief et al	1971	6	NG	South Africa	173
	Derise and Ritchey	1974	175	10–90	United States	175
	Lakomaa and Rytömaa	1977	104	NG	Finland	180
					Median	180
Sr	Steadman et al	1958	12	20–29	United States	570
			12	< 50	United States	330
			12	NG	United States	120
	Lundberg et al	1965b	10	14–25	United States	64
	Söremark and Samsahl	1962a	15	14–16	Sweden	70
	Retief et al	1971	6	NG	South Africa	94
	Derise and Ritchey	1974	175	10–90	United States	180
	Frostell et al	1977	1	NG	Sweden	200
					Median	150
Ba	Retief et al	1971	6	NG	South Africa	129
Fe	Steadman et al	1959	12/23	NG	United States	80
	Söremark and Lundberg	1964b	15	14–16	Sweden	110

Table 3-14 (continued)

Element	Author	Year	n	Age	Origin	Concentration ppm
	Retief et al	1971	6	NG	South Africa	93
	Derise and Ritchey	1974	175	10–90	United States	43
	Lakomaa and Rytömaa	1977	104	NG	Finland	2
						Median 80
Al	Steadman et al	1959	12/23	NG	United States	80
	Retief et al	1971	6	NG	South Africa	69
	Derise and Ritchey	1974	175	10–90	United States	65
	Lakomaa and Rytömaa	1977	104	NG	Finland	150
						Median 75
Br	Söremark and Samsahl	1962a	15	14–16	Sweden	4
	Retief et al	1971	6	Nk	South Africa	114
						Median 59
Pb	Steadman et al	1959	12/23	NG	United States	37 (Est)
	Steadman et al	1959	12/23	NG	United States	60
	Shapiro	1972	NG	NG	United States	52
	Shapiro	1972	NG	NG	United States	43
	Malik and Fremlin	1974	7	NG	United Kingdom	33
	Derise and Ritchey	1974	175	10–90	United States	43
						Median 43
Rb	Söremark and Lundberg	1964b	15	14–16	Sweden	6
	Lundberg et al	1965b	10	14–25	United States	69
						Median 35

Table 3-14 (continued)

Element	Author	Year	n	Age	Origin	Concentration ppm
Cu	Steadman et al	1959	12/23	NG	United States	6
	Steadman et al	1959	12/23	NG	United States	53
	Söremark and Samsahl	1962a	15	14–16	Sweden	0.2
	Derise and Ritchey	1974	175	10–90	United States	7
	Lakomaa	1977	104	NG	Finland	1
					Median	6
I	Derise and Ritchey	1974	175	10–90	United States	4
W	Söremark and Samsahl	1962a	15	14–16	Sweden	3
Sn	Steadman et al	1959	12/23	NG	United States	0.3
	Steadman et al	1959	12/23	NG	United States	3
					Median	1
Ag	Steadman et al	1959	12/23	NG	United States	17
	Steadman et al	1959	12/23	NG	United States	1
	Söremark and Lundberg	1964b	15	14–16	Sweden	0.005
	Retief et al	1971	6	NG	South Africa	2
					Median	1.5
Co	Söremark and Lundberg	1964b	15	14–16	Sweden	3×10^{-4}
	Retief et al	1971	6	NG	South Africa	1
	Derise and Ritchey	1974	175	10–90	United States	32
					Median	1
Cr	Söremark and Lundberg	1964b	15	14–16	Sweden	0.005
	Retief et al	1971	6	NG	South Africa	2
					Median	1

Table 3-14 (continued)

Element	Author	Year	n	Age	Origin	Concentration ppm	
Mn	Steadman et al	1959	12/23	NG	United States		60
	Steadman et al	1959	12/23	NG	United States		31
	Lundberg et al	1965	10	14–25	United States		0.6
	Battisone et al	1967	106	8–30	United States		0.3
	Söremark and Samsahl	1962a	15	14–16	Sweden		0.2
	Retief et al	1971	6	NG	South Africa		0.6
	Derise and Ritchey	1974	175	10–90	United States		6
	Lakomaa and Rytömaa	1977	104	NG	Finland		1
						Median	0.8
Sb	Retief et al	1971	6	NG	South Africa		0.7
Se	Hadjimarkos and Bonhurst	1959	85	> 20–< 50	United States		0.5
	Derise and Ritchey	1974	175	10–90	United States		0.3
						Median	0.4
Au	Söremark and Samsahl	1962a	15	14–16	Sweden		0.03
	Lundberg et al	1965b	10	14–25	United States		ND
	Retief et al	1971	6	NG	South Africa		0.07
						Median	0.03
Cd	Lundberg et al	1965b	10	14–25	United States		ND
Pt	Söremark and Lundberg	1964b	15	14–16	Sweden		ND

NG = not given.
ND = not detected.

Table 3-15
Summary of Reports on the Concentrations of Trace Elements in Dentin of Primary Teeth

Author (Year)	n	Method	Location	Element	Concentration
Lakomaa and Rytömaa, 1977	25	NA	Finland	Na	5500
				Cl	830
				Zn	145
				Cu	1.6
				Mn	1.9
				Fe	4
				Al	210
				K	310
				Mg	9600
Hadjimarkos and Bonhurst, 1959	29	Chemical	United States	Se	2
Shapiro, 1972	9	V	United States	Pb	55
	20	V	United States	Pb	17
	5	V	Iceland	Pb	5
Shapiro, 1973	NG	V	United States	Pb	27
Needleman, 1974	174	V	United States	Pb	198
	304	V	United States	Pb	2
	71	V	United States	Pb	136
De la Burde and Shapiro, 1975	32	V	United States	Pb	195
	36	V	United States	Pb	103
Fosse and Berg Justensen, 1978	2233	AA	Norway	Pb	3
				Median Pb Concentration	56

NA = neutron activation; V = voltametry; AA = atomic absorption spectrophotometry.
NG = not given.

the element occurs at low concentrations (< 6 ppm). Steadman et al (1959), however, reported unusually high levels of the element in their largest analyses of ancient North American Indian teeth, in particular Pueblo Indians, where all six dentin layers had Mn in excess of 400 ppm. Indian Knoll whole dentin averaged 31 ppm, and Mn in enamel of these teeth was also much higher than that found in other studies. In general, dentin and enamel Mn levels are similar (Derise and Ritchey, 1974; Lakomaa and Rytömaa, 1977; and others).

The few data on Sb and Se indicate that they occur in concentrations less than 1 ppm, which is slightly less than enamel (Retief et al, 1971; Hadjimarkos and Bonhurst, 1959; Derise and Ritchey, 1974).

Elements occurring at concentrations between 1 and 10 ppm Concentrations of Cu occur at about 6 ppm which is the same as in enamel. An exception to this is recorded by Steadman et al (1958) who found Cu levels of 50 ppm in various dentin layers and whole dentin of ancient North American Knoll Indian teeth, whereas in the same study Pueblo Indian teeth have 6 ppm Cu. Presumably this indicates a geographic influence on Cu. Both groups of teeth had higher Cu levels in the outer dentin layers. No increase in dentin Cu has been reported with the age of tooth, but one study has identified relatively lower Cu concentration in dentin compared with enamel (Derise and Ritchey, 1974).

In dentin I, Sn, and W have received scant attention but appear to occur at about similar levels in dentin or in enamel. Reports show I, 6 ppm; W, 0.2 ppm; and Sn, 1.5 ppm (Derise and Ritchey, 1974); Söremark and Samsahl, 1962a; Steadman et al, 1959).

The concentration of Ag varies from 0.005 to 17 ppm with a median value estimated at 1.5 ppm. On the available evidence Ag is higher in the outer dentin layer and occurs in much the same concentration range as in enamel. No doubt these wide variations can be partially attributed to the widespread use of Ag amalgam (Söremark and Lundberg, 1964; Steadman et al, 1959; Retief et al, 1971).

The mean values of Co have been reported from 3×10^{-4} ppm to 32 ppm. No data are available to indicate any change of concentration with depth in age or between dentin and enamel (Söremark and Lundberg, 1964; Derise and Ritchey, 1974). Low levels of Cr are present in dentin at about 1 ppm which is similar to enamel (Söremark and Samsahl, 1962a; Retief et al, 1971).

Elements occurring at concentrations between 10 and 100 ppm Concentrations of Fe in dentin appear to occur similarly in enamel although an early study, which analyzed dentin by layer, reported higher Fe levels in the outer layer of dentin than that of enamel. No change in dentin Fe was reported with age (Derise and Ritchey, 1974; Lakomaa and Rytömaa, 1977; Retief et al, 1971; Steadman et al, 1959).

With a median value of 75 ppm Al appears to be lower in dentin than in enamel which has a median value of 35 ppm. However, in three of the four comparative studies (Retief et al, 1971; Derise and Ritchey, 1974; Lakomaa and Rytömaa, 1971) of enamel and dentin, the Al concentrations in dentin were higher than in enamel. However, Steadman et al (1959) reported converse findings and also observed much higher levels of Al in the outer dentin layer whereas Derise et al (1974) noted no change in dentin Al with age.

Concentration of Br averaged, in only two reports, as 59 ppm; however, the studies individually reported 4 ppm (Söremark and Samsahl, 1962a) and 114 ppm (Retief et al, 1971). The latter reported the only comparable data for enamel and dentin at 34 ppm and 114 ppm respectively.

In whole dentin of permanent teeth Pb is about 40 ppm in several studies. Mean values reported show less variability than for primary dentin although the latter included subjects with known exposure to high environmental Pb. In enamel and dentin Pb at similar levels increased with age at the amelodentinal junction and particularly in secondary dentin, no doubt again reflecting environmental exposure. Concentration ranges between individual teeth are of the order 15 to 97 ppm (Steadman et al, 1959; Shapiro, 1972; Malik and Fremlin, 1974; Derise and Ritchey, 1974).

It has been reported that Rb is found in unerupted teeth at 69 ppm by analytic technique. Both of these mean values fall within the mean concentrations reported for enamel (Lundberg et al, 1965b; Söremark and Lundberg, 1964b).

Elements occurring at concentrations between 100 and 1000 ppm

The halogen, Cl, at 600 ppm in dentin, is much less common than in enamel (3200 ppm). Four comparative studies (Retief, 1971; Derise et al, 1974; Frostell et al, 1977; Lakomaa and Rytömaa, 1971) have reported lower Cl levels in dentin than enamel. A noncomparative study by Söremark and Samsahl (1962a) reported high Cl levels decreasing with age and also with proximity to the dental pulp.

Steadman et al (1959) reported Si, average 860 ppm, in dentin of ancient North American Pueblo Indians. However, the dentin layer adjacent to cementum contained very high levels of Si of 2500 and 1360 ppm while deeper dentin Si concentrations were of the order of 300 ppm. Chips of whole dentin from much older teeth (Indian Knoll) contained 30 to 50 ppm Si, and in this study dentinal Si was higher than that of enamel. Other studies of Si in enamel suggest enamel levels of 130 to 240 ppm (Losee et al, 1974a and 1974c).

Concentrations of K have been determined with three studies reporting 180 to 200 ppm, a level a little below that found in enamel. No variation of K with age or depth of dentin has been reported.

Five studies (Söremark and Samsahl, 1962a; Retief et al, 1971;

Derise and Ritchey, 1974; Lakomaa and Rytömaa, 1977) have reported Zn mean values of between 160 and 199 ppm. Only that of Steadman et al (1959), on ancient North American Indian teeth, found high levels of Zn at 2200 and 570 ppm with Zn concentrations in the dentin layer adjacent to cementum as high as 5300 ppm. There is a tendency for Zn to increase in whole dentin with age and also for postmortem accumulation of trace elements such as Zn.

There are many reports on Sr in dentin. The median concentration of the various studies is 150 ppm, marginally higher than overall levels reported in enamel. As a calcified tissue element capable of substituting for Ca in apatite and also an element reportedly linked to caries susceptibility, markedly different levels of accumulation of Sr are found in dentin.

Steadman et al (1958) found Sr of interest because of the possibility of its deposition in calcified tissue being influenced by the availability of F. Using a layer-by-layer analysis of dentin obtained from teeth from diverse geographic areas they concluded that dentin and enamel Sr accumulations are similar, and considerable variation in Sr in dentin occurs in teeth from different geographic areas. Most Sr is deposited in dentin prior to eruption. Large geographic variations are seen in tooth dentin from Texas, South Dakota, and Tonga. Other studies report that Sr levels in enamel and dentin are similar and that very little change occurs with age or depth of dentin (Retief et al, 1971; Derise and Ritchey, 1974; Frostell et al, 1977; Steadman et al, 1958).

Another alkaline earth, Ba, has been reported in only one study. Based on analysis of 6 teeth, the concentration (129 ppm) was similar to that found in enamel in the same study (Retief et al, 1971).

Elements occurring at concentrations exceeding 1000 ppm The most common trace element reported in dentin is Mg, reported in several studies (Derise et al, 1974; Frostell et al, 1977; Lakomaa and Rytömaa, 1977, and others) as being in excess of 0.7% by weight of dentin. The median value for Mg in dentin is about three times greater than that in enamel. In dentin Mg appears to increase slightly with age (Derise et al, 1974) and with proximity to the dental pulp (Frostell et al, 1977).

Finally, Na occurs at a median concentration of 5700 ppm in dentin—a level lower than that found in whole enamel. In four comparative studies Na was found at consistently lower concentrations in dentin than in enamel, and similar concentrations increase with age. No marked change in Na levels occur with depth of dentin. The ratio of Na:Mg varies considerably between enamel (2.6) and dentin (0.6).

Primary Dentin

Trace elements in primary dentin have been studied a few times (Table 3-15). The element that has attracted principal interest is Pb (see

Chapter 20 where a much wider range of studies on Pb in teeth is reviewed). This element, because of its toxic and cumulative effect in body tissues, has been investigated in teeth, including dentin, because of its value as an indicator of environmental exposure to Pb. Marked elevations of dentin Pb have been found in teeth from children in high Pb environments (Needleman, 1974). Almost 200 ppm Pb were identified in teeth of Negro children exposed to incidental ingestion of Pb paint in old housing areas, compared with 42 ppm Pb in teeth of Caucasian children from new housing areas where Pb-free paints were used. It has also been observed (Shapiro et al, 1973) that Pb in dentin of American children with Pb poisoning was 55 ppm whereas in healthy children levels were only 17 ppm. Dentin of teeth from Icelandic children showed Pb was as low as 5 ppm. The concentration of Pb in dentin appears to be dependent on the amount of paint or plaster ingested and the period of exposure (De la Burde and Shapiro, 1975). Fosse and Berg Justesen's (1978) major study of Pb in primary teeth from various urban and rural communities in Norway showed that Pb in dentin was more than twice the level of Pb in enamel and that Pb in teeth varied between geographic areas. Shapiro (1972) found the Pb content of enamel, coronal dentin, and root dentin similar but there was a marked variation of Pb in secondary dentin. Reported Pb levels in this study were low compared with other studies, the only comparable low concentration being that found in Icelandic dentin by Shapiro (1973).

A few other elements have been reported in primary dentin by Lakomaa and Rytömaa (1977), namely Na, Cl, Zn and Mn, Fe, Al, K, Mg, and Se; Hadjimarkos and Bonhorst (1959) have contributed data on Se. The concentrations reported agree in general with the level reported in permanent dentin.

TRACE ELEMENTS IN BONE AND CALCULUS

Bone

The growth and development of bone are complex and depend on many factors paramount among which are cellular and calcification mechanisms including metallo-enzyme systems. Trace elements can influence the mineral phase of bone by: 1) incorporation into the apatite lattice and consequently affecting its chemical reactivity; 2) incorporation into a non-apatitic phase, thereby affecting the rate of mineralization, although not becoming incorporated into crystal lattice; and 3) affecting the amorphous to crystalline transition of the mineral phase during bone development.

The structure and function of bone set it apart from enamel and dentin despite the common crystallographic features of its main component, hydroxyapatite. While trace elements may influence the mineral phase of bone, they might also affect the organic phase. For instance Si, probably an essential element, is localized in active growth centers and has a particular significance in the maturing of bone. It is associated with phosphate during calcification and decreases in concentration with maturity (Underwood, 1977). Another example is Zn, which, if deficient, reduces cell proliferation at endochondral growth sites in rats (Bergman et al, 1972).

Elements are not evenly distributed throughout bone; this is explained by differing skeletal turnover rates (Bryant and Loutit, 1964). In adults the annual turnover of the cortical (ivory) bone of limb bones is about 1% whereas that of trabeculated bone is about 8%. The rates increase with decreasing age to 100% turnover in the first year of life.

Of the 26 essential and several toxic trace elements some are known to be specifically associated with physiologic functions during the development and mineralization of bone. For example, Na, K, Rb, Cs, Mg, Zn, Cd, Cr, Cu, Mn, Fe, Co, Ni, Mo, and Al have all been associated with enzyme systems in normal bone cell functions (Vaughan, 1970).

Underwood (1977) reviews in detail actual and potential effects of trace element excesses and deficiencies on bones with respect to role, function, and concentration. Although it is known that environmental and dietary availability of some elements influences their uptake into bone in man and animals with resultant abnormalities (fluorosis, molybdenosis) no direct consequences on dental bone in man have been reported.

Biologic apatite can be appropriately described as a Ca-deficient carbonated apatite that includes trace and variable amounts of foreign elements. Trace element contamination where the metallic ion is doubly charged and has a similar ionic charge probably occurs by substitution of an element in the Ca position in the lattice (and possibly also phosphate), and contamination by surface adsorption onto the crystal also occurs. Although referred to as "in trace amounts" this can mean that almost the complete or a substantial proportion of the total body content of some elements is associated with bone. Most (99%) of Sr, a widely studied bone-seeking element in man and animals, is present in bone; however, the total Sr content of the average man is only about 0.3 g (Underwood, 1977). Skeletal Mg accounts for 70% of total body Mg and is readily mobilized for use elsewhere when required. About 90% of total body Pb and 5% of Mn are found in the skeletal tissues whereas other elements—Cd, Hg, and Zn—appear to remain at low concentrations in bone (Doyle, 1979).

Much more interest has been shown for element content of animal tissues, including bone, than for that of humans because of the increasing availability of toxic elements, such as Pb, Hg, and Cd, in the soil-food-animal chain. This has occurred because of their increased use in agricultural chemicals or as a result of industrial pollution.

Bioavailability of elements in food and water varies considerably and systemic absorption is modified by many factors. For instance, trace element concentrations are low in refined carbohydrate diets and diets high in phytic acid (due to chelation) such as found in unrefined cereals. These diets interfere with the absorption of Fe, Zn, and other elements (Burch and Hahn, 1979; Moynahan, 1979). In general it can be said that information on the trace element content of the human skeleton is scarce, and much of it needs substantiation. It was recently stated that the data base for the normal range of minor elements in bone was inadequate (Hopps and O'Dell, 1981) and also that much of the accumulated data were widely divergent and should, perhaps, be disregarded (Smith et al, 1981). Nevertheless, with these reservations in mind, some perspective on the element concentrations can be obtained from several studies of human bone and a larger number of studies of animal bones.

Human bone Nusbaum et al (1965) described the trace elements in calvarium and rib samples obtained by autopsy from 175 subjects in Los Angeles. Although the study was primarily concerned with Pb levels in bone resulting from air pollution, some 15 other elements were also considered. Three elements, Cd, Ag, and V, were generally not detected but the remaining 12 elements were identified and quantitated (Table 3-16).

Table 3-16
Trace Elements in Human Bones* μg/g Ash (ppm)

			Age[a]	
	Calvarium	Rib	*0–20 yrs*	*81–90 yrs*
Al	33	27	64	28
Ba	30	28	41	33
Cr	34	32	42	28
Co	44	43	49	37
Cu	25	26	27	24
Fe	640	773	680	560
Pb	69	69	54	61
Mn	13	14	15	11
Mo	102	104	109	94
Ni	108	111	148	104
Sr	148	146	167	140
Zn	180	194	198	187
As	6	3	(no data)	

*Summarized from Nusbaum et al, 1965.
[a]Data for calvarium samples. Young and old ages only quoted.

For all 12 elements there was a consistent pattern—in bone, neither Pb nor other elements were much influenced by either age or a heavily polluted environment.

Certain elements are possibly functional components of bone matrix, and others are contaminants according to Becker et al (1968), who studied human tibia and fibia bones from modern Americans and ancient (1300 AD) Peruvians. The data for five samples of modern and ancient bones showed that American samples had ten times the levels of Pb and appreciably higher levels of V than did Peruvian bones, while Cu and Zn were similar for all specimens. No B was detected in modern bones but was found in all Peruvian bones probably because of contamination from the sand in which the skeletons had remained buried for about six centuries. Contamination may also explain the higher levels of Al, Fe, Mn, Si, and Sr found in the Peruvian specimens. Several other studies should be mentioned although the analytic methods used, the type of bone, and age of specimens vary widely.

Alexander and Nusbaum (1962) assessed Zn levels and reported mean values of the order 181 ppm in ash, the range being between 106 and 372 ppm. They observed no variation in element concentration with the age of their specimens or their geographic origin. On the other hand, several studies have shown that the level of Sr in human bones varies according to their geographic origin. It has been proposed that drinking water is likely the source of the variation observed (discussed more fully in Chapter 15). Sowden and Stitch (1957) found an average Sr level of 100 ppm in long bones with a range of 42 to 157 ppm in bone ash, also correlated with the age of the specimens. In young children (3 months) bone Sr was 42 ppm whereas in older people (33 to 74 years) levels were 157 ppm. No attempt was made to examine the geographic origin of the specimens. In the same study Ba was noted at an overall concentration of 7 ppm with a variation between 2 and 17 ppm which, like Sr, was age dependent.

Stitch (1957) separately reported on the occurrence of Al, Cu, and Mn, in 13 to 14 samples of bone analyzed by neutron activation. The mean and range of values reported in ppm of bone ash was: Al, 40 (20–200); Cu, 20 (10–70); Mn 4 (4–20).

Strehlow and Kneip (1969) analyzed more than 40 samples of different bone from one single skeleton of a 62-year-old man with no history of unusual exposure to high Pb or Zn environments. The skeletal distribution of the two elements was described and compared with the concentrations found in teeth obtained from other sources. Findings showed (Table 3-17) a significant variation between bone samples, and that teeth (from other subjects) differed from bone in their Pb content, thereby reflecting the differences in the mineral metabolism of dental and skeletal tissues. Compared with comparable bones analyzed by Nusbaum et al (1965), these values are a little high. However, the concentrations of Zn are similar to those reported by Alhova et al (1977) for cancellous bone from iliac crest specimens of 94 adults. Concentrations ranged between 25 to 100 ppm, with a tendency for maximum Zn level to occur in the fifth decade of life. These investigators proposed a role for Zn in osteoporosis.

Animal bone The most extensive data on trace elements in bone are those accumulated by Curzon (1977) using powdered anorganic bone, prepared from bovine femur shaft by Losee. The bovine samples were used as reference apatite material during extensive study of trace elements in dental enamel. Two independent laboratories used spark source mass spectrometry for the analyses, and the procedure included the use of internal standards. The mean concentrations and standard errors are shown in Table 3-18 together with bone sample data reported by Cutress (1979). In the latter study the objective was to determine trace element content of the outer layer of dental enamel, and the bovine bone was included as a standard reference sample. Curzon's study analyzed for 30 elements, of which 20 were identified and quantitated. Over a period of about two years 23 bone samples were included in these dental studies.

Doyle (1979) expressed concern over the extent of contamination of the environment with toxic elements after reviewing data on As, Hg, Cd, Pb, Se, F, Mg, Zn, Cu, Fe, and Mn in the bones of cattle, sheep, pigs, and chickens. While it is impossible to be definite about the relevance of the data to human bone composition, they do imply that high levels of elements can accumulate in bone and that our knowledge of the subject is very meager. A summary of the concentration of elements found in bone shows (Table 3-19) many variations for elements such as Pb, Zn, and Fe.

Table 3-17
Lead and Zinc Concentrations in Bones and Teeth*

	Lead		Zinc	
	Fresh bone	*Ash*	*Fresh bone*	*Ash*
Teeth	219	301	–	–
Mandible	57	121	71	153
Calvaria	44	88	83	166
Vertebrae	16	156	26	266
Rib	15	123	33	262
Long bone, ends	23	132	33	178
Long bone, shafts	36	98	57	152
Skeletal Average	34	120	53	189

Concentrations in ppm.
*After Strehlow and Kneip, 1969.

However, much of this variation is no doubt attributable to methods of analysis and basis of concentration (ash, wet, dry, marrow-free, state of bone) as well as bone type analyzed.

In summary, the current level of knowledge on trace elements in bone shows that they occur at widely differing concentrations; the concentrations are dependent on the type of element, the type and age of bone, and the availability of the elements from food or environmental sources (Wolf et al, 1973). In dentistry the trace element content of bone has not as yet been demonstrated to influence form, function, or health of relevant bone or skeleton.

Similarly, little (Hals and Selvig, 1977) is known about the trace element composition of cementum. Recent interest in root caries may promote such an interest as this condition has to do with cementum.

Table 3-18
Results of Analyses of Anorganic Bovine Bone by Mass Spectrometry from Three Laboratories Over a Period of Two Years

Element	Curzon Lab 1 (n = 7) Mean ± SE	Curzon Lab 2 (n = 16) Mean ± SE	Cutress Lab 3 (n = 5) Mean	Range
Li	1 ± 0.4	1 ± 0.2	12	1–26
B	2 ± 0.3	2 ± 0.4	3	2–4
F	419 ± 32	382 ± 32	—	—
Al	29 ± 9	11 ± 2	148	108–241
Si	193 ± 90	20 ± 8	20	3–40
S	27 ± 6	220 ± 63	—	—
K	271 ± 22	119 ± 28	—	—
Mn	1 ± 0.4	1 ± 0.5	9	1–26
Fe	28 ± 5	19 ± 4	53	30–119
Ni	2 ± 0.6	1 ± 0.3	12	6–18
Cu	25 ± 3	17 ± 6	—	—
Zn	86 ± 3	81 ± 8	117	61–230
Se	0.3 ± 0.2	0.3 ± 0.1	12	4–21
Br	2 ± 0.3	0.7 ± 0.2	3	1–5
Rb	0.3 ± 0.2	0.04 ± 0.01	14	0.3–0.7
Sr	324 ± 15	337 ± 47	120	101–144
Nb	0.2 ± 0.1	1 ± 0.5	0	0
Mo	2 ± 0.5	3 ± 0.6	0.06	0–0.1
Ag	82 ± 78	0.4 ± 0.2	9	1–19
Cd	1.3 ± 0.6	1 ± 0.2	3	1–8
Sn	5 ± 3	2 ± 0.3	3	4–13
Ba	37 ± 23	237 ± 87	—	—
Pb	5 ± 0.5	9 ± 1	14	10–18

Be, Ti, V, Cr, Co, Zn, Sb—below detection limits.
n = number of samples.
Curzon samples of unknown size.
Cutress samples 1.5 mg.
Concentrations in ppm.

Table 3-19
Summary of Range of Trace Element Concentrations in Bone of Animals*

Element	Range (ppm)	Element	Range (ppm)
Cd	0.01 - 5	Fe	33 - 460
Pb	0.3 - 66	Mn	1 - 12
Mg	0.2 - 0.9 mg %	As	0.004- 0.02
Zn	3 - 370	Hg	0.1
Cu	0.7 - 12	Se	0.04 - 0.3

*After Doyle, 1979—abbreviated data.

Trace Elements in Dental Calculus

Calculus is a biologic apatite which like enamel and dentin is liable to incorporate foreign elements into the crystal lattice during crystallization and, by diffusion, to exchange with elements in the oral environment after mineralization.

The factors influencing the deposition of calculus can at best be only speculative. It is a common observation that the rate, distribution, and volume of calculus formation vary widely from person to person, and although it seems unlikely that trace elements play a role in calculus formation, it should be kept in mind that X-ray diffraction patterns of synthesized hydroxyapatite differ with the concentration of elements available during precipitation (Featherstone et al, 1979; Legeros et al, 1977; Leatherstone and Nelson, 1980) and theoretically many substitutions are possible for Ca or P (Young, 1974). Also the presence or absence of some elements, such as alkaline earths, may affect the crystallization nuclei of apatite or the role of bacterial proliferation and activity on the surface of calculus.

A review of studies on calculus composition reveals that calculus lacks homogeneity with respect to density (Little et al, 1963), distribution in the mouth, and crystal structure (Little and Hazen, 1964; Theilade and Schroeder, 1966; Schroeder and Bambauer, 1966). In addition, the gross distribution varies as to whether it is subgingival or supragingival and also by tooth-to-tooth location. These variations may influence the element incorporation into calculus because of the varying oral biochemical environment. For example, gingival exudate will influence subgingival more than supragingival calculus; proximity to salivary gland ducts will produce differing biochemical environments.

Until more information becomes available on the elements in calculus the conclusions to be reached are more restricted to the distribution variables and heterogeneity of the material. Several studies have produced information on elements in subgingival and supragingival calculus

Table 3-20
Summary of Reports on Trace Elements in Dental Calculus

Author (year)	Samples Method and Source	Element	Concentration ppm	
Söremark and Samsahl, 1962	8 supragingival neutron activation, Sweden	Na	15,000	
		Cl	9,000	
		Zn	255	
		Sr	34	
		Br	12	
		Cu	4	
		Mn	2	
		W	0.1	
		Au	0.01	
Little et al, 1963	200 subgingival flame photometry, United States	Na	20,000 to 71,900	
Little and Hazen, 1964	231 supragingival, marginal deep	Na	16,700	
	flame photometry, United States	Na	23,700	
Lundberg et al, 1966	22 supra-, subgingival		SUPRA-	SUB-
	neutron activation, Sweden	Cl	153	194
		Br	7	9
		Cu	4	6
		Mn	3	0.9
		Na	1783	2552
		Sr	10	24

Author (year)	Samples Method and Source	Element	Concentration ppm		
Grøn et al, 1967	Supra- and subgingival	Mg	7800		12,000
Retief et al, 1973	4 -10 supra-, subgingival neutron activation, South Africa	Sb		0.7	
		Ag		0.2	
		Zn		174	
		Co		0.1	
		Fe		54	
Knuutila et al, 1979	Supra-, subgingival neutron activation	Mg	5800		9100
		Mn	20		11
		Sr	13		22
		Zn	91		487

(Lundberg et al, 1966; Grøn et al, 1967; Retief et al, 1973; Knuutila et al, 1979); calculus from various intraoral locations (Grøn et al, 1967; Little and Hazen, 1964); and density of calculus (Little et al, 1963; Little and Hazen, 1964). Of the small group of elements reported in calculus Mg, Fe, Cu, W, Ag, Co, Sb, and Au occur in concentrations of the same order as those in dentin. The elements Cl, Sr, and Br occur in lower concentrations and Na and Mn at higher concentrations than in dentin. Knowledge about trace elements in calculus is sparse.

Perhaps of most interest is Na which has been reported as high as 7% (Little et al, 1963) of dry weight. In this study the investigators classified the calculus according to period of formation (one week to nine months) and its density as measured by a soft X-ray technique. The most radiolucent calculus (newly formed) has the highest percentage of Na and the most opaque calculus, the lowest concentration, 1.8%. Subgingival and supragingival calculus appears to incorporate higher Na, Mg, Zn, and Sr than subgingival calculus (Lundberg et al, 1966; Grøn et al, 1967; Knuutila et al, 1979). The higher Mg content of subgingival calculus may be explained by the relatively high level of whitlockite in subgingival calculus (Grøn et al, 1967) which is present as Mg whitlockite. The fourfold higher content of Zn subgingival calculus has been attributed to the higher levels of Zn in serum than in saliva (Knuutila, 1979). Apart from Na and Cl, Zn is the next most common trace element in calculus.

A summary of trace elements and their concentrations reported in calculus is shown in Table 3-20.

REFERENCES

Alexander GV, Nusbaum RE: Zinc in bone. *Nature* 1962;195:903.

Alhova HO, Puittimen J, Nokso-Koivisto VM: Content of human cancellous bone. *Acta Orthop* 1977;48:1–4.

Attramadal A, Jonsen J: Heavy trace elements in ancient Norwegian teeth. *Acta Odont Scand* 1978;36:97–101.

Battisone GC, Fieldman MH, Reba RC: The manganese content of human enamel in dentin. *Arch Oral Biol* 1967;12:1115–1122.

Becker RD, Spadaro JA, Berg EW: The trace elements of human bone. *J Bone Joint Surg* 1968;50A:326–334.

Bergman B, Friberg V, Lohmander S, et al: The importance of zinc to cell proliferation in endochondral growth sites in the white rat. *Scand J Dent Res* 1972;80:486–492.

Brudevold F: Changes in enamel with age. Univ. Roch. Dent. Res. Fellowship Prog. Proc. 25th yr. Celebration 1930–35. pp 185–192, June 1957.

Brudevold F, Aasenden R, Srinivasian BN, Bakhos Y: Lead in enamel and saliva, dental caries and the use of enamel biopsies for measuring past exposure to Pb. *J Dent Res* 1977;56:1165–1171.

Brudevold F, Reda A, Aasenden R, et al: Determination of trace elements in surface enamel of human teeth by a new biopsy procedure. *Arch Oral Biol* 1975;20:667–673.

Brudevold F, Söremark R: Chemistry of the inorganic phase of enamel, in Miles AEW (ed): *Structural and Chemical Organization of Teeth* vol 2. New York, Academic Press, 1967.

Brudevold F, Steadman LT: A study of copper in human enamel. *J Dent Res* 1955;34:209–216.

Brudevold F, Steadman LT: A study of tin in enamel. *J Dent Res* 1956a;35:749.

Brudevold F, Steadman LT: The distribution of Pb in human enamel. *J Dent Res* 1956b;35:430.

Brudevold F, Steadman LT, Smith FA: Inorganic and organic compounds of tooth structure. *Ann NY Acad Sci* 1960;85:110–132.

Brudevold F, Steadman LT, Spinelli MA, et al: A study of zinc in human teeth. *Arch Oral Biol* 1963;8:135–144.

Bryant FJ, Loutit JF: The entry of strontium-90 into human bone. *Proc Roy Soc* 1964;B159:449–465.

Burch RE, Hahn HK: Trace elements in human nutrition. *Med Clin North Am* 1979;63:1057–1067.

Burnett GW, Zenewitz JA: Studies of the composition of teeth. VIII. The composition of human teeth. *J Dent Res* 1958;37:590–600.

Calonius PEB, Visäpää A: The inorganic constituents of human teeth and bone examined by X-ray emission spectrography. *Arch Oral Biol* 1965;10:9–13.

Curzon MEJ: *Trace Element Composition of Human Enamel and Dental Caries.* PhD Thesis, University of London, 1977.

Curzon MEJ, Crocker DC: Relationships of trace elements in human tooth enamel to caries. *Arch Oral Biol* 1978;23:637–653.

Curzon MEJ, Losee FL: Dental caries and trace element composition of whole human enamel: Eastern United States. *JADA* 1977a;94:1146–1155.

Curzon MEJ, Losee FL: Strontium content of enamel and dental caries. *Caries Res* 1977b;11:321–326.

Curzon MEJ, Losee FL, Macalister AD: Trace elements in the enamel of teeth from New Zealand and U.S.A. *NZ Dent J* 1975;71:80–83.

Curzon MEJ, Spector PC, Losee FL: Dental caries related to Cd and Pb in whole human dental enamel, in Hemphill DD (ed): *Trace Substances and Environmental Health* vol 9. Columbia, MO, University of Missouri, 1977; pp 23–29.

Cutress TW: The inorganic composition and solubility of dental enamel in several population groups. *Arch Oral Biol* 1972;17:93–109.

Cutress TW: A preliminary study of the micro-element composition of the outer layer of dental enamel. *Caries Res* 1979;13:73–79.

De la Burde FA, Shapiro IM: Dental lead, blood Pb and pica in urban children. *Arch Environ Health* 1975;30:281–284.

Derise NL, Ritchey SJ, Furr AK: Mineral composition of normal human enamel and dentin and the relation of composition to dental caries. I. Macrominerals comparison of methods of analysis. *J Dent Res* 1974;53:847–852.

Derise NL, Ritchey SJ: Mineral composition of normal human enamel and dentin. II. Microminerals. *J Dent Res* 1974;53:853–858.

Doyle JJ: Toxic and essential elements in bone—A review. *J Animal Sci* 1979;49: 482–497.

Featherstone JDB, McGrath MP, Smith MW: The effect of trace elements in some New Zealand water supplies on synthetic apatite structure. *NZ Dent J* 1979;75:206–211.

Featherstone JDB, Nelson DA: The effect of fluoride, zinc, strontium, magnesium and iron on the crystal-structural disorder in synthetic carbonated apatites. *Aust J Chem* 1980;33:2363–2368.

Fosse G, Berg Justesen NB: Lead in deciduous teeth of Norwegian children. *Arch Environ Health* 1978;13:166–175.

Frostell G, Larsson SJ, Lodding A, et al: SIMS study of element concentration profiles in enamel and dentin. *Scand J Dent Res* 1977;85:18–21.

Grøn P, van Campen GJ, Lindstrom I: Human dental calculus. Inorganic chemical and crystallographic composition. *Arch Oral Biol* 1967;12:829–837.

Haataja J, Pohto P, Kleemola-Kujala E: On the macrominerals of deciduous teeth. *Saom Hammaslaak Toim* 1972;68:67–72.

Habercam JW, Keil JE, Riegart JR, Croft HW: Lead content of human blood, hair, and deciduous teeth: Correlation with environmental factors and growth. *J Dent Res* 1974;53:1160–1163.

Hadjimarkos DM, Bonhorst CW: The selenium content of human teeth. *Oral Surg* 1959;12:113–116.

Hals E, Selvig KA: Correlated electron probe analysis and microradiography of carious and normal dental cementum. *Caries Res* 1977;11:62–75.

Hardwick JL, Martin CJ: A pilot study using mass spectrometry for the estimation of the trace element content of dental tissues. *Helv Odont Acta* 1967;11:63–70.

Helli A, Haavikko K: Macro and micromineral levels in deciduous teeth from different geographical areas correlated with caries prevalence. *Proc Finn Dent Soc* 1977;73:87–98.

Hogdahl OT, Melsom S, Bowen JT: Neutron activation analysis of lanthanide elements in seawater, in Hood DW (ed): *Trace Inorganics in Water. Advances in Chemistry Series* No. 73. Washington, DC, American Chemical Society, 1968.

Hopps HC, O'Dell BL: Introduction and Conclusions to Workshop on "Research Needed to Improve Data on the Mineral Content of Human Tissues." *Fed Proc* 1981;40:2112.

Knuutila M, Lappalainen R, Kontturi-Narli U: Concentrations of Ca, Mg, Mn, Sr and Zn in supra and subgingival calculus. *Scand Dent J* 1979;87:192–196.

Lakomaa EL, Rytomaa I: Mineral composition of enamel and dentin of first and permanent teeth in Finland. *Scand J Dent Res* 1977;85:89–95.

Legeros RZ, Miravite MA, Quirologico GB, Curzon MEJ: The effect of some trace elements on the lattice parameters of human and synthetic apatites. *Calc Tiss Res* 1977;22:362–367.

Little MF, Casciani CA, Rowley J: Dental calculus composition. I. Supragingival calculus: Ash, calcium, phosphorous, sodium and density. *J Dent Res* 1963;42:78–86.

Little MF, Hazen SP: Dental calculus composition 2. Subgingival calculus, Ash, Ca, P, and Na. *J Dent Res* 1964;43:645–651.

Little MF, Steadman LT: Chemical and physical properties of altered and sound enamel IV. Trace element composition. *Arch Oral Biol* 1966;11:273–278.

Losee FL, Cutress TW, Brown R: Trace elements in human dental enamel. *Trac Sub Environ Health* 1974a;7:19–24.

Losee FL, Cutress TW, Brown R: Natural elements of periodic table in human dental enamel. *Caries Res* 1974b;8:123–134.

Losee FL, Curzon MEJ, Little MF: Trace element concentration in human enamel. *Arch Oral Biol* 1974c;19:467–470.

Lundberg M, Söremark R, Thilander H: The concentration of some elements in the enamel of unerupted (impacted) human teeth. *Odont Revy* 1965a;16: 8–11.

Lundberg M, Söremark R, Thilander H: Gamma-ray spectrometric analysis of some elements in coronal dentine of unerupted (impacted) human teeth. *Odont Revy* 1965b;16:97–100.

Lundberg M, Söremark R, Thilander H: Analysis of some elements in supra and subgingival calculus. *J Periodontol* 1966;1:245–249.

Malik SR, Fremlin JH: A study of Pb distribution in human teeth using charged particle activation analysis. *Caries Res* 1974;8:283–292.

Manley RS, Bibby BG: Substances capable of decreasing the acid solubility of tooth enamel. *J Dent Res* 1949;28:160–272.

Moynahan EJ: Trace elements in man. *Phil Trans Roy Soc Land* 1979;288:65–79.

Naujoks R, Schade H, Zelinka F: Chemical composition of different areas of the enamel of deciduous and permanent teeth. *Caries Res* 1967;1:137–143.

Needleman HL: Subclinical lead exposure in Philadelphia schoolchildren. Identification by dentine lead analysis. *N Engl J Med* 1974;290:245.

Needleman HL, Tuncay OC, Shapiro IM: Lead levels in deciduous teeth of urban and suburban American children. *Nature* 1972;235:111–112.

Nixon GS, Smith H: Estimation of arsenic in teeth by activation analysis. *J Dent Res* 1960;39:514–516.

Nixon GS, Smith H: Estimation of Cu in human enamel by activation analysis. *J Dent Res* 1962;41:1013–1016.

Nixon GS, Paxton GD, Smith H: Estimation of Hg in human enamel by activation analysis. *J Dent Res* 1965;44:654–665.

Nixon GS, Livingston HD, Smith H: Estimation of Mn in human enamel by activation analysis. *Arch Oral Biol* 1966;11:247–252.

Nixon GS, Livingston HD, Smith H: Estimation of antimony in human enamel by activation analysis. *Caries Res* 1967;1:327–332.

Nixon GS, Myers VB: Estimation of selenium in human dental enamel by activation analysis. *Caries Res* 1970;4:179–187.

Nixon GS, Helsby CA: The relationship between Sr in drinking water supplies and human tooth enamel. *Arch Oral Biol* 1976;21:691–695.

Nusbaum RE, Butt EM, Gilmour TC, et al: Relation of air pollutants to trace metals in bone. *Arch Environ Health* 1965;10:227–232.

Oehme M, Lund W: The determination of copper, lead, cadmium and zinc in human teeth by anodic stripping voltametry. *Anal Chim Acta* 1978;100:389–398.

Oshio H: The Cd, Zn and Pb content of deciduous teeth from two different geographic areas of Japan. *J Dent Health* 1973;23:203–222.

Pinchin MJ, Newham J, Thompson RPJ: Lead, copper and cadmium in teeth of normal and mentally retarded children. *Clin Chim Acta* 1978;85:89–94.

Proud M: A study of the lead levels in deciduous teeth in Leicester children. *J Dent Res* 1976;55:D109 (Abs).

Rasmussen EG: Antimony, arsenic, bromine and mercury in enamel from human teeth. *Scand J Dent Res* 1974;82:562–565.

Retief DH, Cleaton-Jones PE, Turkstra J, de Wit WJ: The quantitative analysis of 16 elements in normal human enamel and dentine by neutron activation analysis and high resolution gamma-spectrometry. *Arch Oral Biol* 1971;16:1257–1267.

Retief DH, Cleaton-Jones PE, Turkstra J, Smit HJ: The quantitative analysis of Sb, Ag, Zn, Co, and Fe in human dental calculus by neutron activation analysis and high resolution γ-spectrometry. *J Periodontol Res* 1973;8: 263–267.

Retief DH, Scanes S, Cleaton-Jones PE: The quantitative analysis of selenium in sound human enamel by neutron activation analysis. *Arch Oral Biol* 1974; 19:517–524.

Retief DH, Turkstra J, Cleaton-Jones PE, et al: Mineral composition of enamel from population groups with high and low caries incidence. *IADR Abstr* #302, *J Dent Res* 1978;57:150.

Retief DH, Cleaton-Jones PE, Turkstra J, Beukes PJL: Se content of tooth enamel obtained from two South African ethnic groups. *J Dent Res* 1976;55: 701.

Riabulik VS, Maksimouskii YM, Barer GM: A study of tooth and saliva composition by neutron activation analytical methods. *Stomatologua* (Moskow) 1972;51:62–64.

Schamschula RG, et al: WHO *Study of Dental Caries Etiology in Papua-New Guinea.* WHO Pub. No. 40, Geneva, 1978.

Schroeder HE, Bambauer HU: Stages of calcium phosphate crystallisation during calculus formation. *Arch Oral Biol* 1966;11:1–14.

Shapiro MI: The Pb content of human deciduous and permanent teeth. *Environ Res* 1972;5:467–470.

Shapiro MI: Lead levels in dentine and circumpulpal dentine of deciduous teeth of normal and Pb poisoned children. *Clin Chim Acta* 1973;46:119–123.

Shearer TR, Johnson JR, DeSart DJ: Cadmium gradient in human and bovine enamel. *J Dent Res* 1980;59:1072.

Shrestha BM, Mundorff SA, Bibby BG: Enamel dissolution I effects of various agents and titanium tetrafluoride. *J Dent Res* 1977;51:1561–1566.

Smith JC, et al: Evaluation of published data pertaining to mineral composition of human tissue, in "Workshop on Research Needed to Improve Data on Mineral Content of Human Tissues." *Fed Proc* 1981;40:2120–2125.

Söremark R, Samsahl K: Gamma-ray spectrometric analysis of elements in normal human enamel. *Arch Oral Biol* 1961;6:275–283.

Söremark R, Samsahl K: Gamma-ray spectrometric analysis of elements in normal human dentin. *J Dent Res* 1962a;41:603–606.

Söremark R, Samsahl K: Analysis of inorganic constituents in dental calculus by means of neutron activation and gamma-ray spectrometry. *J Dent Res* 1962b;41:596.

Söremark R, Lundberg M: Gamma-ray spectrometric analysis of the concentrations of Cr, Ag, Fe, Co, Rb and Pt in normal human enamel. *Acta Odont Scand* 1964a;22:225–259.

Söremark R, Lundberg M: Analysis of the concentrations of Cr, Ag, Fe, Co, Pt and Rb in normal human dentine. *Odont Revy* 1964b;15:285–289.

Sowden EM, Stitch SR: Trace elements in human tissue. *Biochem J* 1957;67: 104–109.

Spector PC, Curzon MEJ: Relationship of Sr in drinking water and surface enamel. *J Dent Res* 1978;57:55–58.

Stack MV, Burkitt AD, Nickless SA: Trace metals in teeth at birth. *Bull Environ Conc and Toxic* 1976;16:764–766.

Steadman LT, Brudevold F, Smith FA: Distribution of Sr in teeth from different geographic areas. *JADA* 1958;57:340–344.

Steadman LT, Brudevold F, Smith FA, et al: Trace elements in ancient Indian teeth. *J Dent Res* 1959;38:285–292.

Steinnes E, Dahm S, Furseth R: Concentration of rare earths in dentine and enamel (A pilot study). *Acta Odont Scand* 1974;32:125–129.

Strehlow CD, Kneip TJ: The distribution of lead and zinc in the human skeleton. *Am Ind Hyg Assoc J* 1969;30:372–378.

Theilade J, Schroeder HE: Recent results in dental calculus research. *Int Dent J* 1966;16:205–221.

Underwood EJ: *Trace Elements in Human and Mammal Nutrition,* ed 4. New York, Academic Press, 1977.

Vaughan JM: *The Physiology of Bone.* Oxford, Clarendon Press, 1970, p 99.

Vrbic V, Stupar J: Dental caries and the concentration of aluminum and strontium in enamel. *Caries Res* 1980;14:141–147.

Weatherell JA, Weidmann SM, Hamm SM: Sampling of enamel particles by means of strong acids for density measurements. *Arch Oral Biol* 1966;11: 107–111.
Wolf N, Gedalia I, Yariv S, Zuckermann H: The Sr content of bones and teeth of human foetuses. *Arch Oral Biol* 1973;18:233–238.
Young RA: Implications of atomic substitution and other structural details in apatites. *J Dent Res* 1974;53:193–203.

4 Trace Elements in Saliva

T. W. Cutress

No broad screening of saliva for total trace element content has been attempted although the techniques to do so are available. Saliva presents major problems in evaluation because of its heterogeneous and variable nature which includes physiologic as well as biochemical factors. The composition and secretion of saliva and their influence on oral health have been the subject of numerous reviews and symposia that adequately emphasize the complexity of the subject (Shannon et al, 1974).

Saliva Sampling

Saliva is secreted from four glandular sources—parotid, submandibular, sublingual, and mucosal. With few exceptions interest in salivary composition has been restricted to the major elements and a few trace elements which contribute to the electrolyte balance. Different glands not only secrete saliva of differing composition, but also their contribution to the total saliva output varies with flow rate. Submandibular, parotid, and sublingual glands contribute less than 60% of total saliva production; the remainder is derived from the small mucosal glands. Saliva, together with the oral flora, diet, and dental enamel, must be considered as a potentially important factor in the cause of caries, because the oral environment is pervaded at all times by a salivary phase which influences both dental plaque and the enamel surface. It is well known that damage to glands resulting in diminished salivary flow is commonly associated with an increased incidence of caries.

Studies on Whole, or Mixed, Saliva

Studies on whole (mixed) saliva overcome the problems of variations in saliva composition from different glands and have advantages because they deal with a more representative salivary environment contributing to oral biology. However, mixed saliva is nevertheless subject to changed composition with varying flow from contributing sources, particularly as the contribution of each gland varies with flow rate, time of day, and type of stimuli.

In 1961 Afonsky published a comprehensive review on saliva. At that time observations on the constituents of saliva rather than major components were predominantly concerned with Na, K, Cl, and Mg with Fe, Cu, Co, S, Br, Mo, and Ni receiving brief mention.

Afonsky concluded:

1. The concentration of Na in unstimulated saliva varied from 1 to 65 mg %, but the concentration is increased on stimulation. It was also suggested that Na is the predominant cation in the buffer system for stimulated saliva but that no relationship between oral health and Na concentration in saliva has been demonstrated. Salivary levels of Na are not affected by serum Na levels.
2. The concentration of K in unstimulated saliva varied between 30 to 95 mg %; it varied greatly between individuals and was not much affected by stimulation. Salivary levels of K are not affected by serum K levels. Thus, K is the major cation of the buffer system of unstimulated saliva.
3. The concentration of Cl in unstimulated saliva varied between 30 and 145 mg %. Levels increased in stimulation but were not influenced by blood circulation of Cl.
4. The concentration of Mg was possibly 0.1 to 0.7 mg %, but techniques available for analysis gave rise to doubt on the range of concentration.

5. The concentration of Fe varied from 0 to 0.6 mg/g; Cu from 10 to 48 μg % with wide individual variations; Co from 0 to 13 μg %; Br from 0.2 to 7.1 mg/1.
6. The concentration of S varied from 3 to 20 mg and was probably associated with salivary thiocyanates.

Trace Element Concentrations in Saliva

Of the ten reports on trace element concentrations in saliva (Table 4-1) only three (Dreizen and Levy, 1970; Cutress, 1972; Schamschula et al, 1978) looked for substantial numbers of elements. Dreizen and Levy (1970), in a study primarily concerned with comparing the effects of varying stimuli on the trace elemental composition of mixed saliva, observed that Fe was the only element consistently identified in all saliva samples. Most samples contained Sn while Mo and Cd were present least often. It was concluded that the salivary glands produce secretions which differ in trace element composition according to the type and intensity of the stimulus applied.

Cutress (1972) studied the flow rates and trace element composition of saliva from trisomy 21 and other mentally retarded persons. Trisomic subjects were found to have slow rates of salivary secretion, but trace element concentrations differed little for mixed or parotid salivas between the two groups being studied or between the two types of saliva. Nevertheless, the output of some elements over a time period of one hour differed considerably because of variations in flow rate of the saliva. For example, trisomy 21 subjects on average secreted 74 μg Mg per hour compared with 154 μg for non-trisomic subjects. Similarly trisomics secreted 35 μg compared with 14 mg K, and 34 μg compared with 9 μg Fe.

Mixed saliva from pregnant women has been reported (Kayavis and Papanayotou, 1976) high in Mg which may be explained by the increase in alkaline phosphates in the saliva. Interest in this subject may aid in furthering knowledge of eclampsia.

A study of some minor elements in the saliva (parotid) of smokers and nonsmokers reported similar levels of Cl, Na, and Mg with Cl and Na concentrations showing diurnal variations, with increase during sleeping hours. It appears, however, that K is influenced by nicotine action being higher in smokers' parotid saliva. No diurnal variation was reported (Dogan et al, 1971).

Lear and Grøn (1968), apart from observing an inverse relationship between saliva flow rate and Mg concentration, suggested that the higher Mg in resting parotid saliva produces higher Mg in calculus forming on teeth adjacent to Stensen's duct. They also proposed that low Mg levels found on mandibular teeth could be explained by the lower Mg content of unstimulated submandibular saliva.

Table 4-1
Summary of All Reports on Trace Elements in Saliva

Author (year)	Element	Saliva Type	No.	Location	Concentration μg/100 ml
Dreizen et al, 1952	Cu	Mixed	48	United States	10–48 μg/100
	Co		37		0–13 μg/100
	Ni		21		ND
	Mo		21		ND
Gow, 1965	Mg	Mixed	183	Australia	356–506
Dawes, 1967	Mg	Parotid	10	United Kingdom	377
Lear and Grøn, 1968	Mg	Parotid		United States	318
Dreizen and Levy, 1970	Zn	Mixed	30	United States	< 13.5
	Cd				< 1.5
	Pb				35
	Fe				17.1
	Mo				< 2.4
	Cu				1.9
	Sn				64.8
	Ni				< 1.1
	Al				10.5
	Mn				1.7
	Cr				8.6
	Ce				< 4.3
	Sr				1.3
	Ba				< 3.2

Table 4-1 (continued)

Author (year)	Element	Saliva Type	No.	Location	Concentration μg/100 ml	
Dogon et al, 1971					Nonsmoker	Smoker
					ppm	
	Cl				816–1136	710–1278
	Mg				8– 13	8– 13
	Na				552– 666	437– 827
	K				547– 665	743– 899
Brudevold et al, 1977	Pb	Mixed	251	United States	44	
					Mixed	Parotid
Cutress, 1972	Na	Mixed and parotid	31	New Zealand	20,800	12,500
	K				69,800	82,800
	Mg				360	200
	Mn				2	2
	Si				160	290
	Al				160	98
	Mo				0.3	0.3
	Cu				18	48
	Fe				60	15
	Ti				5	7
	Ba				6	3

Table 4-1 (continued)

Author (year)	Element	Saliva Type	No.	Location	Concentration μg/100 ml	
	Sr				1	2
	Cr				1	1
	Zn				60	60
Kayavis and Papanayotou, 1976	Mg	Mixed	354	Greece	Pregnant	630
					Nonpregnant	500
Schamschula et al, 1978	Sr	Mixed	283	Papua-New Guinea	9	
	Li				0.07	
	K				96,300	
	Mg				722	
	Zn				26	
	Pb				2	
	Cu				ND	
	Ba				8	

Some studies have shown that some salivary constituents such as Ca or protein (Dawes, 1967, 1969) vary with individual salivary flow. It has been shown that Mg varies inversely with flow rate (Dawes, 1967; Lear and Grøn, 1968; Gow, 1965). However, it should be noted that these variations apply to changing rates of flow in individuals; it cannot be assumed that persons with low salivary flow rates have, for example, higher salivary Mg than persons with high flow rates (Cutress, 1972).

The element Pb is present only at low levels in saliva and is not much influenced by Pb poisoning. Brudevold et al (1977) observed that saliva Pb levels were only about one-tenth that in blood, and while blood Pb levels were much increased in poisoning cases, saliva levels were only slightly raised. Saliva Pb levels reported by Brudevold et al are much lower than those found by Dreizen and Levy (1970) and Schamschula et al (1978).

TRACE ELEMENTS IN SALIVA AND DENTAL CARIES

Schamschula et al (1978), like Dreizen and Levy (1970) and Cutress (1972), studied unusual population groups in an attempt to identify potential salivary effect on caries susceptibility and resistance. For seven trace elements analyzed in salivas collected from four villages in Papua-New Guinea it was concluded that, with the exception of Pb, all trace elements showed an inverse trend between concentration and caries experience. Elements not detected in any samples were Sr, Cr, and Mn.

The most interesting finding was related to Li, which showed a strong inverse relationship to caries experience but was present in saliva in very low concentrations (0.0007 ppm)—a finding supported by similar correlations found in enamel and plaque. Substantiation of the suggested relationship of Li with caries in future studies is awaited with much interest. From these three studies and a few studies analyzing for single or few elements (Dreizen et al, 1952; Gow, 1965), a comparison for the few elements in mixed saliva common to two or more studies can be made (Table 4-2). Agreement is good between respective investigators for Zn, Pb, Mo, Cu, Al, Mn, Sr, Ba, Mg, and K. This increases confidence in the prospects of future saliva studies providing significant and comparable observations on the trace element content of saliva.

SUMMARY

A review of the literature shows how little research has been done on trace elements in saliva. Because saliva is constantly bathing the teeth it plays an important role in the prevention or remineralization of caries.

Table 4-2
Trace Elements in Saliva (Comparison of Selected Studies)

Element	Dreizen et al 1952	Gow 1965	Dreizen and Levy 1970	Cutress 1972	Schamschula et al 1978
Zn			0.17	0.6	0.26
Cd			0.02		
Pb			0.04		0.02
Fe			< 0.25	1	
Mo	ND		0.02	0.001	
Cu	0.1–0.5		0.01	0.1	
Sn			0.48		
Ni	ND		0.01		
Al			0.1	0.9	
Mn			0.02	0.0	
Cr			0.35	0.01	
Cs			< 0.04		
Sr			0.01	0.01	0.09
Ba			< 0.03	0.1	0.08
Na				208	
K				698	963
Mg		3.6–5.1		4	7
Si				1	
Ti				0.04	
Li					0.0007
Co	0–0.1				

ND = not done.

The data we have show that a limited number of elements (Zn, Pb, Fe, Mo, Cu, Al, Mn, Cr, Sr, Ba, K, and Mg) have been found in saliva. In most cases concentrations are very low. In only one study has the trace element content of saliva been related to caries. This study showed a strong inverse relationship of caries to Li, a finding that remains to be substantiated.

REFERENCES

Afonsky I: *Saliva: Survey of the Literature. A Review.* Birmingham, University of Alabama Press, 1961.

Brudevold F, Aasenden R, Srinivasian BN, et al: Lead in enamel and saliva, dental caries and the use of enamel biopsies for measuring past exposure to Pb. *J Dent Res* 1977;56:1165–1171.

Cutress TW: Comparison of flow-rate and pH of mixed and parotid salivas from trisomic 21 and other mentally retarded subjects. *Arch Oral Biol* 1972;17: 1081–1094.

Dawes C: The secretion of Mg and Ca in human parotid saliva. *Caries Res* 1967; 1:333–342.

Dawes C: The effects of flow-rate and duration of stimulation on the concentrations of protein and the main electrolytes in human parotid saliva. *Arch Oral Biol* 1969;14:277–294.

Dogon IL, Amdur BH, Bell K: Observations on trace element variation of some inorganic constituents of human parotid saliva in smokers and non-smokers. *Arch Oral Biol* 1971;16:95–105.

Dreizen S, Spies HA, Spies BS: Copper and cobalt levels in human saliva and dental caries. *J Dent Res* 1952;31:137–142.

Dreizen S, Levy BM: Comparative concentrations of selected trace metals in human and marmoset saliva. *Arch Oral Biol* 1970;15:179–188.

Gow BS: Analysis of metals in saliva by atomic absorption spectroscopy. II. Magnesium. *J Dent Res* 1965;44:890–894.

Kayavis J, Papanoyotou P: Salivary Mg levels in pregnant women. *J Dent Res* 1976;55:706–708.

Lear RD, Grøn P: Mg in human saliva. *Arch Oral Biol* 1968;13:1311–1319.

Schamschula RG, Adkins BL, Charlton KG, et al: *WHO Study of Dental Caries Etiology in Papua-New Guinea.* Geneva, WHO Publication #40, 1978.

Shannon IL, Suddick RP, Dowd EJ: Saliva: Composition and secretion, in *Monographs in Oral Science* vol. 2, Basel, S. Kause, 1974.

5 Trace Elements in Dental Plaque

J. L. Hardwick

The term *dental plaque* will be applied to those soft deposits that adhere to the surface of teeth exposed to the oral environment and that can be removed for analysis, avoiding as far as possible contamination with saliva, calculus, or food debris. Dental plaque consists almost entirely of deposits acquired after eruption but in a few sites may include small amounts of the remains of the original formative tissues of enamel.

Unless collected within a few hours after the removal of the previous plaque, it will contain dense concentrations of micro-organisms of many types lying within a matrix of organic and inorganic matter. This matrix is usually formed of products produced by the breakdown by plaque bacteria of saliva, gingival fluid, desquamated epithelial cells, and food debris collecting on the surfaces of the teeth. Once formed, plaque will remain in sheltered areas almost indefinitely but in other areas much of it will be removed by mastication or tooth cleaning. Therefore, within the same mouth it is present in different stages of maturity; on occlusal surfaces, apart from the fissures, little plaque is present and, where present, much will be recently formed whereas it may never be removed in well-protected areas.

The bacteria use the food debris, saliva, gingival fluid, and desquamated cells for their growth and reproduction. In the absence of food, the residues of enzymic and bacterial action will consist largely of modified mucosubstances and relatively insoluble calcium phosphate-protein complexes. However, in the presence of some sugars (especially sucrose) derived from food deposits, certain types of plaque bacteria produce large amounts of high molecular weight, sticky, extracellular material, mainly polymers of glucose. Such polymers greatly increase the bulk of the plaque and therefore its thickness, and will alter significantly the proportions of its carbohydrate or protein moieties. It is believed that these polysaccharide polymers restrict diffusion through the plaque, thus favoring anaerobic conditions and increased production of acid metabolites by the plaque bacteria. They also form a reservoir of carbohydrate which, together with reservoirs of carbohydrates within the bacterial cells, will enable the period of acid production to be prolonged.

About 80% to 90% of plaque consists of water with dissolved salts. As the plaque is often only a few microns thick, it would be expected that dissolved inorganic salts and free ions would quickly diffuse to the plaque boundaries where they would be lost to the saliva. However, the passage of such ions will be restricted by the presence of the polysaccharide polymers and insoluble inorganic matter. In addition, the proteins of the plaque matrix will act as poly-electrolytes and, depending largely on the pH of the environment, will selectively bond or restrict the movement of charged particles, including trace elements.

The extracellular calcium and phosphate content of the plaque (and of the saliva from which it is largely formed) is high and at neutral pH is believed usually to be supersaturated with calcium and orthophosphates. Some of the calcium and phosphate ions will be bonded to proteins but some are likely to precipitate slowly, initially in a poorly structured, amorphous form. Such amorphous calcium phosphates are also likely selectively to bond with other ions, including trace elements, slowing their loss from the plaque; with a fall in pH some of these bound ions may be liberated again. Eventually, if left undisturbed, the amorphous calcium phosphates may become crystalline as plaque is transformed to calculus.

This brief consideration of the composition and nature of dental plaque indicates that it is a most heterogeneous material. Its composition is likely to vary considerably between different sites in the same mouth or even on the same tooth. Indeed, dental plaque in a mouth consists of a conglomeration of many small ecosystems. The composition of these ecosystems will change rapidly according to the availability of the substrates from which they are being derived, variations in pH, degree of anaerobiasis, and their maturity.

Problems of Trace Element Analysis of Plaque

In this section the difficulties associated with the trace element analysis of dental plaque and the interpretation of data in relation to the cause of dental caries will be discussed.

The size of the sample of plaque for analysis Although plaque covers most surfaces of erupted teeth, it is usually only a few microns thick. In a mouth it may often not be possible to collect more than 10 mg plaque from all surfaces of all teeth. Since many trace elements produce biological effects at ppm levels and sometimes below, highly sensitive analytical techniques must be employed. Until the late 1950s analytical techniques for many elements of the required sensitivity were not available. For instance, with fluorides an isotope-dilution method (Kudahl et al, 1962) sensitive to near ng levels had to be developed specifically for analysis of plaque to show that fluorides were present in plaque in much higher concentrations than in saliva (Hardwick, 1963; Kudahl, 1963). It was discarded when an adequate, cheaper, and less sensitive method of chemical analysis (Singer and Armstrong, 1959) became available. Despite the rapid advances in microanalytical techniques the content of a few elements can still be estimated with reasonable precision only in plaque samples pooled from several teeth.

The heterogeneity of dental plaque The trace element content of plaque is likely to vary substantially between sites on the same tooth and on the same tooth from day to day. As it is often possible to analyze only samples pooled from several teeth—and sometimes even pooled samples collected from the same mouth over several days—such samples can only provide a rough guide to the general levels of trace elements throughout the mouth. It is therefore not surprising to find that trace element contents of dental plaque from different mouths tend to show a greater variance about the mean than do most biologic tissues.

Contamination It is obvious that microanalysis to the low detection limits required demands scrupulous care to avoid adventitious contamination of the samples at every stage of their collection and subsequent analysis. This is particularly difficult with plaque sampling when contamination with saliva and food debris is a constant problem. The instruments used to collect plaque are a further source of contamination if metal or uncoated instruments are used.

The relevance of the analytical data Most trace elements produce their biologic effects only when present in the tissues in a specific chemical form or physical state. For instance, for some trace elements in Group D (see below) it is clear that only elements in an ionic or ionizable form are likely to be incorporated into the apatite lattice. They will not be available for incorporation if they are very strongly complexed or bonded to another ion or to organic matter.

Some analytical techniques estimate the total content of the actual element (e.g., atomic absorption spectrophotometry; spark source mass spectrometry). Other techniques estimate only the content of an element in a specific chemical form or physical state. A particularly good example of this type of technique is the F′ electrode which estimates F′ activity. If the analysis of a sample is conducted as usually recommended with adjustment of the ionic strength and of the pH to about 5.5 of both the samples and controls, it will estimate this at pH 5.5. However, if the sample has been ashed or diffused before the analysis with the F′ electrode, it will estimate the F′ activity of the solution in which the ash (or diffusate) is dissolved. After undertaking the necessary calculations the estimate of the F content of the original sample will usually be the sum of its ionic F and F which was ionizable under the ashing or diffusing conditions.

Before attempting to interpret the relevance of the analytical trace element estimations obtained on a tissue to a biologic process it is essential to understand what has been estimated. The total content of a trace element may have little relevance if it is biologically active only in a specific chemical form or physical state and if it is present in the sample only in inactive forms. Fortunately, enough is known about which chemical forms or physical states are likely to be active for most elements in biologic tissues to indicate the most appropriate method of analysis. In the more homogeneous dental enamel the choice will be easier than in a dynamic and heterogeneous biologic material such as dental plaque.

Interactions between trace elements The action of a trace element in a biologic tissue may often be modified by the presence of another element which potentiates or inhibits its action. Occasionally two or more trace elements can perform essentially the same role and therefore can act as replacements for each other. Several interactions are well recognized in trace elements influencing enzyme activity and may be relevant to the activities of plaque micro-organisms.

The presence of certain trace elements in Group D (see below) in plaque fluid is also likely to have an important influence on the rate of solution under acid conditions of the inorganic moiety of the underlying hard dental tissues and therefore on caries. In plaque fluid only ions in trace amounts which can either replace or bond strongly with ions present in the apatite lattice of enamel are likely to influence the carious process in this way.

It is theoretically possible that the same trace element (eg, F′) may have an effect both on the enzymic activities of the plaque micro-organisms and on the rate of solution of the underlying hard tissues.

The complexity of the possible interactions between trace elements makes it desirable that investigations on the relation of caries experience to trace element content of dental plaque be undertaken simultaneously on several of the more likely trace elements in the same plaque sample or on aliquots from the same plaque sample.

Trace Element Content of Dental Plaque Related to Dental Caries

The mechanism which primarily initiates the carious process is located in the dental plaque although its effects are observed in the adjacent hard dental tissues. It is well recognized that plaque micro-organisms produce the acid metabolites which decalcify the underlying hard dental tissues, the essential process of dental caries.

Although the relation of plaque micro-organisms to caries has been studied extensively, until recently the trace element content of plaque has received little attention. Yet it was known that some trace elements such as F produce significant effects on the growth and metabolism of several bacteria, and that the presence of others in plaque might influence the rate of solution of the underlying enamel at low plaque pH. This lack of research on trace elements in plaque apart from F was largely due to the difficulties inherent in trace element analysis of plaque samples.

The significance of elements present in trace amounts in plaque with reference to the carious process must be assessed. Elements present in trace amounts may be classified into four groups according to the role they play.

In assessing the possible significance of an element it will be necessary to attempt to classify the element into one or more of these four groups to obtain some indication as to whether it is likely to have any effect and, if so, at what concentration the potentiating or inhibiting action is likely to occur. It is therefore desirable to comment on the relevance of this classification to caries initiation and progress.

Group A Those with no known biologic role, probably the largest group. The composition of substances entering the mouth from other parts of the body and from food, drinks, dental restorations, air, culinary or other instruments as well as from drugs, medicines, tooth pastes, and cosmetics is diverse. At one time or another in life all elements will probably have been present in the oral cavity and may enter the dental plaque. Over half the elements of the periodic table have been detected, and there is every reason to assume that others may be present.

Most of the elements in plaque are unlikely to have an effect on the initiation or progress of caries.

Group B Elements necessary for enzymic activities (essential to the health of the cell due to their connection with its enzyme processes). These are usually present as metallo-enzymes, firmly associated with and in a precise proportion to the protein of the enzyme or in a less rigid and specific association with an enzyme which they activate. This group includes Fe, Cu, Zn, Mo, I, Co, and Mn. Sometimes one trace element may replace another in the enzyme system without loss of the enzymic activity.

This group will chiefly affect the growth, reproduction, and metabolism of the plaque bacteria. In dental caries they will act largely by favoring some types of micro-organisms at the expense of others with which they are in competition for the available substrates. A few are likely to have an effect on caries.

Group C Those which inhibit essential enzyme processes of cells. Many trace elements, which in low concentrations are essential to the enzyme activity of cells, are injurious to their growth and metabolism in much higher concentrations. In such cases, the relation between the activities of the cells and the concentration of the trace element will not be linear, thus making this relation more difficult to demonstrate.

An inhibitory effect of the F′ on the formation of acid metabolites from carbohydrate substrates by plaque bacteria or bacteria in salivary sediments was demonstrated about 30 years ago. Other trace elements may have a similar action.

Group D A miscellaneous group of essential elements whose metabolic role is often not well understood, usually connected with some electrochemical, nutrient, or structural function.

When caries is active, metabolites produced by the micro-organisms of the dental plaque cause solution of the main inorganic structural units, the apatite cells, of the underlying dental tissues. The unit hydroxyapatite cell is often assumed to have the formula $Ca_{10}(OH)_2(PO_4)_6$, but each of its constituent ions may be replaced by a few other specific ions without destroying the crystal lattice. Such apparently minor substitutions may affect its characteristics and, most importantly in caries, its solubility or rate of solution.

The effect of the substitution of hydroxyl ions by F′ on enamel solubility is well recognized. In this substitution the F′ has a similar charge and a closely similar ionic radius to the hydroxyl ion. This is not so with most other substitutions which therefore often require some rearrangement and/or substitution of other ions within the lattice to maintain charge balance and approximately the same spatial arrangement of ions within the lattice. These substitutions can only occur to a limited extent but may have a major effect on the solubility of the apatite and the initiation and progress of caries.

These effects are likely to be especially significant at a pH near the so-called critical pH of 5.0 to 5.5 often found in fermenting plaque. At the level of Ca and phosphate ion activities present in dental plaque, and at pH values above this range, most apatites will be less soluble than either CaF_2 or dicalcium orthophosphate (as brushite $CaHPO_4 2H_2O$), whereas below this range the reverse is usually true, even for fluorapatites. Therefore, after a fall in pH to this critical level, as the pH slowly rises again the presence of a few specific elements in ionic form in the plaque may well determine whether precipitation occurs. This can produce apatite cells in the plaque fluid or on the surface of the enamel crystals (with the reformation of any apatite cells dissolved by the earlier drop in pH), or as amorphous or crystalline CaF_2 or dicalcium phosphate. Furthermore, if a trace ion such as F reduces the free energy of the apatite and is available in an ionic form in plaque, any apatite cells reprecipitated on the enamel crystals are likely to have a higher content of that trace element and to be less soluble than those which were dissolved earlier. This phenomenon has been shown to occur in vitro with F under conditions of fluctuating pH (Ramsey et al, 1973).

Finally, any element present in plaque fluid that bonds strongly with one of the ions present in enamel apatite, at or below the critical pH, will reduce the activity in the plaque of that ion and thereby increase the rate of solution of the surface cells of the enamel apatite crystals. It is sometimes assumed that the rate of solution of the inorganic constituents of enamel in the slow decalcification that occurs in caries is determined solely by the inherent solubility of these cells. Their rate of solution in vivo, especially around the so-called critical pH, will be determined not only by their solubility but also by the concentrations of the free ions, of which they are composed, in the dental plaque. It is therefore clear that some elements in Group D when present in plaque in trace amounts may significantly affect caries.

TRACE ELEMENTS IN PLAQUE

Investigations on Trace Elements Detected in Dental Plaque

Because of possible interactions among trace elements in dental plaque, it is important to have information on as many of the elements present in a single plaque sample as it is possible to obtain. Three investigations have been undertaken primarily for this purpose, two using spark source mass spectrometric methods (Hardwick and Martin, 1967; Swift, 1967) and one X-ray spectroscopy (Puttnam et al, 1966). Mass spectrometry theoretically permits detection in a single sample of all elements from mass number 8 to 240, and X-ray spectroscopy elements with atomic numbers above 12. Neither method would detect Li, now claimed to have

Table 5-1
Summary of Reports on Concentrations (dry weight) of Elements in Dental Plaque

Element	Study		
	Hardwick and Martin 1967	*Swift 1967*	*Puttnam 1966*
Major Elements (%)			
K	Present	about 5	1.1
P	Present	about 3	0.45
Ca	Present	about 3	0.22
Na	Present	about 1	NG
Minor Elements (%)			
Mg	Present	0.3	NG
Cl	Present	0.3	NG
S	Present	NG	NG
Trace Elements (ppm)			
Si	(100)	70	NG
Fe	90[a]	230	92
Zn	80	30	66
Mn	60	70	4
Rb	50	35	NG
Br	40	3	NG
Cu	25	80	35
Pt	(12)	NG	NG
Al	11	35	NG
Sn	10	5	16
Ni	9	NG	8
Ti	9	NG	NG
Cr	8	7	10
W	NG	8	NG
Mo	6	0.8	2
Pb	5	6	7
Te	NG	5	NG
Sb	4	NG	NG
Sr	2.5	2	NG
Ag	1.2	0.3	9
Ba	1.1	2	NG
Ge	1.0	NG	NG
Au	1.0	NG	NG
F	1.0	NG	NG
Co	0.9	0.7	6
B	0.8	NG	NG
V	0.8	2	NG
I	0.7	2	NG
Zr	0.6	0.8	NG
Ru	0.6	NG	NG
In	0.5	NG	NG
Se	0.5	30	NG
La	0.4	NG	NG

Table 5-1 (continued)

Element	Study		
	Hardwick and Martin 1967	*Swift 1967*	*Puttnam 1966*
As	0.3	3	NG
Y	0.2	NG	NG
Sc	0.2	NG	NG
Nb	NG	0.1	NG
Samples	Mean of 2 pooled from 40 subjects	One sample	Mean of up to 14 pooled from 5 subjects

( ) = Concentration probably high due to contamination. NG = Detection not mentioned in publication.
[a]Significant contamination from the electrode die was not detected in graphite electrode "blanks."

significant associations with dental caries. Nor would mass spectrometry detect elements used in the preparation of the electrodes such as carbon or rhenium. A summary of these investigations is shown in Table 5-1.

In the study of Hardwick and Martin their reported ashed weights have been converted to dry weight, assuming that during drying the weight of plaque is reduced by 83%. These authors did not claim that their analyses were more than semiquantitative at this stage of the development of their technique: in Table 5-1 the best estimate is given for each element.

With few exceptions, there is general agreement about the concentrations of the elements detected in the three studies; recognizing the heterogeneity of dental plaque and the difficulties in collection of plaque free from contamination with saliva, food debris, blood, and calculus are accepted.

Several points are noteworthy about the data in Table 5-1. Puttnam's estimates of the concentrations of major elements are lower than those of most other investigators. The concentration of F is considerably lower than that observed in most other investigations but no special precautions were taken by Hardwick and Martin to avoid its loss due to its volatility during ashing. The presence of about half the elements has been detected in dental plaque samples. As expected, elements in the alkaline earth group or with multiple valences are particularly well represented. The high concentrations of K and Rb probably chiefly reflect the bacterial cell content of plaque.

The elements detected and their order in a concentration scale, show reasonably close agreement with those in the hard dental tissues (cf, Table 3-13, or Hardwick and Martin, 1967). This may indicate some similarities in chemical or physical characteristics of the biologic materials despite their entirely different origins.

Although the presence of most elements considered to have a possible association with dental caries has been detected, there is no evidence that the remainder play a significant role in dental caries. It is probable that they are adventitious contaminants from the oral cavity. Both methods estimate the total concentration of each element in the sample but do not show to what extent each element is, in a chemical or physical form, likely to influence the carious process. The importance of this has already been discussed.

Trace Elements Other Than Fluoride in Dental Plaque

It will be shown later that, according to the two main widely accepted theories of the action of fluoride (one an effect on bacterial metabolism and the other an effect on the solution rate of the underlying enamel), its action is probably mediated by free F′ and that a reservoir of bound ionizable F is present in plaque. Although trace elements other than F in plaque may influence caries by a direct action on bacterial metabolism, or the rate of solution of the enamel, it is also possible that some may act indirectly by affecting the availability of free F′ or their storage in bound form. The possibility of such interactions between trace elements, not only with F, must always be considered.

Few investigations have been undertaken on the content of trace elements other than F in plaque, and in the knowledge of the author only two on their relation to caries experience. In view of the exceptionally extensive nature of one of these studies, its general outline should be described briefly.

The Papua-New Guinea study This study has been reported in nearly 20 publications and comprehensively in Schamschula et al (1978). The subjects of this study were 301 12- to 24-year-old natives who had lived all their lives in 16 villages in part of the drainage area of the Sepik River. To analyze the data obtained, several villages were grouped together in four clusters of those villages close to each other and whose environment and inhabitants had similar important characteristics. Soil parameters varied between the groups of villages, and most of the food used was produced locally and included little sucrose. The subjects did not practice oral hygiene, and their exposure to F was exceptionally low. Sugar consumption, oral hygiene, and exposure to F are well recognized as major factors affecting the incidence of caries. As sugar consumption and oral hygiene would have minimal effects on caries incidence observed in these subjects and as their exposure to F, although differing between the groups, was also low, it was likely that effects from other less dramatic factors affecting caries, such as trace elements, would be recognized more easily than in most other societies. Furthermore, although overall the mean cumulative caries experience was low compared with that observed in most advanced societies, it ranged from nil in one group of villages to a DMFT of 6.4 in the group of villages with the highest caries experience and was still higher in individual villages and individual subjects.

For each subject, data on an exceptionally large number of variables which might influence caries incidence were collected, including among others the concentrations of several major and trace elements present in plaque (see Chapter 2). Indeed two of the primary aims of the study were to test hypotheses that the concentrations of any of 10 elements in plaque, saliva, and surface enamel or of any of 16 elements in soil or food produced in village gardens were related to the caries experience of their inhabitants. The data obtained were largely analyzed by advanced statistical methods designed to enable detection of interactions between elements. The microanalytic techniques used were similar to those in Australian investigations (Agus et al, 1976).

Major elements in plaque Because of possible interactions between trace elements and other constituents of plaque, the significance of trace elements cannot be assessed adequately without knowledge of the major elements present. Many of the investigations on major elements have used analytic methods which estimate the total content of an element, disregarding its chemical form and physical state.

Table 5-2 summarizes the findings of a selection of typical and extensive investigations and indicates the levels of some major elements. Hydrogen and carbon have been omitted from the table because their concentration as elements gives little understanding of the role they play. Similar considerations apply to O, N, and S while Cl and Na are present in concentrations of about 1000 ppm or above.

The data in Table 5-2 show reasonable agreement for P, Mg, and K but less so for Ca. The variability in the reported Ca concentrations may be associated with the maturity of the plaque samples analyzed. Grøn et al (1969) in further investigations on freshly forming buccal plaque, grown under somewhat artificial conditions, showed that the Ca content rose steadily from $0.53 \pm 0.44\%$ at day 1 to $1.55 \pm 1.10\%$ at day 4: P_i and Mg did not show a similar change with degree of maturity.

Tatevoissian and Gould (1976) have also studied the concentrations of some major elements in plaque fluid. The following concentrations (mMol/l) were found:

K	61.5 ± 13.5	P_i	14.2 ± 3.1
Na	35.1 ± 9.0	Mg	3.7 ± 1.1
Ca	6.5 ± 2.1		

Table 5-2
Concentrations (% dry weight) of Calcium, Phosphorus, Magnesium, and Potassium in Dental Plaque

	Mean Concentrations (± SD)				
Author (year)	*No.*	*Calcium*	*Phosphorus*	*Magnesium*	*Potassium*
Dawes et al, 1962	16–85	0.45 ± 0.21	1.65 ± 0.42	0.17 ± 0.33	2.30 ± 0.36
Grøn et al, 1969	8	1.68 ± 0.62	1.26 ± 0.24 (P_1)	0.15 ± 0.08	NG
Ashley, 1975	33	0.55	1.38	NG	NG
Schamschula et al, 1977	72	0.72 ± 0.68	1.42 ± 0.38	0.19 ± 0.04	2.75 ± 0.46
Schamschula et al, 1978	293	2.44 ± 1.51	1.69 ± 0.70	0.16 ± 0.07	1.76 ± 0.58
Tatevoissian and Gould, 1979 (plaque residue)	56–79	0.68 ± 0.28	1.15 ± 0.2	0.22 ± 0.19	2.57 ± 0.38

NG = not given

With the exception of Na in gingival fluid (88.8 ± 31.4), these mean levels were usually several times as high as those found in saliva or gingival fluid. Ionic Ca activity measurements indicated, as would be expected, that the solubility product of apatite at neutral pH was grossly exceeded. It must therefore be assumed that the Ca and inorganic phosphate are not present to a substantial extent as free ions but must be present largely either as "complexed or protein-bound Ca" or as the initial stages of the formation of so-called "amorphous" poorly structured Ca and phosphate-containing precipitates, as postulated by Hardwick (1971a).

In the Papua-New Guinea studies Schamschula et al (1978) found that the Ca content, as dry weight, of samples of plaque was nearly three times as great as it was in the solids in stimulated saliva. (For F it was nearly 20 times as great.) A similar relation did not apply for several of the other elements, and some showed a reverse relation. In this study Ca concentrations were unusually high but were not considered to be affected by betel nut chewing, a habit of some of the subjects. It may be noteworthy that the subjects did not practice oral hygiene and thus their plaque was mature.

Minor and trace elements in plaque Table 5-3 summarizes the findings of six studies in which concentrations of five trace elements in plaque were studied. Beighton et al (1977) have also analyzed plaque for Cd and found a median value of 1.0 ppm with a range from 0.1 to 190 ppm.

Where the same elements have been studied in several investigations, there is a wide range in estimates. Where the standard deviations are reported, they are often larger than the mean, also indicating a wide range with a skewed distribution.

Interrelations Between Elements in Plaque; Relations of Individual Elements to Cumulative Caries Experience

Two extensive studies, undertaken in Papua-New Guinea and in Australia (Schamschula et al, 1978 and 1978a), have analyzed interrelations between the plaque concentrations of ten elements and the relation between their concentrations and the cumulative caries experience of the subjects from whom the plaque was collected. The age of the subjects ranged from 12 to 24 and 9.7 to 13.0 years respectively.

In Table 5-4 a matrix of product-moment correlation coefficients between certain elements is shown. Remarkably high correlation coefficients for the relation between Ca and P concentrations and Ca:P ratios were observed in both studies. This suggests that Ca and P are in some way associated with each other, possibly as a calcium phosphate. Furthermore, the correlation coefficients of both these elements with F,

Table 5-3
Summary of Reports on Concentrations of Five Trace Elements in Dental Plaque

		Mean ± SD (ppm, dry weight)				
Author (year)	*No.*	*Copper*	*Lead*	*Lithium*	*Strontium*	*Zinc*
Puttnam, 1966	8–14	35	7	NG	NG	66
Swift, 1967	1	80	6	NG	2	30
Hardwick and Martin, 1967	2	25	5	NG	2.5	80
Schamschula et al, 1977	72	22.3 ± 23.4	7.9 ± 7.4	0.13 ± 0.13	2.6 ± 2.2	103.5 ± 45.3
Beighton et al, 1977	15	12.0[b]	17.1[b]	NG	NG	NG
Schamschula et al, 1978	~290	9.84 ± 7.57	3.28 ± 2.43	0.31 ± 0.46	20.4 ± 17.4	120.2 ± 86.5

[a] = semiquantitative; best estimate
[b] = median value
NG = not given

Table 5-4
Matrix Showing the Significant Interrelations Between Elements in Plaque for the Papua-New Guinea and Australian Studies

Australian Study	Papua-New Guinea Study											Australian Study
	Ca	P	Ca:P	Sr	Mg	F	K	Li	Zn	Cu	Pb	
Ca		0.82	0.76	0.79	0.43	0.40	–	0.19[a]	–	–	–	Ca
P	0.75		0.33	0.62	0.63	0.29	0.30	0.24	0.20	0.16[a]	0.12[a]	P
Ca:P	0.96	0.60		0.64	–	0.41	– 0.28	–	–	0.15[a]	– 0.12[a]	Ca:P
Sr	0.81	0.71	–	–	0.35	0.28	–	0.19[a]	–	–	–	Sr
Mg	0.40	0.64	–	–	–	–	0.55	0.25	0.26	0.29	0.12[a]	Mg
F	0.51	0.40	–	–	–	–	–	–	–	0.11[a]	–	F
K	–	0.51	–	–	0.67	–	–	0.15[a]	0.17[a]	0.22	0.11[a]	K
Li	–	–	–	–	–	–	–	–	0.21	0.22	0.20	Li
Zn	–	–	–	–	–	–	–	–	–	0.26	0.17[a]	Zn
Cu	–	–	–	–	–	–	–	–	0.33[a]	–	0.28	Cu
Pb	–	–	–	–	–	–	–	–	0.44	–	–	Pb
	Ca	P	Ca:P	Sr	Mg	F	K	Li	Zn	Cu	Pb	

[a] $p = < 0.05$; all others $p = < 0.001$
Data from Schamschula et al, 1978, 1978a.

while not as large, are still highly significant and tend to support the hypothesis that part of the calcium phosphate may be present as an apatite. The two alkaline earth metals, Sr and Mg, in general show high correlation coefficients with Ca and P and, in the case of Sr, also with F; it may be relevant that Ca, Sr, and Mg all form relatively insoluble F and orthophosphate salts.

Table 5-5 shows those significant product-moment correlation coefficients for each element against the cumulative caries experience of the subjects from whom the plaque was collected. Once again high correlation coefficients are in general shown for Ca, P, and Ca:P ratios. Before these two studies, Ashley (1975) had shown a similar inverse relation for both Ca and P not only against cumulative caries experience but also for caries increments over a two-year period in British school children. Several earlier investigations had indicated similar trends. On theoretical grounds such an inverse relation is to be expected as the rate of solution of the hard dental tissues under the acid conditions of fermenting plaque will depend both on the solubility of those tissues and on the concentration of free ions, such as Ca and PO_4 in plaque. A similar inverse relation is shown for Sr in both studies and for Mg in the Australian studies.

Both studies show the expected inverse relation between plaque F content and cumulative caries experience but the correlation coefficients are lower than for several other elements. In view of the generally recognized value of F in preventing caries, this apparently surprising observation calls for comment. If the action of F is primarily or entirely

Table 5-5
Summary of Significant Product-Moment Correlation Coefficients Between Concentrations (dry weight) of Elements in Plaque and Caries Experience (DMFT)

Study	Correlation		Significance $p <$
Papua-New Guinea	Ca:P	− 0.32	0.001
	Ca	− 0.22	0.001
	Pb	+ 0.20	0.001
	F	− 0.19	0.001
	K	+ 0.15	0.01
	Sr	− 0.15	0.01
Australian	P	− 0.41	0.001
	Mg	− 0.39	0.001
	K	− 0.30	0.01
	F	− 0.28	0.01
	Ca	− 0.25	0.05
	Sr	− 0.23	0.05

After Schamschula et al, 1977a, 1978a.

due to an inhibition of the metabolism of plaque bacteria, such a low correlation coefficient would be surprising. However, if the F is bound to minute, poorly structured Ca-containing granules in resting plaque and is released in free ionic form under the acid conditions of fermenting plaque, it would not be expected to show as high a correlation coefficient as Ca and probably P. Even the presence of trace amounts of F in apatites would reduce their solubility and in plaque fluid would also encourage the transformation of $CaPO_4$ species, such as brushite, to relatively insoluble apatite. Therefore, a rise in plaque F levels would tend to increase the amount of minute Ca- and PO_4-containing granules present in resting plaque. During the fermentative stage these granules, as they went into solution, would release into the plaque fluid higher concentrations of Ca, PO_4, and F ions, reducing the rate of solution of the underlying hard dental tissues. The effects caused by the release of the free Ca and PO_4 ions will be more apparent than by release of F ions because there are many more hydroxyl ions (which can replace F ions in the apatite lattice) than F in most biologic apatites. The Ca and PO_4 cannot be replaced by other ions to the same great extent. The relatively low correlation coefficient for F should therefore not be regarded as an indication that F is not as highly effective in caries prevention as Ca or PO_4; fluoride would act both directly as F ions and also by increasing the concentrations of the free Ca and PO_4 ions present in fermenting plaque.

A highly significant direct correlation between plaque Pb levels and cumulative experience is shown in the Papua-New Guinea studies. This is of great interest because of the association of Pb with increased dental caries (see Chapter 20).

In the case of K, significant trends in opposite directions are observed in the two studies. It is easy, but of little value without further information, to speculate on the reasons for this anomalous finding. However, it is worth noting that K levels reflect both the bacterial content and the inorganic constituents of enamel and that K is involved in the metabolic processes of bacteria (Luoma, 1964).

Finally, it is interesting to note that Schamschula and his colleagues have shown significant inverse relations of the Li contents of both enamel and saliva with caries experience but only a consistent but non-significant inverse trend between plaque Li levels and caries experience in the Australian study. They also showed a highly significant relation between plaque Li concentrations and plaque quantity. This element, although present in plaque in apparently low concentrations when expressed as ppm, is present in similar numbers as atoms or ions to some minor elements such as Pb or Sr.

These two important studies have clearly confirmed that the plaque Ca and PO_4 concentrations are inversely and highly significantly related to cumulative caries experience. This relation is almost certainly due to a solubility-reducing mechanism. The similar inverse relation between total plaque F levels and caries experience is probably due both to a solubility-reducing effect and to effects on bacterial metabolism. Substantial evidence has been obtained showing that Sr and Mg plaque concentrations are inversely related to caries experience; further research will be needed to elucidate the mechanism responsible. High Pb concentrations in plaque appear to have an adverse effect on caries experience. Several other elements in plaque appear to show associations with cumulative caries experience but the evidence is not conclusive at this stage. In particular, further research is needed on Li and Cu concentrations which show interesting trends usually not statistically significant in these two studies.

Fluoride in Plaque

In the introduction we stated that in general F would not be discussed as it has already been the subject of a number of textbooks and the emphasis of our book was to be trace elements other than F. Nevertheless, on the question of plaque the role of F is so interesting, particularly in relation to trace elements, that to ignore it completely would be negligent.

Theoretically F in plaque can be classified in three forms:

1. Free F′ in the plaque fluid. At neutral pH, and in the presence of high concentrations of Ca and PO_4, F′ can be present in this form as it will either precipitate or diffuse out to the saliva or underlying enamel (Hardwick, 1963a).
2. Ionizable F. At the pHs occurring in resting plaque, F is likely to be present mainly bound either to inorganic constituents, or to organic components of plaque. In fermenting plaque this F may be liberated as F′ and be capable of influencing dissolution of enamel and bacterial metabolism (Jenkins, 1959; Jenkins et al, 1969).
3. Firmly bound F. Some F molecules may not be decomposed even under drastic procedures, such as dry or wet ashing, but are unlikely to be present in significant amounts in dental plaque.

Fluoride content of plaque. Table 5-6 shows the results of most investigations reported in the literature, using chemical microanalytic techniques to estimate the "total" F content. They are likely to indicate the sum of the free ionic and ionizable F. In some investigations they will estimate ionizable F released at pHs well below that which occurs in plaque.

Table 5-6
"Total" Fluoride in Human Dental Plaque (ppm dry weight)

Author (year)	Age of Subjects	Samples Analyzed	Fluoride conc. Mean ± SD (range)	Remarks
Hardwick and Leach, 1963	8–55	30	201 ± 99	Very thick plaque. No F′ therapy.
	13–60	51	393 ± 269	Hospital patients and staff. No F′ therapy.
Mühlemann et al, 1964	24–30	8	164 ± 116	Collected on intra-oral foils for 10 days. F′ free dentifrice.
	24–30	8	702 ± 284	Collected on intra-oral foils for 10 days. 0.2% NaF dentifrice used.
Dawes et al, 1965	11	25[a]	147 ± 100	Low fluoride district.
	11	22[a]	276 ± 129	Drinking water 2.0 ppm F′.
Grøn et al, 1969	Adult	8	25.2 ± 8.3	No tooth brushing for 3 days. Buccal surfaces only.
Birkeland, 1970	12–13	31	52 ± 78	Collection 5 to 6 days after last weekly rinse of 0.2% NaCl.
	12–13	28	85 ± 179	Collection 5 to 6 days after last weekly rinse of 0.2% NaF.
von der Fehr, 1971	About 19	100	561 (5–8,655)	All used F′ toothpaste. No cleaning for 2 days before collection.

Table 5-6 (continued)

Author (year)	Age of Subjects	Samples Analyzed	Fluoride conc. Mean ± SD (range)	Remarks
Birkeland et al, 1971	11–13	8	216 ± 128	4 hours after last fortnightly 0.2% NaF rinse (highest estimate).
	11–13	9	42 ± 31	3 days after last fortnightly rinse of 0.2% NaF (lowest estimate).
	11–13	142	55 ± 53	Test group received 0.2% NaF rinse, the remainder acted as controls. Differences between groups in F′ concentration failed to reach significance statistically.
Agus et al, 1976	9–13	26	13.5 ± 8.3	Drinking water 0.1 ppm F′.
	9–13	21	22.6 ± 16.8	Drinking water fluoridated for 4 years.
	9–13	25	25.6 ± 16.4	Drinking water fluoridated for 16 years.
Jenkins and Edgar, 1977	5	58	21.2 ± 11.8	Low fluoride district.
	5	49	26.5 ± 15.3	Drinking water 1.0 ppm F′.
Schamschula et al, 1978	12–24	229	36.7 ± 30.2	Natives of Papua-New Guinea. Low fluoride district. No oral hygiene.

[a]Mostly pooled samples.

Table 5-7 summarizes the findings of two important recent publications which show the lability of F in both plaque fluid and residues and between F in the ionic form and loosely bound form following F rinses or addition of sugar.

The estimates of the F content of plaque shown in recent investigations are on the whole substantially lower than those of the earlier investigations (before 1971). Most of the former used the ingenious Birkeland microanalytic technique, or modifications of it, on very small samples in which an F electrode was used, on a small amount of fluid. The timing and method of collection may have major effects on plaque F concentration (Hardwick, 1971).

It is therefore concluded that the great variability in the mean F concentrations reported in these 10 studies is not accounted for to a substantial extent by the differing microanalytic techniques used but may be related to the timing and method of plaque collection. They also reflect great and real variations between plaque samples.

The free ionic F of plaque; bound ionizable F It is generally believed that the action of the element F in dental caries, with the probable exception of the mono-fluorophosphate ion, is mediated mainly or entirely as free F ions. The mono-fluorophosphate ion is readily hydrolyzed at certain pH conditions and in the presence of some trace elements. It is possible that its entire action is accounted for by the release of F and PO_4 into the plaque or enamel (Grøn et al, 1971; Eanes, 1976).

In the presence of the high levels of Ca and orthophosphate ions in plaque, at neutral pH such as in resting plaque, high concentrations of free F′ cannot be present over significant periods of time since they will bond with Ca or both ions to form Ca-containing precipitates. They may also bond to organic constituents in the plaque, such as some proteins or other inorganic ions.

Two elegant studies, summarized in Table 5-7, have confirmed the low concentrations of free ions. Tatevoissian and Gould (1978) used their centrifuge technique (1976) to separate the plaque fluid from pooled plaque samples and subsequently did not dilute them. (Dilution will probably release free F′ from any reservoirs of bound fluoride in plaque.) The individual samples of Agus et al (1980) were diluted to the minimum extent but a TISAB buffer was used. When the extreme difficulties of such analyses are taken into account, the findings of the two investigations are in reasonable agreement.

It is clear from these studies that only a small percentage of the total F of resting plaque (Agus et al, 1980: 2.8 ± 2.5%) is present as free F′. Agus et al (1980) also showed, as expected, that some of the bound ionizable F of resting plaque was released as free ions in fermenting plaque with a fall in bound F levels, confirming that the bound F is a source of additional free F ions at times when the plaque pH falls.

Table 5-7
Free Ionic and Ionizable Fluoride Concentrations in Plaque

Author (year)	Age Group	Samples Analyzed Fluoride Content (Mean ± SD)		Remarks
Tatevoissian, 1978 units	18–24	plaque fluid aF^1 (μM) wet	plaque residue $\mu g/g$ dry	No tooth brushing for 24 hours Following 0.02 M NaF rinse
		4930 ± 800	414 ± 32	1 minute after rinse
		40 ± 9	104 ± 7	30 minutes after rinse
		28.5 ± 6.5	76 ± 10	1 hour after rinse
		8.1 ± 2.9	80 ± 28	2 hours after rinse
		2.1 ± 0.7	26 ± 18	3 hours after rinse
		2.0 ± 0.7	36.7 ± 21.5	No rinse
Agus et al, 1980 units as parts/10^6 wet wt.	10–11	plaque fluid F^1	plaque sediment Bound F	
		0.07 ± 0.07	4.96 ± 7.62	Resting plaque mean pH 5.9
		0.16 ± 0.14	3.96 ± 4.50	Fermenting plaque after 15% sucrose addition; mean pH 4.5

NB: Units given and whether wet or dry weight.

Evidence of a relation of plaque F to caries Agus et al (1976) and Schamschula et al (1978) have reported a study on children, aged between 9.7 and 13.0 years, all lifelong residents of one of three Australian towns with differing levels of F in their water supplies. Their findings are summarized in Table 5-8. Cumulative caries experience was inversely and significantly related to plaque concentrations for all the 72 children living and those in the city whose water supply had been fluoridated for four years. In the town with low F water no association was observed. No association was demonstrated between plaque F and other forms of F therapy. It is interesting to note that high and significant product-moment correlation coefficients were demonstrated between F contents and both Ca and P levels in the plaque specimens from all those towns. With the exception of the Ca levels in the plaque of the low F town, Ca and P concentrations were also inversely and significantly associated with cumulative caries experience, an association which is consistent with observations in two earlier studies (Sutfin et al, 1970; Ashley, 1975).

Table 5-8
Findings on Plaque Fluoride and Dental Caries

	Water Supplies of Town		
	0.1 ppm F	*1.0 ppm F 4 years*	*1.0 ppm F 16 years*
No.	26	21	25
Age (years)	11.3 ± 0.5	12.2 ± 0.6	10.8 ± 0.7
DMFT	4.7 ± 2.9	6.6 ± 2.7	2.2 ± 2.0
Plaque F (ppm dry weight)	13.5 ± 0.5	22.6 ± 16.8	25.6 ± 16.4

After Agus et al, 1976.

Schamschula and colleagues (1977a, 1978) in their Papua-New Guinea study found results close to those of the Australian investigations. Consistent trends were observed for an inverse relation between "total" plaque F concentration and cumulative caries experience in the inhabitants of each of the village groups; they were statistically significant in one of the village groups and throughout the whole sample of subjects. Once more, high product-moment correlation coefficients between "total" F concentrations and Ca levels in the plaque samples were demonstrated. The inverse relation between F concentrations in plaque and cumulative caries experience is in stark contrast to the direct relation, observed in the same study, of F concentrations in surface enamel etched from clinically sound labial surfaces of upper canine teeth with the cumulative caries experience of the subjects. The latter, an apparently anomalous finding, is perhaps not unexpected in such sites in view of the investigations by Ramsey et al (1973) and Charlton et al (1974) on conditions encouraging uptake of F onto apatite samples or enamel respectively from their external environments. Indeed there is now sound evidence (Schamschula et al, 1981) showing that the F content of surface enamel confuses the inverse relation between caries experience and the F content of bulk enamel, which has been observed in most investigations.

MECHANISMS OF ACTION OF TRACE ELEMENTS IN PLAQUE

Data on the trace element content are still not sufficiently precise to say definitely how plaque trace elements act on dental caries or periodontal disease. However, it is possible to speculate on several mechanisms of action on caries.

Plaque Fluids

Because many of the trace elements in dental plaque may be present in an ionic form, their availability at the plaque-enamel interface may be important for their role in dental caries. The data presented for F in Table 5-7 indicates that for adequate study of this subject further work is needed on analysis of plaque fluids. As yet analytic techniques are not sufficiently well developed to analyze for many trace elements on the minute quantities of plaque fluid available from individual plaque samples. This will have to be accomplished for a full understanding of the role of trace elements in dental caries.

Effects on Oral Bacteria

The presence of trace elements in plaque may well have an action on the bacteria present in the plaque. Apart from F little is known of such action and it is discussed, where appropriate, in Part III of this book.

The mechanisms by which trace elements may affect bacteria are varied. Limited studies on aggregation, adsorption, acid production, growth, internal and external polysaccharide production, and glycolysis have begun in several research laboratories. It is now well established that F affects bacterial metabolism. Other trace elements, such as Sr and Li, have already been tested and have not shown any dramatic effects on bacterial metabolism (Curzon et al, 1981). In a recent publication Treasure (1981) showed that although F has a significant effect on external polysaccharide production (EPS) Li and Sr had little effect alone. In combination with F these two trace elements tended to reverse the effects of F.

Other work has shown some effects of trace elements. For example, Bowen (1971) found that ions of Ca, Mn, and Mg influenced glucan production by *S. mutans.* Boyd (1978) did not find a strong effect for Mg but showed that Zn more than doubled EPS production by *S. salivarius.* Other workers have also noted effects of trace elements in EPS production (Curzon et al, 1981).

The area in northwest Ohio where there is very low caries prevalence has been described in Chapter 2. In a related study Handelman and Losee (1971) reported increased EPS formation in *S. mutans* grown in a medium prepared with highly mineralized water from northwest Ohio. Later work (Herbison and Handelman, 1975) showed that although a combination of F, Sr, and Li affected hydroxyapatite dissolution these elements did not appear to affect bacterial growth or acid production. Other authors have studied acid production for a number of trace elements.

Effects of Trace Elements on Acid Production by Oral Bacteria

Gallagher and Cutress (1977) have undertaken an interesting and extensive in vitro study of possible effects on acid production by *S. mutans, S. sanguis, S. salivarius, A. naeslundii,* and *A. viscosus* by the addition of 25 trace elements to a complex sucrose-containing medium. The trace elements selected for screening were those found by the authors in the outer 20 μm layer of enamel, and the concentration of the trace elements added to the medium usually tended to span the ranges found in surface enamel. At about the maximum concentrations found in surface enamel F, V, Mn, Ni, Cu, Zn, Se, Ag, Cd, Sb, and Ba significantly reduced acid production by *S. mutans,* the fall in pH usually being 1 to 2 pH units less than in the control cultures. They also hindered growth and plaque formation. Lesser but still significant effects were produced by Be, Si, Ti, and La. With the other four micro-organisms less dramatic effects were observed but F, Mn, Zn, Ag, and Sb inhibited at least one aspect of their metabolism.

The best estimate now available of the mean concentrations in plaque of the trace elements screened indicate that they agree reasonably satisfactorily with some of those used by Gallagher and Cutress with F, Mn, Cu, Cd, and Ni but are lower, and sometimes considerably lower, with the other elements showing an effect on acid production. However, the effect on the fall in pH was also very substantial, and it is probable that concentrations similar to those usually present in plaque would still sometimes affect acid production sufficiently to influence the caries process provided they are in an appropriate chemical and physical state. The evidence obtained indicates that certain elements in addition to F may influence caries by effects on the metabolism of some oral bacteria.

SUMMARY AND CONCLUSIONS

The processes which result in the initiation of a carious lesion take place in dental plaque, the hard dental tissues playing largely a passive role. If a trace element produces its effect on the initiation of the lesion by influencing the activities of the plaque bacteria, the action can only be mediated in the plaque. If, on the other hand, the trace element affects the rate of solution of the underlying hard tissues, the plaque content of that element will usually be equally important as that of the hard dental tissues, and possibly more so. For example, solution of the inorganic moiety of the hard dental tissues under the acid conditions of fermenting plaque will not occur if the plaque becomes saturated at such pH with respect to those ions present in the apatitic lattices of the surface enamel. Indeed, the role of dental plaque is paramount to a thorough understanding of the relation of trace elements to the onset and progression of the carious lesion.

Because of well-recognized interactions between some major, minor, and trace elements, the effects of an individual trace element on the onset of caries may be modified considerably by the presence of other elements. Therefore, in assessing the significance of a specific trace element in plaque, it is highly desirable that the concentration of other major elements (such as Ca and PO_4) and of trace elements (such as F) that are known to inhibit caries should also be estimated. It is possible that the effects of a trace element on the initiation and progress of a carious lesion may arise because it affects the availability in an active form of elements, such as Ca, F, and PO_4, and it is not due to a direct effect of the element on the rate of solution of the hard dental tissues or on the metabolism of the plaque micro-organisms. It is probable that these two theoretical mechanisms are mainly involved in the onset and progression of dental caries.

In the present state of knowledge, effects on the rate of solution of the hard dental tissues appear to be more important than effects on the metabolism of plaque bacteria. Yet in the case of F there is sound evidence that the preventive action of this ion is in part mediated by effects on bacterial metabolism, and there is every reason to believe that a few other trace elements, known to influence the activities of certain specific microorganisms, will also affect the onset and progression of caries. It is desirable that such trace elements should be screened, as has already been done by Gallagher and Cutress, but using concentrations of the elements commonly found in dental plaque, rather than in surface enamel.

Some elements will only affect the onset and progression of carious lesions when they are present in a specific chemical form or physical state. Yet if some of these remained permanently in this active form, they would quickly diffuse out of the plaque. For much of the time they must be present in a bound form and will only be released in an active form when plaque conditions, such as low pH, favor progression of the carious lesion. For example, a major reservoir of ionizable F is present in plaque and appears from recent research (Leach and Appleton, 1981) to be largely in the form of minute Ca-containing precipitates. In the acid conditions of fermenting plaque, F in an active ionic form will be released into the plaque fluid and will be available to affect both the rate of solution of enamel and bacterial metabolism. It would be expected that similar mechanisms will bond temporarily several other trace elements. In interpreting microanalytic data on the trace element content of plaque, it is desirable to investigate the concentration of an element in the active form and whether it is also present in a stored form that would be released in an active form in the acid conditions of fermenting plaque.

There is now substantial evidence that trace elements in addition to F influence the onset and progression of carious lesions. The evidence for a beneficial effect is particularly strong for Sr; inverse associations between Li contents of enamel and of saliva and cumulative caries experience have been observed, but similar and consistent trends in the Li content of plaque have not been statistically significant. Concentrations of Pb in plaque are directly related to cumulative caries experience. Unfortunately, there is little evidence at present to indicate the precise mechanism that may be responsible for these observed associations.

In studying the possible relations between elements in plaque and cumulative caries experience or caries increments in a population, it is more realistic to undertake analyses on specimens of plaque as they normally occur in that population and not on plaque freshly formed after cleaning of tooth surfaces a short time before. There is now strong evidence that plaque contents of at least some elements increase with time as the plaque becomes more mature.

Despite the paramount importance of the role of plaque in investigations on the effects of trace elements on the onset and progression of caries, until recently relatively little research had been undertaken in this field because of the inherent great analytical difficulties. It now appears possible that development of some elegant physical microanalytical techniques will enable concentrations of some trace elements, even within constituents of plaque, to be determined. If so, such developments should greatly increase knowledge and understanding of the mechanisms responsible for any observed relation between the trace element content and the onset and progress of the carious lesion and where reservoirs of "stored" trace elements, if present, are located.

The amount of information on trace elements in plaque is still sparse when compared with the extensive amount known about enamel. Still less is known about plaque fluids where trace elements in ionic form may be important in the demineralization, remineralization process.

REFERENCES

Agus HM, Schamschula RG, Barmes DE, et al: Associations between the total fluoride content of dental plaque and individual caries experience in Australian children. *Commun Dent Oral Epidemiol* 1976;4:210–214.

Agus HM, Un PSH, Cooper MH, et al: Ionized and bound fluoride in resting and fermenting dental plaque and individual human caries experience. *Arch Oral Biol* 1980;25:517–522.

Ashley FP: Calcium and phosphate concentrations of dental plaque related to dental caries in 11-14 year-old male subjects. *Caries Res* 1975;9:351–362.

Beighton D, Fry P, Higgins TN, et al: Determination of Cu, Pb, and Cd concentrations in dental plaque using anodic stripping voltametry. *J Dent Res* (Spec Issue) 1977;56:D191.

Birkeland JM: Direct potentiometric determination of fluoride in soft tooth deposits. *Caries Res* 1970;4:243–255.

Birkeland JM, Jorkjend L, von der Fehr FR: The influence of fluoride rinses on the fluoride content of dental plaque in children. *Caries Res* 1971;5:169–176.

Bowen WH: The effects of Ca, Mg, and Mn on dextran production by a cariogenic streptococcus. *Arch Oral Biol* 1971;16:115–119.

Boyd RF: The effect of some divalent cations on EPS synthesis on *S. salivarius. J Dent Res* 1978;57:380–383.

Charlton G, Blainey B, Schamschula RG: Associations between dental plaque and fluoride in human surface enamel. *Arch Oral Biol* 1974;19:139–143.

Curzon MEJ, Eisenberg AD, Treasure PE: The effect of Sr, Li and F on the in vitro formation and metabolism of dental plaque and on plaque formation and caries development in the rat. Final Report Grant DE 72409. Washington, D.C., National Institute of Dental Research, 1981.

Dawes C, Jenkins GN: Some inorganic constituents of dental plaque and their relation to early calculus formation and caries. *Arch Oral Biol* 1962;7:161–172.

Dawes C, Jenkins GN, Hardwick JL, et al: The relation between fluoride concentrations in the dental plaque and in drinking water. *Br Dent J* 1965;119:164–167.

Eanes ED: The reaction of mono-fluorophosphate with amorphous and apatitic calcium phosphates. *Caries Res* 1976;10:59–71.

Gallagher IHC, Cutress, TW: The effects of trace elements on the growth and fermentation by oral streptococci and actinomyces. *Arch Oral Biol* 1977; 22:555–562.

Grøn P, Yao K, Spinelli M: A study of inorganic constituents in dental plaque *J Dent Res* (Suppl to No. 5) 1969;48:799–805.

Grøn P, Brudevold T, Aasenden R: Mono-fluorophosphate interaction with hydroxyapatite and intact enamel. *Caries Res* 1971;5:202–214.

Handelman SL, Losee FL: Inhibition of enamel solubility in a highly mineralized water. *J Dent Res* 1971;50:1605–1609.

Hardwick JL, Leach SA: The fluoride content of the dental plaque. Proc 9th ORCA Congress, *Arch Oral Biol* (Suppl) 1963;8:151–158.

Hardwick JL: The mechanism of fluorides in lessening susceptibility to dental caries. *Br Dent J* 1963a;114:222–228.

Hardwick JL, Martin CJ: A pilot study using mass spectrometry for the estimation of the trace element content of dental tissues. *Helv Odont Acta* 1967;11: 62–70.

Hardwick JL: Association between plaque fluoride concentrations and other parameters, in McHugh WD (ed): *Dental Plaque.* Edinburgh, Livingstone, 1971, pp 171–178.

Hardwick JL: Die Bedeutung von Fluor in der dentalen Plaque. *Dtsch Stomat* 1971a;21:107–110.

Herbison RJ, Handelman SL: Effect of trace elements on dissolution of hydroxyapatite by cariogenic streptococci. *J Dent Res* 1975;54:1107–1114.

Jenkins GN: The effect of pH on the fluoride inhibition of salivary acid production. *Arch Oral Biol* 1959;1:33–41.

Jenkins GN, Edgar WM, Ferguson DB: The distribution and metabolic effects on human plaque fluorine. *Arch Oral Biol* 1969;14:105–119.

Jenkins GN, Edgar WM: Distribution and forms of F in saliva and plaque. *Caries Res* 1977;11 (Suppl 1): 226–242.

Kudahl JN, Fremlin JH, Hardwick JL: The isotope-dilution method of fluorine micro-estimation, in *Radio-isotopes in the Physical Sciences and Industry.* I.A.E.A. Vienna, 1962, pp 317–323.

Kudahl JN: On the determination of fluoride in small samples. Proc 9th ORCA Congress. *Arch Oral Biol* 1963; (Spec Suppl) 8:53–56.

Leach SA, Appleton J: Ultrastructural investigations by energy-dispersive x-ray microanalysis of some elements in the formation of dental plaque, in Rölla G, Sonju T, Emberry G (eds): *Tooth Surfaces Interactions and Preventive Dentistry.* London, Information Retrieval Ltd, 1981, pp 65–79.

Luoma H: Lability of inorganic phosphate in dental plaque and saliva. *Acta Odont Scand* 1964;22 (Suppl 41).

Mühlemann HR, Schatt A, Schroeder HE: Salivary origin of fluorine in calcified dental plaque. *Helv Odont Acta* 1964;8:128–129.

Puttnam NA, Bradshaw F, Platt P: X-ray spectroscopic determination of the constituent elements of dental plaque, in James PMC, König G, Held HR (eds): *Advances in Fluorine Research and Dental Caries Prevention* vol 4. Oxford, Pergamon Press, 1966, pp 157–162.

Ramsey AC, Duff EJ, Paterson L, et al: The uptake of F′ by hydroxyapatite at varying pH. *Caries Res* 1973;7:231–244.

Schamschula RG, Agus HM, Bunzel M, et al: The concentrations of selected major and trace minerals in human dental plaque. *Arch Oral Biol* 1977; 22: 321- 325.

Schamschula RG, Adkins BL, Barmes DE, et al: Caries experience and the mineral content of plaque in a primitive population in New Guinea. *J Dent Res* (Spec Issue) 1977a;56:C62–C70.

Schamschula RG, Adkins BL, Barmes DE, et al: *WHO Study of Dental Caries Etiology in Papua-New Guinea.* WHO Offset Publication No. 40. Geneva, Switzerland, World Health Organization, 1978.

Schamschula RG, Bunzel M, Agus HM, et al: Plaque minerals and caries experience: Associations and inter-relationships. *J Dent Res* 1978a;57:427–432.

Schamschula RG, Cooper MH, Agus HM, et al: Oral health of Australian children using surface and artesian water supplies. *Commun Dent Oral Epidemiol* 1981;9:27–31.

Singer L, Armstrong WD: Determination of fluoride in blood serum. *Anal Chem* 1959;31:103–111.

Singer L, Jarvey BA, Venkateswarlu P, et al: Fluoride in plaque. *J Dent Res* 1970;49:455.

Sutfin LV, Sweeney EA, Ascoldi W: Calcifying dental plaque and reduced dental caries in permanent molars of children from two Guatemalan villages. *J Dent Res* 1970;49:772–775.

Swift P: A method for the trace element of analysis of dental tissues. *Br Dent J* 1967;123:326–327.

Tatevoissian A, Gould CT: The composition of the aqueous phase in human dental plaque. *Arch Oral Biol* 1976;21:319–323.

Tatevoissian A: Distribution and kinetics of fluoride ions in the free aqueous and residual phases of human dental plaque. *Arch Oral Biol* 1978;23:893–898.

Tatevoissian A, Gould CT: The kinetics of inorganic phosphate in human dental plaque and saliva. *Arch Oral Biol* 1979;24:461–466.

Treasure PE: Effects of F, Li and Sr on extracellular polysaccharide production by *S. mutans* and *A. viscosus. J Dent Res* 1981;60:1601–1610.

von der Fehr FR, Møller IJ, Kirkegaard E, Kold M: Individual variations of fluoride and calcium in plaque. *J Dent Res* 1971;50:718.

PART III
The Relationships of Individual Trace Elements to Dental Disease

In dealing with the information available about individual trace elements several approaches were possible. One was to place the elements in an order based on the amount of information available. Second, we could have dealt with them alphabetically. There was also a third order of the elements based upon their essentiality to life and their toxicity.

This part of the book deals with those elements which are known to be essential for animal metabolism. This section therefore includes such elements as molybdenum (Mo) and selenium (Se) which are well known as important for life and for which there is now considerable evidence implicating them as important in dental disease. We have also emphasized, in separate chapters, several other elements such as manganese (Mn), sulfur (S) and copper (Cu) for which the dental literature is not large but which we feel deserve greater attention. This section finally includes a group of ten elements, all essential or probably essential, such as silicon (Si), for which the dental literature gives almost no information.

A large number of trace elements—the majority of those in the periodic table—are not essential for life and are inactive according to present knowledge. Included in this section are four elements for which we devote separate chapters, strontium (Sr), lithium (Li), barium (Ba), and titanium (Ti). For Sr and Li there is now considerable evidence to show they may be cariostatic or act in some way to inhibit dental caries. There is also slight evidence to suggest they may be essential. The element Ba is given a separate chapter because it is an alkaline earth and therefore has many chemical similarities to Mg, Ca, and Sr which are related to dental disease. In addition, there is some evidence to show Ba may affect enamel composition and dental decay. Because of recent studies showing Ti, as TiF_4, to have an anticaries effect we have given this element a separate chapter. Finally, there is a large group of elements for which there is scant, or no, evidence of any relationship to dental disease.

Two trace elements are nonessential yet also toxic; we have given these prominence in this book by giving them not only separate chapters but also a separate section in this third part of the book. Both lead (Pb) and cadmium (Cd) are serious environmental pollutants for which there is now considerable information in the dental literature.

Part III is specifically designed to give comprehensive reviews on individual trace elements. Our objective is that these chapters will serve as a starting point for further research on their effects on dental disease.

42 Mo

95.94

18-13-1

6 Molybdenum

G. N. Jenkins

Molybdenum, symbol Mo and atomic number 42, is grouped among the transition elements of the periodic table and as a metal. Originally its ores were confused with those of lead, hence its name derived from *molybdos,* Greek for lead. Its essential nature to plant and animal life makes it an important biologic element.

Atoms of Mo can exist in compounds in which its electrovalency is 2, 3, 4, 5, or 6 or its coordinate valency 4, 6, or 8. In solution, molybdate ions (MoO_4) may polymerize with the incorporation of an additional Mo atom to form paramolybdate $(Mo_7O_{24})^{6-}$. Sodium molybdate is a stable salt (Na_2MoO_4), but ammonium molybdate readily loses ammonia to become the paramolybdate:

$$7(NH_4)_2MoO_4 \rightarrow (NH_4)_6Mo_7O_{24} + 4H_2O + 8NH_3$$

This complexity and these particular salts have had a direct bearing on the confusion that has arisen in animal experiments on dental caries;

to be discussed later. Molybdate itself also readily forms with other groups an extraordinary variety of complex ions such as phosphomolybdates, oxyfluoromolybdates, silicomolybdates, and others. Organic complexes are also known, and thus it has been difficult to determine which salt should be used in dental research.

Naturally Occurring Compounds

Among a large number of naturally occurring compounds are molybdenite MoS_2 and Wulfenite Pb, MoO_4. The ore is used widely in industry. Of great interest is the use of Mo-containing phosphate fertilizers which have an effect on agricultural production, and the element is widely incorporated into the food chain. Although there is little strong evidence of normal Mo exerting an anticaries effect in man, if it does so the supplement needed is small, of the order of the amount that prevents Mo deficiency in plants (about 2 oz per acre).

Metabolism of Molybdenum

There are two mammalian enzymes that require Mo as a cofactor. In both cases they are dehydrogenases, and both are molybdoflavoproteins. Xanthine oxidase is found in liver, kidney, and milk, and oxidizes xanthine, hypoxanthine, and some other prolines and aldehydes (Schroeder et al, 1970).

In agriculture and animal husbandry Mo has been used for many years. A toxic state in cattle, teart, has been known since the 1930s, as well as a deficiency state induced by excessive Cu intake (Underwood, 1977). There is a complex interaction of Mo with Cu, both of which antagonize each other in mammalian metabolism. In addition, the Mo-Cu interaction is governed by the presence of S. No doubt these interactions of

Table 6-1
Molybdenum Concentrations in Human Tissues

Tissue	Mean ± SE	Tissue	Mean ± SE
Brain	ND	Kidney	0.37 ± 0.01
Muscle	ND	Liver	1.10 ± 0.05
Lung	0.016 ± 0.003	Bone	ND
Ileum	0.02 ± 0.006	Enamel	7.2 ± 1.35
Heart	0.017 ± 0.003		

Data from Schroeder et al, 1970; Losee et al, 1974.
ND = not detected.
Concentration in $\mu g/g$.

Mo and Cu also play a role in the caries process as both elements have been associated with caries. No research has been done, and the role of the Mo-Cu interaction remains to be investigated.

Molybdenum in Calcified Tissues

In human tissues Mo is found principally in liver, kidney, fat, and blood, although the concentrations are not high. Table 6-1 gives the results of analysis of several selected human tissues which show that Mo content in most tissues, including enamel, is low.

The role of trace elements, including Mo, in enamel and dentin has already been discussed, and not much further elaboration is needed here. The concentrations of Mo in both whole and surface enamel have always been found to be low. It is unlikely that the incorporation of Mo into enamel apatite will have much influence on the dental caries process. Where differences in enamel content of Mo have been found, it is suggested that this is merely a record of environmental differences.

Human Food and Water Sources

Until Schroeder published his work not much was known about the Mo content of foods; we now know that the major food sources are meats, grains, and legumes. In members of the legume family used as human and animal foods, Mo concentrations can reach toxic levels where these plants are grown on soils high in Mo. The element is essential for the enzyme used by the bacteria *Azotobacter,* found in root nodules of legumes, in the fixation of nitrogen. A partial list of food analyses carried out by Schroeder is shown in Table 6-2. The relatively high concentration of Mo in peas is notable.

Table 6-2
Molybdenum Concentrations in Typical Foods Used in United States

Food Group	Item	Molybdenum $\mu g/g$
Organ meats	Beef liver	1.97
Meat	Beef	0.0
Cereals	All Bran	2.75
Milk	Whole	0.20
Vegetables	Carrot	0.08
	Spinach	0.26
	Peas	3.50
	Tomato	0.0

After Schroeder et al, 1970.

A geographic difference in Mo content of plant materials is to be expected since wide variations in soil Mo occur. However, in a country such as the United States where there is extensive use of molybdenized superphosphates as top dressing for food plants, such differences are no longer readily apparent.

DENTAL CARIES

There have been more studies on the relationship of Mo to dental caries than on any other trace element except fluoride. These studies have been carried out in many countries and have comprised epidemiologic surveys (Table 6-3), as well as animal studies using rats and monkeys. The results have been confusing and no clear picture of the relationship, or role, of Mo to dental caries has arisen.

Table 6-3
Summary of Human Studies Reporting a Relationship of Molybdenum to Caries Prevalence

Author	Year	Area of Study	Caries Effect
Adler and Straub	1953	Hungary	Decrease
Ludwig et al	1960	New Zealand	Decrease
Anderson	1966	England	Decrease
Curzon et al	1971	United States	No effect

Man

Several epidemiologic surveys in many parts of the world have indicated a negative association of Mo to dental caries. Early studies arose as the result of baseline dental surveys preparatory to conducting fluoridation studies. In these instances it was found that significant differences already existed between the communities chosen for the fluoridation projects. Thus further research was required to determine why such differences in caries prevalence existed when the water fluoride concentrations were very similar.

Adler and Straub (1953) were among the first to suggest that trace elements other than fluoride may be associated with a low caries prevalence. They observed communities in Hungary where caries was lower than would have been expected considering that the fluoride concentration of the water ranged only from 0.13 to 0.84 ppm. Nagy and Polyik (1955) reported that the water supplies of these communities contained 0.1 ppm Mo and suggested that this constituent caused the low caries rate. This figure is much higher than that of most other water analyses, perhaps because the method of analysis was not very accurate (Waters, cited by Anderson, 1969). This conclusion was made more probable by experiments in rats where 0.1 ppm Mo in their water supplies reduced caries.

In New Zealand, baseline studies on caries before the introduction of fluoridation in the adjacent cities of Napier and Hastings revealed a much lower prevalence in Napier (Ludwig et al, 1960). The water supplies came from the same strata, and the only obvious difference between the two cities was that Napier derived most of its vegetables from land raised from the sea by an earthquake some 30 years previously. Analyses of soils showed no significant differences except a higher pH of the Napier soil (Healy et al, 1961), but spectroscopy of the ash of the vegetables grown in Napier showed much higher concentrations of several trace elements including Mo. It was already known that an alkaline soil favors Mo uptake by plants, a finding which probably explained the higher levels in Napier-grown vegetables. When fed to rats, the ash of vegetables grown in Napier and Hastings reduced caries and the effect of Napier ash was almost as large as that produced by equivalent amounts of Mo, which indicated that this was probably the active element (Ludwig et al, 1962).

The deciduous teeth from natives of Napier and Hastings were analyzed for Mo by a colorimetric method (unfortunately without separating the enamel from the dentin) and the results were 0.069 and 0.046 ppm, respectively (Healy et al, 1963). Permanent teeth had lower concentrations that were the same in both towns (0.034 and 0.032 ppm in Napier and Hastings, respectively), possibly because the teeth had developed either before the earthquake or before the area was used for growing food (only 15 years before the study began). These figures are similar to those of Nixon (1969) who found concentrations between 0.026 and 0.12 ppm in enamel from permanent teeth in Scotland but differ markedly from some other estimates. Losee et al (1974) give a mean figure of 7.2 ppm for Mo in enamel collected from widely distributed sources in the United States and analyzed by spark source mass spectroscopy; Hardwick and Martin (1967) put Mo of both enamel and dentin in the 1 to 10 ppm group of trace elements. Twenty-four samples of urine from the children in Napier and Hastings contained 3.5 and 2.7 μg Mo/g ash, respectively (Healy, 1966 cited by Anderson, 1969), clear evidence of a higher Mo intake in Napier.

Anderson (1964, 1966, 1969) reported the caries incidence among school children in an area of Somerset, England where cattle suffer from teart (Mo poisoning) compared with that in neighboring areas. In all three surveys Anderson found lower caries in the teart area (Table 6-4). If this was produced by Mo, there was a question about the source of this element. The concentration in the water was very low although higher

(0.003 ppm) than in the control area (0.0018 ppm), but Mn was also higher (0.09 ppm compared with a control of 0.01 ppm). Vegetables were not produced commercially in this area, the main food product being milk which was higher in Mo (0.05 ppm) than in milk from the control area (0.03 ppm); both were within the normally accepted range of 0.03 to 0.06 ppm (Underwood, 1977). Some of the milk produced in the teart area might have reached the control areas. The Mo concentration of the whole teeth was higher in the teart area but fluoride concentrations were also.

Table 6-4
Average DMFT in 12-Year-Old Children in Molybdenum and Control Areas in England

	Mo		Control		
	n =	*DMFT*	*n* =	*DMFT*	% Dif
First survey					
boys	132	4.15	276	5.32	22
girls	138	4.59	261	5.93	23
Second Survey	309	4.59	373	5.71	20

Data from Anderson, 1966, 1969.

Spot samples (not 24-hour collections) of urine from children in Anderson's teart and control areas showed a higher Mo output in the former; he estimated approximate daily Mo intakes of between 38 and 76 μg, contrasting with 23 to 46 μg in the control areas. From some other (unpublished) data on milk and water consumption by children in Britain, Anderson calculated the intake from milk and water at 24 μg and 14 μg in the teart and control areas, respectively.

Although Anderson was cautious in suggesting that Mo was responsible for the low caries scores, he found no alternate explanation. The difference in the Mn concentrations in the water was considered to be too small to affect total intake significantly. In the same Mo area, associated with a low decayed, missing and filled teeth (DMFT) index, a higher percentage of children had orange stain on their teeth and a lower percentage had brown stain. Even in areas not high in Mo, extrinsic enamel stains described as "dark" or "brown" were reported by James (1965) and Sutcliffe (1965), respectively, more common in children with low DMFT, values. The nature of the relationship between the stains, caries, and Mo intake is not known.

Curzon et al (1971) found no difference between the DMFT or DMFS (surfaces) of 12- to 14-year-old children in a teart and a control area of California. The Mo concentrations of a limited number of enamel samples were higher in older children from the teart area (> 10 ppm compared with 1) but unerupted and newly erupted teeth were reported to contain no Mo. The enamel from a 50-year-old man was stated to be 5000 ppm, grossly in excess of any other published figures, but the Mo concentration in the milk and water was not consistently higher in the teart area, nor did the urine contain more Mo; this may explain the absence of a caries effect. The puzzling feature of this work is the origin of the higher concentration of Mo in the teeth (more than 1000 times greater than that given by previous authors) without any evidence of a higher intake. The method of analysis, semiquantitative laser microprobe, may have been too crude to give any accurate figures.

In Papua-New Guinea, certain villages were grouped on the basis of their caries prevalence, and the trace elements in the soil and staple diet (sago) analyzed (Barmes, 1969) (see Chapter 2). An important finding was that the soils of the villages with a low caries score were, in general, more alkaline than those with a high caries score and contained much higher concentrations of Ba and Sr. A high soil pH is associated with high uptakes of Mo. Analyses of the sago grown in the groups of villages did not, however, show a consistent difference in Mo concentration, and only half the values showed the expected relation to pH, but the levels reported were very low (from 0.03 to 0.54 ppm). Barmes stated that the reason was "because of the relative abundance of alkalis and alkaline earths in the soil, the detection level for Mo was inadequate."

In a later, thorough series of analyses, trace elements in soil were correlated not only with caries but with many physiologic factors believed to be associated with caries, such as the composition of enamel, plaque, and saliva, and certain dimensions of the teeth (Schamschula et al, 1978). The Mo of soil was not consistently related to caries, since for high caries groups the mean Mo concentrations were 1.5 and 2.0 ppm while for low caries groups the values were 1.7 and 1.0 ppm.

The results of relating trace elements in the soil to the composition of plants have been promised (Schamschula et al, 1978) for the future; the earlier work of Barmes (1969), however, did not suggest that the Mo content of sago was much affected by the composition of soil. These workers have emphasized the distinction between factors concerned with the initiation and progress of caries (what these authors define as the severity score). Out of the 2088 simple correlations calculated, 17% were significant at the <0.05 level or better, but, as the authors themselves point out, about one-third of these would arise by chance. With this proviso in mind, Schamschula et al found that the concentration of total Mo of soils had a positive correlation of 0.5 with the mean severity score of caries, contrary to the expected trend.

The birthplace of 360 recruits for the United States Navy who had been found to be caries-free was traced over a ten-year period; a high proportion of them were clustered into three small areas of northwest

Ohio, northeast South Carolina, and west central Florida, all farming communities where the inhabitants consumed local produce (Losee and Adkins, 1969). Analysis of the water supplies from the Ohio and South Carolina areas revealed unusually high concentrations of B, Li, Mo, Sr (with Mo especially high in the Ohio region). However, in South Carolina, which had moderately high Sr and Mo concentrations in the water, more green vegetables are eaten than in Ohio, and these concentrate Sr and Mo. In both areas, Sr and Mo intakes would be expected to be higher than for the United States.

Since the discovery that waterborne fluoride had a powerful anticaries effect, it was assumed by some workers that water would be an important source of other trace elements. Compared with foods, however, water is a negligible source of most trace elements. Losee and Adkins (1976) studied the effect of cooking vegetables in water high in F, Sr, and Mo from northwest Ohio and found that the vegetables took up Sr, Mo, F, and Li but that other trace elements were washed out of the vegetables (Table 2-3). Water may, therefore, be important in providing certain trace elements which are concentrated by vegetables.

The epidemiologic evidence as a whole is suggestive rather than conclusive. Lower caries incidence associated with Mo in Hungarian and American communities could be explained by other factors; in the former, few other factors were studied, and in the latter it was realized that other trace elements were also involved. The most convincing evidence was provided by an animal experiment (see below) in which the anticaries effect of the ash of the vegetables was shown to be almost equal to its Mo content. On the other hand, the Mo of the ash would be in a different (more available) form than that of the vegetables so results were biased toward a false positive.

Animals

Many experiments studying dental caries in animals related to Mo have shown no clear picture. This may be partially because of the complexity of the chemistry of Mo, its many salts, and lack of purity of many of these salts. The commercial salt ammonium molybdate is an uncertain mixture but in animal experiments this and the sodium salt have been used indiscriminately, with confusing results.

After it was found in the Hungarian study that Mo was associated with low caries, an animal experiment was carried out in which three generations of rats on a coarse-corn diet (therefore liable to incur fracture caries) were given 0.1 ppm of Mo in their water both before and after the eruption of their teeth (Adler and Porcsalmy, 1961). The caries was reduced in the groups receiving Mo both before and after eruption compared with controls, but the experiment was difficult to interpret because the caries scores in the control animals (first generation: 4.5) rose in the second and third generations to 6.6 and 11.1 respectively. Clearly, some important factor that influenced caries was not controlled.

Shaw and Griffiths (1961) compared the effect of similar ranges of concentrations of ammonium molybdate and ammonium paramolybdate on caries in rats although there is no mention of testing the purity of additives. The molybdate produced considerable reduction in caries except at the highest dosage used (0.1% in the diet); this led to a large but nonsignificant increase in caries which they attributed to a general toxic reaction shown by low weight gains. Ammonium paramolybdate in approximately equimolecular dosage produced no reduction in caries and a much lower general toxicity with the higher dose. This experiment, the only direct comparison of the effects of molybdate and paramolybdate, clearly indicates a greater biologic activity for the former. Büttner (1961) also found that ammonium paramolybdate had no effect in reducing caries. Thus it seems likely that negative results were obtained in some other experiments because the inactive paramolybdate had been inadvertently fed or had been formed from molybdate during the experiment.

The ash of green beans grown in Napier when used as a dietary supplement to rats effected a reduction in caries (Table 6-5). Similarly the solids from samples of water from the low caries area of northwest Ohio reduced caries in rats (Losee and Adkins, 1976) to a greater extent than could be accounted for by its fluoride content. The trace elements detected (F, B, Li, Mo, and Sr) were not tested separately, however, so these results provide no evidence for an effect of Mo, although its concentration (0.1 ppm) was the same as that reported by Adler and Polyik and higher than most other published figures.

In the majority of the rat experiments a reduction in caries was reported, but some negative results were obtained. In many experiments

Table 6-5
Effect on Rat Caries of Vegetable Ash from Napier and Hastings

Diet	Number of Rats	Mean Number of Cavities	Mean Caries Severity Score	Mo Content of Diet (ppm)
Control	17	6.4	16.6	0.05
Control + Napier ash	20	3.5	7.4	1.60
Control + Hastings ash	20	5.3	13.9	0.20
Control + Hastings ash + molybdenum	20	4.5	9.5	1.35

Data from Ludwig et al, 1962.

the dosage of Mo was so high that nonspecific effects (eg, a reduced intake of the cariogenic diet arising from an unpleasant taste, or general malaise following toxic reactions) may explain the reduction in caries. Experiments done until 1967 were analyzed and set in tabular form by Jenkins (1967a). Results have not materially changed since then. Jenkins pointed out that Mo interacts with several other constituents of the diet (eg, Cu, W, Mn, and sulphate) (Suttle, 1974) and that the concentration in the basal diets of these substances and of Mo itself had not been taken into account in any of the experiments, an explanation for some of the contradictory results.

Bowen (1973) found that 2 ppm of molybdate, as the sodium salt, in the drinking water of monkeys (*Macaca fascicularis*) had no effect on caries when given either before or after mineralization of their teeth, and it did not supplement the marked reduction observed with 2 ppm F (as NaF). The effect of Mo supplements during the development of the teeth (separately from postnatal effects) has been tested in animal experiments by four groups of workers and, except for one (Kruger, 1959) who injected a realistic dose of 7 μg Mo/kg of body weight, the results have been negative. Shaw and Griffiths (1961) stated that prenatal ammonium paramolybdate tended to increase caries in rats on the large dose of 1 mg per day (although the data in their Table 8 is confusing and difficult to understand). Van Reen et al (1962) found no effect, and Hunt and Navia (1975) reported that Mo (500 μg/10g body weight, fed by esophageal intubation) increased caries and abolished or reversed the reduction of caries by fluoride. The dose was massive, equal to 3.5 g Mo for a 70 kg adult man. The question of a developmental effect of Mo on caries has not, therefore, been adequately studied but the only experiment in which a reasonable dose was used gave a significant reduction.

MECHANISMS OF ACTION

Although an effect of Mo on caries in man or a specific effect in experimental animals cannot be regarded as established, indications are sufficiently strong to justify some experiments on possible modes of action. The possibility that Mo affects caries would be increased if any biologic properties of Mo suggested such modes of action. The three most obvious actions would be an effect on the solubility of enamel, on the metabolism of plaque bacteria, or on the shape of the tooth and its capacity for retaining plaque.

Effects on Enamel Solubility

The presence of Mo in whole and surface enamel was discussed in Chapter 3. By its incorporation into enamel, either during development or by posteruptive uptake, the physical-chemical characteristics of the enamel might be affected.

The addition of 1 and 10 ppm of Mo (as paramolybdate) to acetate buffer in which the rate of solubility of enamel or calcium phosphate (as a model for enamel) was tested (Jenkins, 1967c) showed a small reduction (measured as the amount of phosphate dissolving). As a standard of reference, 1 and 10 ppm of F were also tested under the same conditions. The effect of Mo was about one-third that of F but, unlike F, Mo did not bind to the enamel as, after washing with water, its solubility was the same as that of the control. If, however, the calcium phosphate was washed with acetate buffer previously saturated with calcium phosphate (which would resemble the effect in vivo of washing with saliva) the reduced solubility was still detectable. Mo and F were tested together and at the 1 ppm level had an additive effect; at 10 ppm, the mixture had the same effect as F alone. In a similar experiment with lactate buffer Mo had no effect, probably because it forms a complex ion with lactate (Jenkins, 1967c).

Sodium molybdate had approximately one-tenth the activity of paramolybdate in reducing the solubility of apatite. McLundie et al (1968) tested, in an artificial mouth, the effect of sodium molybdate on the solubility of the enamel in whole teeth in a Tris acid maleate buffer at pH 6.0. They reported that Mo *increased* solubility, which they explained as a result of the formation at this pH of a complex between Ca and molybdate. A study of the amounts of Ca and phosphate dissolving in the buffer used by McLundie et al in the presence of a range of concentrations of molybdate or paramolybdate and at pH 5 and 6 did not support this explanation. In no case did the dissolving Ca exceed that of the control (Jenkins, unpublished data).

There are no data on the effect of solubility of Mo incorporated into apatite or adsorbed onto its surface, but the low concentrations of Mo reported in enamel and dentin (less than 0.1 ppm [Healy and Ludwig, 1963] or 0.026 to 0.12 ppm [Nixon, 1969]) do not seem likely to influence solubility. As a standard of comparison, differences in the F concentration in enamel of several hundred ppm result in only a small reduction in solubility.

Ericsson (1966 and Ericsson and Söremark (1968)) injected ^{99}Mo into pregnant mice and found that Mo passed through the placenta readily and was taken up by the fetal mineralizing bones. When Mo and F were given simultaneously, the F in the intestine showed a tendency, of borderline significance, to be higher in the group receiving F. This implies that Mo was delaying or reducing the absorption of F, a conclusion contrary to that of some other workers.

Injection of ^{99}Mo into 8-day-old rats showed that Mo was taken up temporarily by bone, dentin, and even more by enamel during the final stage of mineralization (the matrix did not take it up) (Bawden and Hammarström, 1976). One day after injection the bones had lost most of the Mo but the concentration in enamel increased; after four days only the enamel and dentin contained Mo. The soft tissues took up low concentrations immediately after the injection but by four days it had disappeared except in the cortical part of the kidney. Rapidly mineralizing enamel had the highest affinity for Mo of all tissues and retained it most firmly. In tissue culture Mo was taken up by the mineralizing enamel of rat molars but not by the soft tissues (Rahkamo and Tuompo, 1974). Bertrand (personal communication, 1975), on the other hand, reported that Mo could bind to odontoblasts and the organic matrix of enamel and dentin.

Effects on Bacterial Acid Production

The effect of Mo, as both molybdate and paramolybdate, was tested on acid production by saliva incubated with sugar (Jenkins, 1967c). Neither form had any influence on saliva when the initial pH was about 7, but if the saliva was adjusted to pH 5.0 before the incubation began, a further fall in pH was inhibited slightly but significantly by paramolybdate at Mo concentrations exceeding 8 ppm. In the molybdate form, Mo had no effect at either initial pH. This result is not in agreement with that of Bowen and Eastoe (1967) who reported that 30 ppm of Mo (as molybdate) were adequate to produce significant inhibition of pH change of monkey plaque in vivo after the application of sucrose.

Coulter and Russell (1974) found that the yield of *S. mutans* in pure culture was greatly increased by up to 500 ppm of Mo (as molybdate) but that the final pH of the culture was unaffected; they did not test paramolybdate. These workers also reported that lactate utilization by *Veillonella parvula* appeared to be increased by Mo (as molybdate) in concentrations of 10 ppm or even as low as 4 ppm under certain conditions. Increased utilization of lactate by this organism might account for the lower pH by mixed salivary organisms in the presence of Mo reported by Jenkins (1967c). However, Coulter and Russell (1974) estimated lactate by the enzyme method; and it is conceivable that their lower concentrations in the presence of Mo (interpreted as increased utilization) could arise by an inhibition of the enzyme by Mo, which they did not check.

Gallagher and Cutress (1977) tested the effects of many trace elements, at the concentrations that have been reported to exist in enamel, on acid production and growth of a number of oral organisms in pure culture. This work seems to be based on the assumption that the element bound into enamel would influence bacterial metabolism. Since there were uncertainties about the concentration of Mo in enamel, *S. mutans* was tested over an arbitrary range of 10 to 1000 ppm of Mo. Gallagher and Cutress found that 1000 ppm paramolybdate inhibited growth and pH drop by this organism and that molybdate increased growth over the range of 10 to 1000 (thus confirming the results of Coulter and Russell) but also stimulated acid production, in contradiction to Coulter and Russell's findings. Herbison and Handelman (1975) tested the effect on hydroxyapatite dissolution by cariogenic streptococci of several trace elements at the concentrations at which they occur in water at Rossburg, one of the areas of Ohio associated with an unusually high proportion of caries-free people. The only elements which exerted an effect were Sr and F; Mo (at 0.12 ppm) had no effect. It is, however, naive to expect the concentration in water to be effective. The concentrations accumulated in enamel or in plaque are those more likely to influence solubility or acid production.

The concentration of Mo in saliva and plaque is uncertain and the extent to which it is influenced by dietary intake and the chemical forms in which it occurs are completely unknown. It is therefore not possible to say whether Mo occurs in sufficient concentration to exert an effect on enamel solubility or plaque acid production. The published figures—2 ppm (Puttnam et al, 1965) and 10 to 100 ppm of plaque ash (Hardwick and Martin, 1967), probably equivalent to about 0.15 to 1.5 ppm wet weight—are much lower than the concentrations shown in vitro to exert significant inhibition. An effect on plaque metabolism would be effective if the Mo were inside the bacterial cells whereas a posteruptive effect on enamel solubility presupposes that the Mo is in the aqueous phase of plaque. As the Mo is presumably distributed between the two phases and much of it bound organically or present in inorganic forms devoid of inhibitory effect, the active concentration available is almost certainly too low to be effective.

Effect on the Morphology of the Tooth

Kruger (1959) included ammonium molybdate in his study of injecting a number of trace elements into young rats to examine the morphology of their molars and their resistance to caries. In his work he reported that out of six dimensions measured, only the mesial fissures differed significantly ($p < 0.05$) from the controls but even this limited effect was much smaller than that of F or B. This difference was noted only when the Mo was injected 7 to 10 days after birth. No significant differences were found when the injection was made either earlier (three to six days) or later (11 to 14 days). In a later, multivariate analysis of the same data, Adkins and Kruger (1966) found a significant effect on dentin thickness only for a Mo + F combination.

Cooper and Ludwig (1965) compared certain dimensions of children's teeth in the New Zealand cities of Hastings (fluoridated at 1 ppm),

Napier (with local vegetables high in Mo), and Palmerston North (control). The bigonial width, mesio-distal and bucco-lingual diameters (indication of the size of the molars) and the cusp height and buccal convexity (related to the shape) were all significantly lower ($p < 0.01$) at Hastings and Napier than at Palmerston North. These results indicate an approximately equal effect of F and (presumably) Mo on the morphology of the molars. The effect of F in reducing caries is almost certainly accounted for by the combined actions on solubility, remineralization, inhibition of acid production, and morphology. Although the relative contribution of these effects is impossible to assess, it is most unlikely that any one single action makes a large contribution. But with Mo, the admittedly inadequate evidence suggests that an effect on morphology very similar to that of F is the only action authenticated in man. It seems unlikely that this effect could influence caries markedly. However, there is the possibility of interaction with F.

Interaction Between Molybdenum and Fluoride

Stookey et al (1961) found that the effects of molybdenum and fluoride were additive, whereas Büttner (1961) concluded that Mo itself (in the form of paramolybdate) had no action on caries but enhanced the effect of fluoride in young rats whose mothers had received both F and Mo during pregnancy and lactation and who had themselves continued to receive F and Mo for a further 120 days. Van Reen et al (1962) were unable to detect any reduction of caries by sodium molybdate either prenatally or postnatally, nor did sodium molybdate enhance the limited protection afforded by F in the particular strain of rats used (it reduced the severity of caries but not the number of lesions). A summary of the percentage reduction in caries in rats is shown in Table 6-6. A possible explanation of the contradictory results arises from the experiments of Stookey et al (1962) whose data suggested that the Mo:F ratio of the additives governed the interaction. If this ratio was below 0.4 there was no apparent synergistic effect on caries, and the effect increased as the ratio approached unity. One implication of these results is that the action of Mo supplements alone might be to reduce caries by enhancing the effect of the F inevitably present in the basal diet. The contradictory results on caries might be explained by variations in the F concentration of the diet.

Crane (1960) reported that Mo increased the absorption of F from the stomach, and Stookey and Muhler (1960) found greater retention of F in both hard and soft tissues when 50 ppm of Mo were fed. Büttner (1961) found no evidence of greater F concentration in bones of rats receiving Mo but he used paramolybdate which has been found to be less active than molybdate. Stookey et al (1962) reported in experiments on

Table 6-6
Percentage Reduction of Caries in Rats by Molybdenum and Fluoride Reported by Three Groups of Authors

Group	Stookey et al 1961	Büttner 1961	Van Reen et al 1962
Mo alone	18	0	"no highly
F alone	32	51	significant
Mo + F	52	76 (25 ppm)	interaction"
		90 (50 ppm)	

fasting, anesthetized rats that, in old animals, Mo increased absorption of F from the duodenum but reduced it in young animals; in the stomach absorption was reduced in old animals by Mo but it either increased slightly or was not affected in young rats. In the lower intestine, F absorption was increased by Mo irrespective of age. These experiments do not appear to have been repeated by other workers so the results cannot be regarded as established. In any case they would seem to be of doubtful relevance because absorption of F from the alimentary canal is almost complete on a normal diet; there is, therefore, little scope for effects of an increased absorption except that a more rapid absorption might lead to sharper peaks in blood concentration and possibly to greater uptake by the tissues. Ericsson (1966) found no evidence of greater uptake of F by bone when ^{99}Mo was given simultaneously by stomach tube. The whole question of an interaction between Mo and F in caries and metabolism remains open.

CLINICAL TRIALS

There have as yet been no clinical trials to test an effect of most of the trace elements discussed in this book. This may be because of the difficulties of carrying out such trials without first undertaking extensive toxicity trials. Nevertheless this has been done for Mo.

A clinical trial of topically applied sodium molybdate on caries in the deciduous teeth of 31 children was reported by Bertrand (1973). The solution containing 1% ionized Mo (pH 7.53) was applied several times, over a few months, on a sponge to teeth with incipient lesions, and the negative pole of an electrolysis apparatus was placed on the sponge for three minutes at 5mA. Untreated teeth on the opposite side of the mouth with similar lesions served as controls. After extraction the teeth were studied radiographically, and Mo was determined by spectrophotometry. A reduction in caries was reported on the treated side, including some reduction in size of the lesions compared with the controls. Mo was reported to have been taken up by the dentinal protein and odontoblasts but not by enamel or cementum. No other supportive work appears to have been published, and the possible mechanism has not been explained.

SUMMARY

Several epidemiologic studies have identified Mo as a possible anti-cariogenic element. In no case has there been shown a direct causative effect; the studies remain equivocal and the influence of other trace elements may possibly have an effect.

The animal experiments have not solved the question of the anti-caries potential of Mo, and a decisive experiment has not yet been carried out. Such an experiment would consist of providing Mo in various forms and over a physiologic range of doses on cariogenic diets of known Mo content given by a feeding machine with controlled concentrations of the other trace elements interacting with Mo.

If Mo does reduce caries, its mechanism must be regarded as still unknown. The failure to discover any property of Mo in the concentrations present in the oral environment, which could explain a caries-reducing action, casts further doubt on the validity of the epidemiologic data which are contradictory and inconclusive.

REFERENCES

Adkins BL, Kruger BJ: Statistical evaluations of a multiple response experiment: Alterations to the morphology of rat molars. *J Dent Res* 1966;45:1205–1213.

Adler P, Straub J: A water-borne caries-protective agent other than fluorine. *Acta Med Hung* 1953;4:221–224.

Adler P, Porcsalmy I: Nevere Versuche uber den Karies-Protektives Effect des in Trinkwasser Enthatenen Molybdans. *Arch Oral Biol* 1961;4:193–198.

Anderson RJ: Dental caries prevalence in teart pasture areas of Great Britain, in Hardwick JL, Held HR, König KG (eds): *Advances in Fluorine Research and Dental Caries Prevention,* vol 3. Basel, Karger, 1964, pp 165–169.

Anderson RJ: Dental caries prevalence in relation to trace elements. *Br Dent J* 1966;120:271–275.

Anderson RJ: The relationship between dental conditions and the trace element molybdenum. *Caries Res* 1969;3:75–87.

Barmes DR: Caries etiology in Septik villages—Trace elements, micronutrient and macronutrient content of soil and food. *Caries Res* 1969;3:44–59.

Bawden JW, Hammerström LE: Autoradiography of ^{99}Mo in developing rat teeth and bone. *Scand J Dent Res* 1976;84:168–174.

Bertrand G: Clinical study of the cariostatic effect of molybdenum on children. *Rev Odonto-Stomat Midi* (France) 1973;31:27–42.

Bowen WH: The effect of fluoride and molybdate on caries activity in monkeys (Macaca Fascicularis). *Br Dent J* 1973;135:489–493.

Bowen WH, Eastoe JE: The effect of sugar solution containing fluoride and molybdate ions on the pH of plaque in monkeys. *Caries Res* 1967;1:130–136.

Büttner W: Effects of some trace elements on fluoride retention and dental caries. *Arch Oral Biol* 1961;6:40–49.

Cooper VK, Ludwig TG: The effect of fluoride and of soil trace elements on the permanent molars in man. *N Z Dent J* 1965;61:33–49.

Coulter WA, Russell C: Effect of molybdenum on the growth and metabolism of *Veillonella Parrula* and *Streptococcus Mutans*. *J Dent Res* 1974;53:1445–1449.

Crane DB: The effect of molybdenum on fluoride absorption. *J Dent Res* 1960; 39:704 (abs).

Curzon MEJ, Kubota J, Bibby BG: Environmental effects of molybdenum on caries. *J Dent Res* 1971;50:74–77.

Ericsson Y: The distribution of simultaneously administered F^{18} and Mo^{99} in the rat, in Hardwick JL, Held HR, Konig KG (eds): *Advances in Fluoride Research and Dental Caries Prevention,* vol 4, 1966, pp 221–222.

Ericsson Y, Söremark R: Placental transfer of molybdenum and its possible caries-preventive effect. *Caries Res* 1968;2:262–265.

Gallagher IHC, Cutress TW: The effect of trace elements on the growth and fermentation by oral streptococci and actinomyces. *Arch Oral Biol* 1977;22: 555–562.

Hardwick JL, Martin CJ: A pilot study using mass spectrometry for the estimation of the trace element content of dental tissues. *Helv Odont Acta* 1967; 11:62–70.

Healy WB, Ludwig TG, Losee FL: Soils and dental caries in Hawkes Bay, New Zealand. *Soil Science* 1961;92:359–366.

Healy WB, Ludwig TG: Molybdenum content of teeth from different soil areas. *IADR Abs* 1963;379:130.

Herbison RJ, Handelman SL: Effect of trace elements on dissolution of hydroxyapatite by cariogenic streptococci. *J Dent Res* 1975;54:1107–1114.

Hunt CE, Navia JM: Pre-eruptive effects of Mo, B, Sr, and F on dental caries in the rat. *Arch Oral Biol* 1975;20:497–501.

James PMC: Dental caries prevalence in relation to calculus, debris and extrinsic dental staining, in Hardwick JL, Held HR, Konig KG (eds): *Advances in Fluorine Research and Dental Caries Prevention,* vol 3, 1965, pp 153–158.

Jenkins GN: Molybdenum and dental caries. *Br Dent J* 1967a;122:435–441.

Jenkins GN: Molybdenum and dental caries. *Br Dent J* 1967b;122:500–503.

Jenkins GN: Molybdenum and dental caries. *Br Dent J* 1967c;122:545–550.

Kruger BJ: The effect of trace elements on experimental dental caries in the albino rat. *University of Queensland Papers* 1959;1:3–28.

Kruger BJ: Interaction of fluoride and molybdenum on dental morphology in the rat. *J Dent Res* 1966;45:714–725.

Losee FL, Adkins BL: A study of the mineral environment of caries-resistant navy recruits. *Caries Res* 1969;3:23–31.

Losee FL, Cutress TW, Brown R: Natural elements of the periodic table in human dental enamel. *Caries Res* 1974;8:123–134.

Losee FL, Curzon MEJ, Little MF: Trace element concentrations in human enamel. *Arch Oral Biol* 1974;19:467–470.

Losee FL, Adkins BL: Effect of water solids on dental caries in the rat. *Caries Res* 1976;10:332–336.

Ludwig TG, Denby GC, Struthers WH: Dental health: I. Caries prevalence amongst dentists' children. *N Z Dent J* 1960;56:174–177.

Ludwig TG, Malthus RS, Healy WB: An association between soil conditions and dental caries in rats. *Nature* 1962;194:456–458.

McLundie AC, Shepherd JB, Mobbs DRA: Studies on the effects of various ions on enamel solubility. *Arch Oral Biol* 1968;13:1321–1330.

Nagy Z, Polyik E: A devavanyai ivoviz specialis vizsgalata femnyomokra. *Fogorvisi Szemle* 1955;48:154–156.

Nixon GS: Trace element content of the hard dental tissues and dental plaque. *Caries Res* 1969;3:60–74.

Puttnam NA, Bradshaw F, Platt P: X-ray spectroscopic determination of the constituent elements of dental plaque, in Hardwick JL, König KG, Held HR (eds): *Advances in Fluorine Research and Dental Caries Prevention,* vol 4, 1965, pp 157–162.

Rahkamo A, Tuompo H: Effect of molybdenum on the developing rat tooth in tissue culture. *Proc Finn Dent Soc* 1974;70:141–146.

Schamschula RG, Adkins BL, Barmes DE, et al: *WHO Study of Dental Caries Etiology in Papua-New Guinea.* Geneva, WHO, 1978.

Schroeder HA, Balassa JJ, Tipton IH: Essential trace elements in man: Molybdenum. *J Chron Dis* 1970;23:481–499.

Shaw JH, Griffiths D: Developmental and post-developmental influences on incidence of experimental dental caries resulting from dietary supplementation by various elements. *Arch Oral Biol* 1961;5:301–322.

Stookey GK, Muhler JC: Effect of molybdenum on fluoride retention in the rat. *J Dent Res* 1960;39:671, (Abs).

Stookey GK, Roberts RA, Muhler JC: Synergistic effect of molybdenum and fluoride on dental caries in the rat. *Proc Soc Exp Biol Med* 1961;109:702–705.

Stookey GK, Crane DB, Muhler JC: Effect of molybdenum on fluoride absorption. *Proc Soc Exp Biol Med* 1962;109:580–583 (abs).

Stookey GK, Muhler JC: Further studies upon the synergistic effect on Molybdenum and fluoride upon experimental caries. *J Dent Res* 1964;43:865–866, (abs).

Sutcliffe P: A clinical study of mandibular incisor caries, in Hardwick JL, Held HR, König KG (eds): *Advances in Fluorine Research and Dental Caries Prevention,* vol 3, 1965, pp 159–163.

Suttle NF: Recent studies of the copper-molybdenum antagonism. *Proc Nutr Soc* 1974;33:299–305.

Underwood EJ: *Trace Elements in Human and Animal Nutrition,* ed 4. New York, Academic Press, 1977.

Van Reen R, Ostrom CA, Berzinskas VJ: Studies of the possible cariostatic effect of sodium molybdate. *Arch Oral Biol* 1962;7:351–356.

34 **Se**

78.96

8-18-6

7 Selenium

T. R. Shearer

A number of excellent reviews on selenium have been published in the last 15 years (Allaway, 1973; Burk, 1976; National Research Council, 1976; Underwood, 1977). This section will emphasize some of the most important aspects of selenium as they relate to dental caries.

Selenium (Se), atomic number 34, is relatively rare, comprising approximately 10^{-4} % of the earth's crust. It is in group VI of the periodic table along with sulfur and tellurium. Commonly found compounds of biologic interest include:

Selenites Na_2SeO_3

Selenomethionine

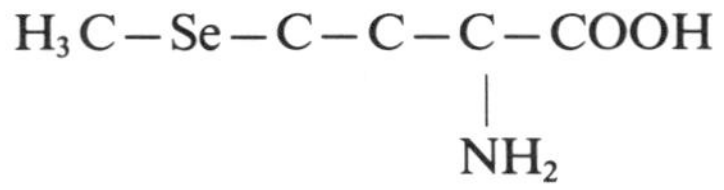

$$H_3C-Se-C-C-\underset{\displaystyle NH_2}{\underset{|}{C}}-COOH$$

Selenocystine

$$\begin{array}{l} Se-CH_2-\overset{\displaystyle NH_2}{\overset{|}{C}}-COOH \\ \,| \\ Se-CH_2-\underset{\displaystyle NH_2}{\underset{|}{C}}-COOH \end{array}$$

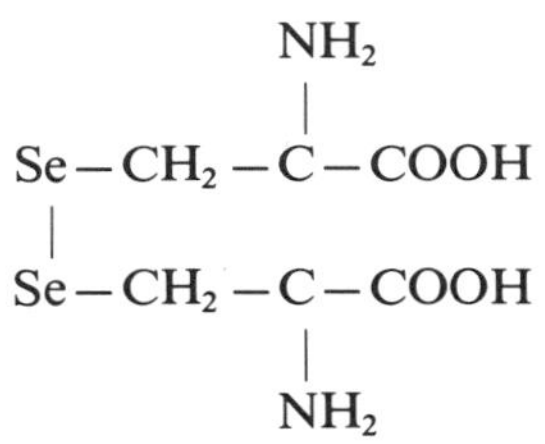

Selenates	Na_2SeO_4
Selenotrisulfides	protein – S – Se – S – protein

Metabolism

In animals Se is metabolized along some of the same metabolic pathways as sulfur (S). The amounts and chemical forms of Se ultimately incorporated into the dental tissues of animals may be influenced by the intermediary metabolism of the specific Se compound ingested. Selenomethionine follows the pathways of methionine metabolism. It can be actively transported across cell membranes, synthesized into proteins via ribosomal template assembly by deceiving the methionine-activating enzyme, and enzymatically converted to a variety of S analogue intermediates. These intermediates include selenocysteine, which can be incorporated into proteins as a selenotrisulfide, or oxidized to selenite, which can also form a proteinaceous selenotrisulfide. On the other hand, ingested selenite is not actively transported across cell membranes; most mammals lack the enzymes necessary for the direct synthesis of selenoamino acids from selenite, and selenite is thought to be incorporated into tissue proteins initially through selenotrisulfide formation (Shearer, 1975). These possible differences in the intermediary metabolism of organic and inorganic selenium compounds should be noted since most published studies on the effect of Se on dental caries in experimental animals have used inorganic selenite, while most of the Se consumed by man is probably proteinaceous, such as selenomethionine found in grains.

Table 7-1
Selenium Content of Human Tissues

Tissue	Selenium	Tissue	Selenium
	ppm		ppm
Nails	1.14	Blood	0.20
Liver	0.43	Urine	0.05
Muscle	0.37	Saliva	0.03
Skin	0.27	Milk	0.02

Data from Burk, 1976; Hadjimarkos and Shearer, 1973; Shearer and Hadjimarkos, 1975.

The Se concentration in human tissues has been determined (Table 7-1), and tissues containing high amounts of S proteins have an affinity for Se. The level of Se in human fingernails and toenails was one of the highest levels reported in man (Hadjimarkos and Shearer, 1973), and this is probably because the protein component of nails is keratin, containing 3.2% S and 12% cysteine. For this reason, the Se content of human nails has been proposed as a new index for epidemiologic studies concerning selenium and dental caries.

Selenium Deficiency and Toxicity

Se is an essential nutrient in the diet of a wide variety of animals including man at about 0.1 ppm. Deficiency diseases include exudative diathesis and pancreatic fibrosis in poultry, white muscle disease (muscular dystrophy) in lambs and calves, hepatosis dietetica in pigs, and liver necrosis in rats. Because of the economic losses from these Se deficiency diseases in farm animals, the Food and Drug Administration recently allowed the use of Se as a feed additive. Se supplements are now sold in health food stores. This means that more Se will enter the food chain where it could conceivably influence dental health.

The exact biochemical function of Se in the prevention of Se deficiency diseases is unknown. Theories center around the association of Se with tissue proteins (Burk, 1976; Luoma et al, 1971). Glutathione peroxidase was recently discovered to be a selenoprotein (National Research Council, 1976), and this enzyme protects tissues from oxidative damage.

Since Se is one of the most biologically active nutrients, the range between toxic doses and essential dietary levels of it may be narrow. Selenosis in rats occurs at about 3 to 5 ppm Se where growth and food consumption are depressed. Chronic Se poisoning in cattle and sheep produces diseases called alkali disease and blind staggers from consumption of grain and forage grasses (5 to 40 ppm Se) or seleniferous weeds ($>$ 800 ppm Se) (Rosenfield and Beath, 1964). The biochemical mechanism for this toxicity is not understood, but some investigators show that Se combines with sulfhydryl groups of enzymes and thereby inhibits their function (Donaldson, 1977). Thus, the biochemical mechanisms underlying the essentiality of Se and its toxicity seem to involve an interaction between Se and tissue proteins.

Human Food Sources and Requirements

The estimated safe and adequate daily intake of Se for adults is 50 to 200 μg which man receives almost entirely from the foods he eats rather than from drinking water. As most of the Se in foods is probably of an organic proteinaceous form, the Se content of foods tends to be highest in foods containing high amounts of protein (Table 7-2).

Typical diets in the United States supply about 132 μg/day of Se, and there is about twice as much consumed as needed to prevent Se deficiency disease (Levander, 1976). The upper level of intake where Se excess may be dangerous to human health is estimated at 200 μg/day, and

Table 7-2
Selenium Levels in Typical Foods (United States)

Food Group	Item	Selenium[a] ppm, wet	Protein %
Organ meat	Beef liver	0.43	24
Meat	Beef	0.34	25
Bread	White	0.28	9
Cereals	Corn flakes	0.03	8
Milk	Whole	0.01	4
Vegetable	Carrot	0.02	1
Fruit	Orange	0.01	1

[a]Data from Morris and Levander, 1970.

certain population groups in Japan and Venezuela may be at risk for selenium toxicity because of the high Se content of foods.

We recently reported a geographic variation in Se content of breast milk across the United States which was roughly related to the Se content of forage crops (Shearer and Hadjimarkos, 1975) (Figure 7-1 and Table 7-3). The Se content of breast milk, as well as its content of other trace elements, is of interest to the dental profession because breast milk is ingested during the period of tooth development. Similar geographic variations for tissue Se have been noted for blood and urinary Se levels (Underwood, 1977).

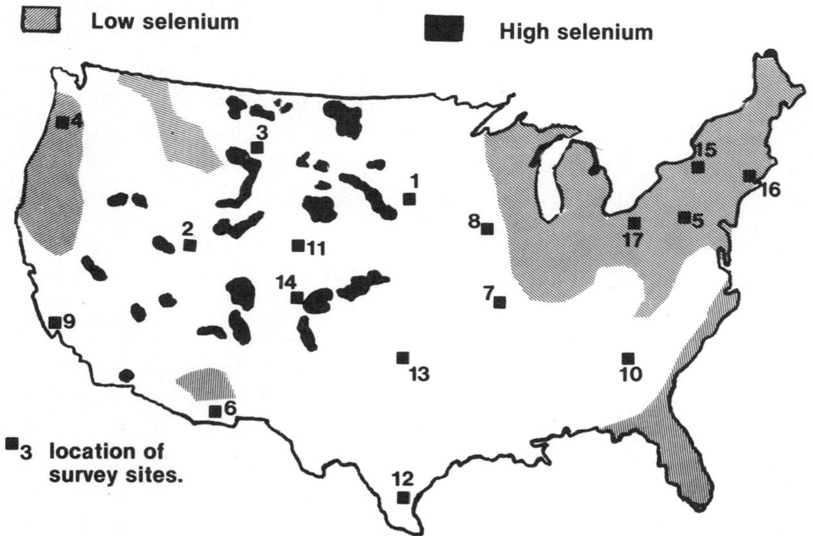

Figure 7-1 Cities in United States where breast milk samples were collected for selenium analysis. Results are given in Table 7-3.

Table 7-3
Selenium Concentrations in Human Mature Milk from 17 Cities in the United States

City	n	Selenium Concentration ppm	
		Mean	± *SE*
1. Sioux Falls, SD	15	0.028	0.003
2. Salt Lake City, UT	14	0.022	0.003
3. Billings, MT	10	0.021	0.001
4. Portland, OR	15	0.021	0.003
5. State College, PA	15	0.021	0.001
6. Douglas AZ	14	0.020	0.001
7. Rolla, MO	15	0.020	0.001
8. Iowa City, IO	15	0.020	0.001
9. Santa Barbara, CA	14	0.018	0.002
10. Athens, GA	11	0.018	0.001
11. Cheyenne, WY	14	0.016	0.001
12. Corpus Christi, TX	15	0.016	0.001
13. Norman, OK	14	0.016	0.001
14. Pueblo, CO	15	0.015	0.001
15. Syracuse, NY	15	0.015	0.001
16. Bristol, CT	15	0.015	0.001
17. Akron, OH	15	0.013	0.001
Overall	241	0.018	0.0004

Paradoxical Effects of Selenium

On the basis of early reports of tumor development in rodents administered high levels of Se, the element was classified as a carcinogen, and it has also been shown to be mutagenic in cultured fibroblasts, a usual property of most if not all carcinogenic agents (Lo et al, 1978). Another group of scientists cite epidemiologic evidence that adequate amounts of Se protect against neoplastic disease in man (Schrauzer et al, 1977). Both toxicity and deficiency of Se cause cataract formation and liver necrosis in rats (Shearer et al, 1980). These paradoxical biologic effects of Se have not been explained. The effect of Se on dental caries likewise at times seems paradoxical; some studies show that if Se is ingested during tooth development, caries is promoted, while other studies give equivocal results.

CARIES

Man

At least nine studies have been published over the last 43 years concerning the relationship between Se and dental caries in man (Table 7-4).

Six of these studies have shown that increased amounts of Se in the diet are associated with dental caries; two studies (Cadell and Cousins, 1960; Retief et al, 1976) conducted in areas of low selenium showed no relation between Se and caries; one recent study (Curzon and Crocker, 1978) found a negative association between caries and the Se content of teeth.

Early emphasis of a positive association between Se and dental caries was reported by the Public Health Service indicating that "bad teeth" were a frequent finding in rural populations living in seleniferous areas of the United States (Smith and Westfall, 1936, 1937). Three independent epidemiologic studies on children in four western states of the United States showed that consumption of somewhat increased amounts of dietary Se during the period of tooth development increased the prevalence of dental caries (Hadjimarkos et al, 1952; Ludwig and Bibby, 1969; Tank and Storvick, 1960). The actual upper limit of dietary Se intake in man above which dental caries may occur is unknown. However, in one of the studies mentioned above, there was at least a tenfold difference in the Se content of milk (0.005 vs 0.067 ppm Se) and eggs (0.056 vs 0.502 ppm Se) between the geographic areas where high and low caries rates were found and a twofold difference in the Se content of 24-hour urine specimens (0.037 vs 0.074 ppm Se) (Hadjimarkos and Bonhorst, 1961).

It should be noted that in the Wyoming study (Tank and Storvick, 1960), not only was there a positive relation between increased Se intake and caries, but also there was evidence to indicate that Se reduced the cariostatic effects of fluoride. Caries rates were 41% higher in areas that were high in both Se and fluoride compared with areas low in Se and high in fluoride.

A study in three population groups of the Chernovitsi region of the Soviet Union showed a direct relationship between caries susceptibility and the concentration of Se in teeth, which reflected the Se levels of the local soils (Suchlov et al, 1973). Not only is this study important because it confirms the other epidemiologic studies on the cariogenic nature of Se, but it is one of only a limited number of reports to study the levels of Se in teeth associated with dental caries. In an epidemiologic study of primitive village communities in Papua-New Guinea, evidence was also found, along with many other associations, of a direct relation between caries prevalence and Se in staple foodstuffs such as sago, sweet potato, and Chinese taro (Barmes, 1969). This study is particularly interesting because the main cariogenic challenge found in many of the other epidemiologic studies performed on more advanced civilizations—refined sugar—was not present. It is possible that some of the more subtle influences on caries, such as Se and other trace elements, may have a more pronounced influence in such an environment.

Two studies (Cadell and Cousins, 1960; Retief et al, 1976) that showed no relationship between Se and dental caries were conducted in

Table 7-4
Summary of Studies in Man Relating Selenium to Dental Caries

Author (year)	Study Conditions	Effect on Caries
Smith and Westfall, 1936, 1937	Health survey in seleniferous areas of United States	Increased Caries, "Bad teeth" frequent finding
Hadjimarkos et al, 1952	Oregon children	Increased caries related to Se in foods and increased Se excretion in urine
Tank and Storvick, 1960	Wyoming children	Selenium reduced cariostatic effects of fluoride
Ludwig and Bibby, 1969	Western states children	Increased caries in high selenium areas
Suchlov et al, 1973	Russian population groups	Increased caries directly related to Se in teeth and soil
Barmes, 1969	New Guinea villages	Variable with a direct relation between caries and Se in foods
Cadell and Cousins, 1960	New Zealand children	No difference in caries in two population groups in low Se areas
Retief et al, 1976	South African children	Se not a factor in caries development in teeth having extremely low Se content
Curzon and Crocker, 1978	Teeth from United States and New Zealand	Negative correlation between DMFT and Se in teeth possibly related to varied background of teeth

areas where Se intake was apparently quite minimal. The mean urinary concentrations of the two groups of New Zealand children showing differences in caries rates were 0.021 and 0.030 ppm Se (Cadell and Cousins, 1960), which are considered below normal. Retief et al (1976) found markedly different caries rates between a group of black South African children (DMFT = 2.5, males; 2.1, females) and white South African children (DMFT = 8.7, males; 11.1, females). An attempt was made to relate these caries rates to the Se content of enamel for these teeth: whites = 0.01 ppm Se and blacks 0.08 ppm Se. These are some of the lowest values for enamel Se ever reported (Table 7-4) and indicate that neither group of children was probably exposed to detrimental amounts of Se.

Curzon and Crocker (1978) recently analyzed whole enamel samples from 451 caries-free premolar teeth collected from 19 states in the United States and Dunedin, New Zealand. They found a negative association between the concentrations of Se in enamel and the caries rate of the tooth donors. These teeth represented a broad spectrum of background including some teeth collected from the Se-deficient areas within New Zealand. It should be remembered that the concentrations of Se found in mature teeth may not necessarily represent the concentrations of Se present during tooth development when the element is thought to exert its effects. Se is associated with the proteins of teeth, and during the maturation process, matrix proteins are lost (Shearer, 1975).

Animals

No less than 12 separate articles have been published reporting the effect of Se on dental caries in various experimental animals (Table 7-5). Bowen (1972) provided evidence for the cariogenic nature of developmental Se in monkeys (*Macaca irus*). He found almost twice as many caries in the monkeys receiving Se at 1 to 2 ppm (16.0 lesions per animal) compared with the control monkeys (8.4 lesions per animal). Caries were also formed faster in the Se group (6.7 months) compared with the controls (21.0 months). In those teeth already formed before the start of the experiment, Se appeared moderately anticariogenic. Drawbacks to this otherwise important experiment were the use of only three animals in the Se group while the control group contained seven animals, and the toxicity of Se to the monkeys.

Using pregnant rats, Büttner (1963) gave 2.3 or 4.6 ppm Se as sodium selenite in drinking water before birth and continued this regimen for 120 days for the pups. He found a statistically significant increase in caries in both Se groups, and the increase in caries was greater for the higher Se group. This developmental effect of Se occurred even

Table 7-5
Summary of Studies Relating Selenium and Caries in Experimental Animals

Author (year)	Animal	Selenium Concentration	Given in	Effect on Caries
Studies with Se Given During Tooth Development				
Bowen, 1972	Monkey	1–2 ppm	Water (5 years)	Increase in caries in only those teeth undergoing development during the experiment
Büttner, 1963	Rat	2.3, 4.5 ppm	Water	Statistically significant increase in caries, proportionate to amount of dietary selenium, decreased weight gain
Navia et al, 1968	Rat	4.0 ppm	Water or food	Increase in sulcal lesions only when Se present in drinking water
Britton et al, 1980	Rat (pregnant)	0.8 ppm	Water	Perinatal Se cariostatic to offspring
Shearer (unpublished observations), 1981	Rat	0.3 mg Se/kg, 2.4 ppm	Injection and water	Cariostasis
Studies Where Selenium Was Started After Tooth Development				
Bowen, 1972	Monkey	1–2 ppm	Water (5 years)	Cariostatic on teeth already developed before start of experiment
Wheatcroft et al, 1951	Rat	1.0, 0.5, or 0.2 mg Se/kg, 100 days	Intraperitoneal injection	Slight trend toward increase in caries
Tempestine, 1962	Rat		Intraperitoneal injection	Small increase in caries
Pappalardo, 1959,	Rat		Stomach tube	Small increase in caries
1962	Rat		Stomach tube	Decreased effectiveness of fluoride

Table 7-5 (continued)

Author (year)	Animal	Selenium Concentration	Given in	Effect on Caries
Muhlemann and König, 1964	Rat	3.3 ppm	Water (20 days)	Inhibition of caries, decreased weight gain
Muhler and Shafer, 1957	Rat	4.5–14.3 ppm	Diet (140 days)	No effect, decreased weight gain
Claycomb et al, 1965	Rat	4.6 ppm	Diet (100 days)	No effect, very low incidence of caries in all groups
Meissner, 1976	Rat	0.5–4.6 ppm	Water	No effect on caries; no effect on cariostatic effects of topical fluoride, Se was toxic
Kaqueles et al, 1977	Rat	a) 0.2–0.4 μg b) 2–4 mg	Ingestions	Cariogenic effect in high Se group; cariostatic effect in low Se group

though the Se groups gained less weight and undoubtedly ate less of the cariogenic diet than did the animals in the control group. A disturbing feature of Büttner's experiment was that there were different proportions of male and female rats in his different test groups. The sexes grow at different rates and hence consume different amounts of cariogenic diet. Navia (1968) fed 4 ppm Se, as Na_2SeO_3, in the drinking water or food of rats from birth until 50 days of age. The only effect of Se was to cause a statistically significant increase (+ 12%) in sulcal caries when Se was present in the drinking water. Again, this occurred despite the fact that the Se animals showed decreased consumption of the cariogenic diet.

Using the modern rat caries model, our laboratory presented Se to rats during pregnancy and lactation (Britton et al, 1980), and to weanling rats (Shearer, unpublished data, 1981), and measured caries after three weeks of feeding a sucrose diet. In both cases Se reduced the frequency of caries. In experiments where Se was delivered after tooth formation the element has been found to have a variable effect; this has occurred in studies in which Se was delivered in the diet by injection or by stomach tube (Table 7-5). Some of these experiments have been reviewed previously (Hadjimarkos, 1973).

Unfortunately, many of the experimental caries studies have failed to control two problems associated with Se:

1. Relatively small amounts of Se are toxic and cause a decrease in consumption of cariogenic diets. This makes direct comparison of caries rates between ad libitum control groups and Se test groups difficult. Recently one of these postdevelopmental experiments (Meissner, 1976) involved 100 rats which received doses of sodium Se ranging from 1 to 10 ppm Se. Some groups even received topical brushing with sodium fluoride paste. There was no effect of Se on caries or on the cariostatic effect of topical fluoride. The higher dosages of Se were so toxic that 22% of the animals died; obviously there must have been decreases in the amount of diet consumed.
2. Se is taken up into teeth during tooth development, which for the first and second molar teeth of the rat occurs in utero and during lactation. Unlike fluoride, the postdevelopmental uptake of Se is quite minimal (Shearer, 1975).

Recently Kaqueles et al (1977) reported an interesting paradoxical effect of selenate on caries development in the rat. Injection of high doses of selenate promoted caries while low doses prevented caries. While it is possible that low doses of Se may be required for optimal calcification (and caries resistance) and high doses may be detrimental to enamel development and may promote caries, there is an obvious lack of consistent results obtained in the rodent experiments.

MECHANISM OF ACTION

Selenium in the Teeth of Man

The uptake and distribution of Se in human teeth has been measured (Table 7-6). Se has been found in almost all samples of human enamel and dentin analyzed. Compared with other elements such as fluoride, enamel Se levels are low. The reported levels of Se in human teeth are influenced by analytical methodology and possibly geographic and ethnic origin.

In a Russian study (Suchlov et al, 1973) different concentrations of Se were measured in whole teeth collected from patients residing in localities having different soil Se levels. Relatively low Se was found in the forest-plains (0.35 ppm Se) and intermediate levels in the steppes (1.26 ppm Se); high levels were found in the mountains (4.57 ppm Se). The soil in the steppes contained approximately 1.4 times more Se than that in the forest-plains, and soils in the mountains contained 1.6 times more Se than that in the steppes. Availability of Se in the soil determines its concentration in the diet, which in turn may explain some of the geographic variations in enamel Se.

Ethnic origin also may influence the enamel Se level. Retief et al (1976) reported that enamel Se levels from black South Africans were significantly higher (6.7 times) than enamel Se levels in white South Africans.

The influence of age on the concentration of Se in human teeth has also been investigated because such data might provide insight into the mechanism of uptake of Se by human teeth. Both the American and the Russian studies (Table 7-6) reported that deciduous enamel and dentin contain higher concentrations of Se than do permanent teeth. This indicated that the diet during childhood probably supplies important amounts of Se for the teeth. In permanent teeth, Hadjimarkos and Bonhorst (1959) could find no definite pattern regarding the Se content of enamel or dentin with advancing age. Such data would indicate that there is no appreciable postdevelopmental uptake of Se by enamel.

Selenium Uptake into the Teeth of Experimental Animals

Further insight into the mechanism of uptake of Se into the teeth has been provided by work using experimental animals. Until now, the main

Table 7-6
Variation in the Reported Concentrations of Selenium in Human Dental Enamel

Country	Author(s)	Enamel Selenium ppm	Comment
South Africa	Retief et al, 1976	0.01, 0.08	Blacks higher than whites
Greece	Hadjimarkos and Bonhorst, 1962	0.03[a]	Ancient higher than modern
England	Nixon and Myers, 1970	0.87	
United States	Curzon and Crocker, 1978;	1.47	No trend with age; deciduous higher than permanent
	Hadjimarkos and Bonhorst, 1961;	0.84[b]	
	Losee et al, 1974;	0.27	
	Johnson and Shearer, 1979	0.05	No gradient from surface to DEJ
Russia	Mamedova, 1965; Suchlov et al, 1973	3.24, 10.19[c]	Whole teeth also high in 2nd study; deciduous lower than permanent

[a]Dentin = 0.13 ppm Se.
[b]Dentin = 0.52 ppm Se.
[c]Dentin = 1.56–3.01 ppm Se.
DEJ = dentino-enamel junction.

species studied has been the rat, and all experiments have used the radioactive isotope of selenium, ^{75}Se, because of the relatively small amounts taken up by teeth. Early studies (Claycomb et al, 1960) showed that the highest concentrations taken up into whole teeth were in the pulp and extravascular areas of the teeth. Much lower concentrations of the Se isotope were found in the mineralized fraction of the tooth. Thomassen and Leicester (1964) then made the significant observation that if rats were injected with radioactive Se during the period of active calcification of the molar tooth (the suckling period for the first and second molars), a prolonged and constant amount of Se remained in the teeth even up to 184 days after the last injection.

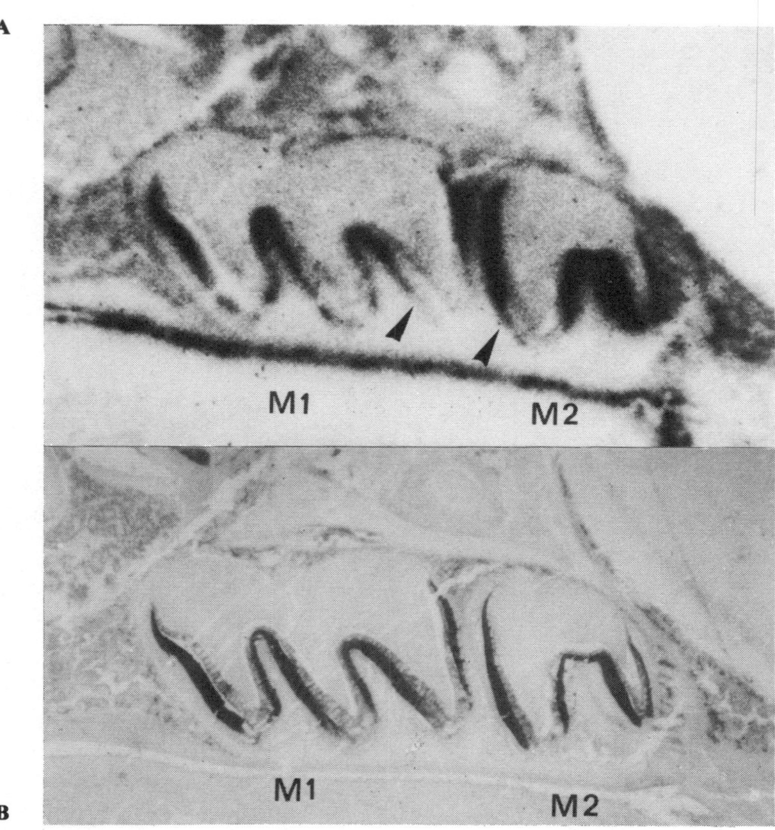

Figure 7-2 **A** H + E stained section of the first (M1) and second (M2) maxillary molar teeth from an 8-day-old rat. Used by permission of Bawden and Hammerström, 1977. **B** Corresponding autoradiogram of first and second molar teeth of 8-day-old rat injected with ^{75}Se. Note the uptake of ^{75}Se only in enamel undergoing matrix formation (black areas at arrows). Used by permission of Bawden and Hammerström, 1977.

Using an elegant autoradiographic technique, Bawden and Hammarström (1977) showed that the uptake of ^{75}Se in the 8-day-old rat pup occurs mainly in the newly formed enamel matrix or the secretory ameloblast of the developoing first and second molar (Figure 7-2). Note the lack of Se uptake into the cuspal areas of the tooth which would have already calcified before administration of the isotope. These results were confirmed in vitro where it was shown that normal cellular function was necessary for Se uptake by ameloblasts (Crisp et al, 1979). Surprisingly, we recently found that the amount of Se taken up by the developing molar teeth of the rat pup is not well correlated with maternal Se intake during pregnancy (Shearer and Britton, 1977). It may be that in the rat proportionate uptake of Se into molar teeth only occurs during the period of tooth development after birth (suckling period).

Recently, we further characterized the mechanism for Se uptake into teeth by providing radioactive ^{75}Se in the drinking water of pregnant rats and analyzing the various fractions of the molar teeth from the pups born to these mothers (Table 7-7) (Shearer, 1975). In such studies, it is extremely important to clean the teeth of all soft tissue since the soft tissues usually contain appreciable quantities of ^{75}Se. We found that developing molar teeth take up much more dietary selenium than do fully erupted, mature molar teeth. The concentration of ^{75}Se was always higher in dentin than enamel. In developing molar teeth, significantly more dietary Se is taken from organic Se (^{75}Se-selenomethionine) than from inorganic selenium ($Na_2{}^{75}SeO_3$). This is important because probably most

Table 7-7
Summary of Studies on ^{75}Se in Molar Teeth of Rats

Dental Tissue	Concentration of ^{75}Se (Mean ± SE)	
	^{75}Se-selenomethionine Drinking water	*$Na_2{}^{75}SeO_3$ Drinking water*
		% Dose/100 g Tissue
Whole tooth powder (mature)	0.29 ± 0.08	0.56 ± 0.02[b]
Whole tooth powder (developing)	2.21 ± 0.08	1.24 ± 0.08[b]
enamel	0.98 ± 0.04	0.70 ± 0.04[b]
dentin	2.27 ± 0.05	1.25 ± 0.08[b]
	% of total ^{75}Se located in protein fraction[a]	
Enamel	79.4 ± 3.8	80.4 ± 1.3
Dentin	94.6 ± 0.4	95.1 ± 0.9

[a]Calculated from: $\frac{^{75}\text{Se remaining after HCl dialysis}}{\text{Se in intact tissue}} \times 100.$

[b]Means within same horizontal columns were significantly different at $p < 0.01$.
Data from Shearer and Hadjimarkos, 1973; Shearer, 1975.

of the selenium consumed by man is in the form of proteinaceous organic Se rather than inorganic Se.

In both enamel and dentin from the developing molar tooth, the vast majority of the Se is nondialyzable in 0.1N HCl. This is evidence that the Se in enamel and dentin is located in the protein fraction of these dental tissues. Over a period of time, there is a movement of Se from the nonprotein fraction to the protein fraction of the teeth (Petrovich and Podorozhnaya, 1974). In soft tissues most of the Se is primarily located in the protein fraction (Underwood, 1977). Dialysis of ^{75}Se-labeled enamel and dentinal proteins has shown that there may be at least three fractions of Se in teeth: proteinaceous selenotrisulfides (protein–cysteine–S–Se–S–cysteine–protein), proteinaceous stable Se (possibly selenocystine and selenomethionine), and loosely bound Se (Shearer, 1975).

Interaction of Selenium with Other Trace Minerals

Fluoride A number of studies have reported that trace minerals such as Mo, V, B, and Cd interact with fluoride to alter its concentrations in dental tissues or its cariostatic effectiveness (Navia, 1970; Shearer et al, 1980). At the enamel surface trace minerals may complex with fluoride and link it to the enamel surface or in some other manner increase or decrease the binding of fluoride to the tooth surface (Shearer and Suttie, 1977). Trace minerals may affect the absorption or the general systemic metabolism of fluoride and thereby alter the availability of fluoride for uptake by the developing tooth. Another factor which may determine the type of interaction which may occur between a trace mineral and fluoride in the enamel is the compartmentalization of the two elements within the dental tissue. For example, Se exists in the organic fraction of enamel and dentin, while fluoride is found in the mineral fracton (Shearer and Ridlington, 1976).

It was important to establish if there was an interaction between Se and fluoride because a previous report indicated that ingestion of Se appeared to negate the cariostatic effectiveness of fluoride (Tank and Storvick, 1960) (see section on caries in man). Büttner (1963) found that Se had no effect on the uptake of fluoride by the rat femur. Büttner's studies should be viewed with caution as his diet contained no added fluoride and the levels of Se used reduced food consumption and hence F intake.

Hadjimarkos (1967) improved the experimental design of these fluoride-Se interaction experiments by pair-feeding both the food and water of a control group of rats receiving 50 ppm F drinking water to an experimental group of rats receiving drinking water containing 50 ppm F plus 3

Table 7-8
Comparative Toxicity of Inorganic and Organic Selenium Compounds in the Presence of Fluoride

Drinking Water	Water Consumption	Body Weight Gain
	mL/day	g/32 days
Controls		
Distilled water	31.0 ± 3.1[a]	242 ± 10
50 ppm F	31.4 ± 1.5	230 ± 8
50 ppm F + 1 ppm Se		
SeO_3	28.0 ± 1.7	217 ± 11
SeO_4	27.8 ± 1.3	226 ± 9
SeMet	31.0 ± 1.7	237 ± 9
SeCys	29.3 ± 1.6	225 ± 7
50 ppm F + 3 ppm Se		
SeO_3	23.1 ± 1.4[b,d]	211 ± 7
SeO_4	23.0 ± 1.1[b,d]	205 ± 7
SeMet	26.7 ± 1.0	211 ± 4
SeCys	23.9 ± 0.7	196 ± 8[c]

[a]Mean ± SE (n = 5 or 6). Within each vertical column, means with superscripts were significantly different ($p < 0.05$) from the following means:
[b]Both control groups
[c]Distilled water control group
[d]1 ppm SeMet group. There were no other statistically significant differences between any two means. The least significant difference (LSD) between any two means for water consumption was 7.71 ml/day, and for body weight gains the LSD was 38 g/32 days.
Data from Shearer and Ridlington, 1976.

ppm Se as Na_2SeO_3. After 28 days, no differences were found in the amount of fluoride taken up into the femurs of both groups.

In a more extensive study, we measured the effect of four different organic and inorganic Se compounds (Table 7-8) on the uptake of fluoride by the teeth and bones of weanling rats (Shearer and Ridlington, 1976). However, the presence of fluoride did not appear to enhance the symptoms of Se toxicity. Likewise, none of the Se compounds enhanced the uptake of fluoride into liver and kidney.

Neither 1 nor 3 ppm Se, as any of the four Se compounds tested, caused statistically significant changes in the uptake of fluoride into the rat femurs or onto the mature molar teeth (Table 7-9). Because of the experimental design, the fluoride uptake onto the molar teeth of the rats represents both the developmental (third molar) and the postdevelopmental (first and second molar) uptake of fluoride onto enamel and onto a small amount of underlying dentin. Clinically important amounts of fluoride are taken up by the developing human enamel.

Table 7-9
Effect of Dietary Selenium on the Uptake of Fluoride by Hard Tissues

Drinking Water	Femur Fluoride	Molar Fluoride
	ppm	ppm
Controls		
Distilled water	482 ± 12[a]	48 ± 4
50 ppm F	2158 ± 89[b]	217 ± 29[b]
50 ppm F + 1 ppm Se		
SeO_3	1849 ± 143[b]	182 ± 30[b]
SeO_4	1896 ± 211[b]	243 ± 37[b]
SeMet	2065 ± 122[b]	210 ± 28[b]
SeCys	2173 ± 146[b]	189 ± 15[b]
50 ppm F + 3 ppm Se		
SeO_3	1722 ± 54[b]	130 ± 15
SeO_4	1764 ± 102[b]	182 ± 36[b]
SeMet	1884 ± 81[b]	160 ± 23[b]
SeCys	2019 ± 103[b]	182 ± 26[b]

[a]Mean ± SE (n = 5 or 6)
[b]Within the vertical columns, means with superscripts were significantly different ($p < 0.05$) from the means of the distilled water control group. There were no other statistically significant differences between any two means. The least significant difference (LSD) between any two means for femur fluoride was 492 ppm F, and for enamel the LSD was 122 ppm F.
Data from Shearer and Ridlington, 1976.

Thus, at present, none of the studies has shown that Se and fluoride at physiologic levels interact directly in dental tissues. These results could be due to the great disparity in the concentrations of these elements in enamel or the location of fluoride and Se principally in two chemically distinct compartments of the body: fluoride in the mineralized fraction of hard tissues as fluoroapatite, and Se in the protein fraction of soft tissues as selenoamino acids and selenotrisulfides.

Trace elements besides fluoride There are many documented examples of the interaction in soft tissues of Se with a variety of other trace minerals, including Te, Hg, Cd, and Cu (Hill, 1975). Se is particularly noted for its ability to ameliorate the symptoms of heavy metal toxicities resulting from excess Cd and Hg. It has been suggested that trace element antagonism may be due to competition between Se and other trace elements for active protein-binding sites. Many other trace elements besides fluoride have been shown to affect the development of caries, and studies should be performed to test whether Se alters the uptake of these elements into enamel. It is possible that Se may affect caries not by interaction with fluoride but by interaction with some other trace element.

Effect of Selenium on Enamel Mineralization

A number of studies have shown that the administration of Se during the period of tooth formation results in structural changes in the tooth enamel (Table 7-10). Such data indicate that large doses of Se are detrimental to calcification. Bowen (1972) put Se as sodium selenate into the drinking water of monkeys for five years and reported that the premolars and the second molar teeth, which were forming during the period of Se administration, had a chalky yellow appearance. Those teeth which had been calcified before the start of the experiment were unaffected.

Using an artificial mouth (McLundie et al, 1968) showed that the addition of 10 or 15 ppm Se as sodium selenite to a pH 6 buffer caused an increased dissolution of human enamel. They suggested that this effect may have been due to calcium selenite chelate formation. However, the physiologic significance of this study may be doubted since human saliva from Portland, Oregon contains only 10 to 15 parts per billion (μg/l) of Se (Hadjimarkos and Shearer, 1971). In experimental animals injected with ^{109}Se, Se rapidly appears in the saliva and salivary glands (Claycomb et al, 1960). Since saliva continuously bathes the teeth it would be pertinent to study the kinetics of Se excretion in saliva from patients consuming the now widely available commercial Se supplements.

In an in vitro experiment, Luoma et al (1971) measured the effect of 17 ppm Se as sodium selenite on acid production in a sucrose inoculum medium containing a cariogenic streptococcus, K-1. This level of Se caused a 15% reduction in the drop in pH by these micro-organisms. Again, the physiologic significance of these changes is unknown especially since the concentration of Se in plaque in man has not been measured. Since Se is an essential element for the growth of a number of micro-organisms, its role in the growth of oral micro-organisms needs much further research.

Thus, the bulk of the studies in the literature have been concerned with the effect of Se on the calcification of the dental tissues. These studies have shown that Se inhibits proper calcification. The underlying biochemical events causing these changes are unknown, but it can be suggested that these effects may be related to the demonstrated uptake of Se into the proteins of enamel during tooth development. Important biochemical events have been attributed to the presence of Se in the proteins of other tissues. For example, the dietary essentiality of Se is thought to be partially explained by its uptake into the enzyme glutathione peroxidase (National Research Council, 1976). Conversely, Se is known to inhibit a number of enzymes, including succinic dehydrogenase, which have been shown to be present in the ameloblast. Future experiments should be directed toward determining the exact alterations occurring in the mineralization process in developing teeth exposed to elevated Se.

Table 7-10
Summary of Published Studies on the Effects of Selenium on Mineralization of Dental Tissues

Author (year)	Conditions	Result
English, 1949	Dog, organic Se compound	Hypoplastic areas on dentin
Irving, 1959	Rat, vitamin E and Se deficiency	Se cures depigmentation of incisor
Kaqueles et al, 1966	Rat, selenate injection	Higher doses reduced calcification
McLundie, 1968	Artificial mouth, 10–15 ppm Se	Increased solubility of enamel in buffer
Eisenmann and Yaeger, 1969	Rat incisor, single injection	Single line of hypomineralization
Newesely, 1970	Literature review	Selenium may inhibit mineralization by changes in protein component of hard tissue
Bowen, 1972	Monkey, 1 and 2 ppm Se, 5 years	Chalky yellow appearance on developing teeth
Anonymous, 1975	Literature review	Studies on effect of selenium on enamel mineralization "need further investigation"

SUMMARY

The trace element Se has been shown to be biologically active and is both essential and toxic to man and other animals. The influence of Se on caries is controversial. Some studies have shown that if somewhat increased amounts of Se are ingested during the period of tooth formation, caries is promoted in man and experimental animals. Many other studies have shown equivocal effects of Se on caries. A working hypothesis for any effects of Se on caries is that Se is incorporated into matrix proteins of enamel where it may influence calcification. With the availability of current improved methodology, such as the rat caries model, Se assays, and isotopes, it seems possible that the mechanism and relative importance of Se in the overall caries process will be assessed in the foreseeable future.

Acknowledgments

The author gratefully acknowledges the support of the USPHS National Institute of Dental Research for some of the studies by the author in this chapter (Grant Nos. DE 03536 and DE 03856).

REFERENCES

Allaway WH: Selenium in the food chain. *Cornell Vet* 1973;63:151–170.

Anonymous. Studies on selenium. *Nutr Rev* 1975;33:138–140.

Barmes DE: Caries etiology in Sepik villages—Trace element micronutrient and macronutrient content of soil and food. *Caries Res* 1969;3:44–59.

Bawden JW, Hammarström LE: Autoradiography of selenium-75 in developing rat teeth and bone. *Caries Res* 1977;11:195–203.

Bowen WH: The effects of selenium and vanadium on caries activity in monkey (M. irus). *J Irish Dent Assoc* 1972;18:83–89.

Britton JL, Shearer TR, DeSart DJ: Cariostasis by moderate doses of selenium in the rat model. *Arch Environ Health* 1980;35:74–76.

Burk RF: Selenium in Man, in Prasad AS, Oberleas D (eds): *Trace Elements in Human Health and Disease.* vol 2. *Essential and Toxic Elements.* Academic Press, New York, 1976, pp 105–133.

Büttner W: Action of trace elements on the metabolism of fluoride. *J Dent Res* 1963;42:453–460.

Cadell PB, Cousins FB: Urinary selenium and dental caries. *Nature* 1960;185: 863–864.

Claycomb, CK, Gatewood DC, Sorenson FM, et al: Presence of Se^{75} in rat saliva after intracardiac injection of radioactive sodium selenite. *J Dent Res* 1960; 39:1264.

Claycomb CK, Summers GW, Jump EB: Effect of dietary selenium on dental caries in Sprague-Dawley rats. *J Dent Res* 1965;44:826.

Crisp FD, Deaton TG, Bawden JW: In vitro study of selenium-75 distribution in developing rat molar enamel. *Caries Res* 1979;13:313–318.

Curzon MEJ, Crocker DC: Relationships of trace elements in human tooth enamel to dental caries. *Arch Oral Biol* 1978;23:647–653.

Donaldson WE: Selenium inhibition of avian fatty acid synthetase complex. *Chem Biol Interactions* 1977;16:313–320.

Eisenmann DR, Yeager JA: Alterations in the formation of rat dentin and enamel induced by various ions. *Arch Oral Biol* 1969;14:1045–1064.

English JA: Experimental effects of thiouracil and selenium on the teeth and jaws of dogs. *J Dent Res* 1949;28:172–194.

Hadjimarkos DM: Selenium-fluoride interaction in relation to dental caries. *Arch Environ Health* 1967;14:881–882.

Hadjimarkos DM: Selenium in relation to dental caries. *Food Cosmet Toxicol* 1973;11:1083–1095.

Hadjimarkos DM, Bonhorst CW: The selenium content of human teeth. *Oral Surg, Oral Med, Oral Pathol* 1959;12:113.

Hadjimarkos DM, Bonhorst CW: The selenium content of eggs, milk, and water in relation to dental caries in children. *J Pediatr* 1961;59:256–259.

Hadjimarkos DM, Bonhorst CW: Fluoride and selenium levels in contemporary and ancient Greek teeth in relation to dental caries. *Nature* 1962;193:177–178.

Hadjimarkos DM, Shearer TR: Selenium concentration in human saliva. *Am J Clin Nutr* 1971;24:1210–1211.

Hadjimarkos DM, Shearer TR: Selenium content of human nails: A new index for epidemiologic studies of dental caries. *J Dent Res* 1973;52:389.

Hadjimarkos DM, Storveck CA, Remmert LF: Selenium and dental caries. *J Pediatr* 1952;40:451–455.

Hill, CH: Interrelationships of selenium with other trace elements. *Fed Proc* 1975;34:2096–2100.

Irving JT: Curative effect of selenium upon the incisor teeth of rats deficient in vitamin E. *Nature* 1959;184:645–646.

Johnson JR, Shearer TR: Selenium uptake into teeth determined by fluorimetry. *J Dent Res* 1979;58:1836–1839.

Kaqueles JC, Maloigne E, Bonifay P: Effects of sodium selenite on caries incidence in the rat. *J Dent Res* 1977;56:69 (abs).

Kaqueles JC, Petrovic A, Shambaugh GE Jr: Effects of sodium selenate on postnatal rat tooth and jaw calcification as detected by tetracycline labeling in vivo. *Anat Rec* 1966;154:365 (abs).

Levander OA: Selenium in foods, in: Proceedings of the Symposium on Selenium-Tellurium in the Environment. Industrial Health Foundation, Pittsburgh, PA, 1976, pp 26–53.

Lo LW, Koropatrick J, Stick: The mutagenicity and cytotoxicity of selenite, "activated" selenite and selenate for normal and DNA repair-deficient human fibroblasts. *Mutat Res* 1978;49:305–312.

Losee FL, Cutress TW, Brown R: Natural elements of the periodic table in human dental enamel. *Caries Res* 1974;8:123–124.

Ludwig TG, Bibby BG: Geographic variations in the prevalence of dental caries in the United States of America. *Caries Res* 1969;3:32–43.

Luoma H, Rants H, Turtola L: The potassium and phosphorous content of a cariogenic streptococcus modified by fluoride and selenium. *Caries Res* 1971;5:96–99.

Mamedova FM: Selenium content in hard human dental tissues in normal conditions, in deep caries, and parodontopathy. *Stomatologia* (Buc) 1965;41:3.

McLundie AC, Shepherd JB, Mobbs DRA: Studies on the effects of various ions on enamel solubility. *Arch Oral Biol* 1968;13:1321–1330.

Meissner W: The effect of selenium and fluorine with respect to caries and toxicity to Osborne-Mendel rats. *Zbl Bakt Hyg,* I Abs Orig 1976;162:330–349.

Morris VC, Levander OA: Selenium content of foods. *J Nutr* 1970;100:1383–1388.

Muhlemann HR, König KG: The effect of some trace elements on experimental fissure caries and on growth in Osborne-Mendel rats. *Helv Odont Acta* (suppl) 1964;79–81.

Muhler JC, Shafer WG: The effect of selenium on the incidence of dental caries in rats. *J Dent Res* 1957;36:895–896.

National Research Council. *Selenium.* Washington, DC, National Academy of Sciences, 1976.

Navia JM: Effect of minerals on dental caries, in Harris RS (ed): *Dietary Chemicals vs Dental Caries.* Washington, American Chemical Society, 1970, pp 123–159.

Navia JM, Menaker L, Seltzer J: Effect of Na_2SeO_3 supplemented in the diet or the water on dental caries of rats. *Fed Proc* 1968;27:2588 (abs).

Newesely H: Factors controlling apatite crystallization, with particular reference to the effect of fluoride and accompanying ions, in Staple PH (ed): *Advances in Oral Biology,* vol 4. 1970, p 11.

Nixon GS, Myers VB: Estimation of selenium in human dental enamel by activation analysis. *Caries Res* 1970;4:179–187.

Pappalardo G: Azione del selenio sulla carie sperimentale del ratto bianco. *Min Stomatolog* 1959;8:748–751.

Pappalardo G: Azione combinato del fluoro e del selenio sulla carie sperimentale del ratto bianco. Proc 6th Congress, ORCA, Pavia 1962, pp. 161–168.

Petrovich YA, Podorozhnaya RP: Selenium turnover in protein and non-protein fractions of calcified tissues of rats of different ages. *Ukrainskyi Biokhinichnyi Zhurnal* 1974;46:639–642.

Retief DH, Cleaton-Jones PE, Turkstra J, et al: Selenium content of tooth enamel obtained from two South African ethnic groups. *J Dent Res* 1976;55:701.

Rosenfeld I, Beath OA: *Selenium, Geobotany, Biochemistry, Toxicity and Nutrition.* New York, Academic Press, 1964.

Schrauzer GM, White DA, Schneider CJ: Cancer mortality correlation studies—III: Statistical associations with dietary selenium intake. *Bioinorg Chem* 1977;7:23–34.

Shearer TR: Developmental and post-developmental uptake of dietary organic and inorganic selenium into the molar teeth of rats. *J Nutr* 1975;105:338–347.

Shearer TR, Britton JL: Protein synthesis in the teeth of rats receiving selenium. *Fed Proc* 1976;35:2058 (abs).

Shearer TR, Britton JL: Relationship of dietary selenium to selenium uptake by teeth. *Fed Proc* 1977;36:4654, (abs).

Shearer TR, Britton JL, DeSart DJ, et al: Influence of cadmium on caries and the cariostatic properties of fluoride in rats. *Arch Environ Health* 1980;35:176–180.

Shearer TR, Hadjimarkos DM: Comparative distributions of ^{75}Se in the hard and soft tissues of mother rats and their pups. *J Nutr* 1973;103:553–559.

Shearer TR, Hadjimarkos DM: Geographic distribution of selenium in human milk. *Arch Environ Health* 1975;30:230–233.

Shearer TR, McCormack DW, DeSart DJ, et al: Histological evaluation of selenium induced cataract. *Exp Eye Res* 1980;31:327–333.

Shearer TR, Ridlington JW: Fluoride-selenium interaction in the hard and soft tissues of the rat. *J Nutr* 1976;106:451–454.

Shearer TR, Suttie JW: Discussion of Paper by Larson, R. Animal studies relating to caries inhibition by fluoride. *Caries Res* 1977;11(Suppl 1):53–57.

Smith MI, Westfall BB: The selenium problem in relation to public health. *Public Health Rep* 1936;51:1446–1505.

Smith MI, Westfall BB. Further field studies on the selenium problem in relation to public health. *Public Health Rep* 1937;52:1375–1384.

Suchlov BP, Katsap IM, Gulgasenko AI: The study of the influence of selenium on the dental caries of the population of Chernovitsi region. *Stomatologia* 1973;52:21–23.

Tank G, Storvick CA: Effect of naturally occurring selenium and vanadium on dental caries. *J Dent Res* 1960;39:473–488.

Tempestine O: Le interferenge tra ormone tiroideo, fluoro ed altri oligo-elementi sulla carie sperimentale del ratto. Proc 6th Congress ORCA, Pavia, 1962, pp 155–158.

Thomassen PR, Leicester JM: Uptake of radioactive beryllium, vanadium, selenium, cerium and yttrium in the tissues and teeth of rats. *J Dent Res* 1964; 43:346–352.

Underwood EJ: Selenium, in *Trace Elements in Human and Animal Nutrition,* ed 4. New York, Academic Press, 1977, p 302.

Wheatcroft MG, English JA, Schlack CA: Effects of selenium on the incidence of dental caries in white rats. *J Dent Res* 1951;30:523 (abs).

23 **V**
50.94
8-11-2

8 Vanadium

M. E. J. Curzon

Vanadium (V), named after the Scandanavian goddess Vanadis, occurs with an average concentration of 150 ppm in the earth's crust distributed throughout many types of rocks. When purified, V is a bright metal whose main industrial use is as an additive to steel. As such it imparts properties ideal for high speed tool steels and for high resistance to corrosion by alkalis and some acids.

Metabolism

A possible physiologic role for V has been reviewed by Schroeder et al (1963), but until Schwarz and Milne in 1971 showed that the element is necessary for growth in the rat, V was not considered an essential trace element for life.

Only between 0.1% and 1% of a dose of V is absorbed from the intestinal tract in man of which 60% is excreted in the urine within 24 hours (Talvite and Wagner, 1954). The highest concentration of V that is retained in the human body has been variously reported to be in the liver, kidneys, bones, and teeth (Talvite and Wagner, 1954; Söremark and Anderson, 1962). Bertrand (1950) recorded V present in various vertebral tissues with a mean overall concentration of 0.1 μg/g. In other studies Schroeder et al (1963) found that lungs contain higher amounts of V than any other tissue, the concentrations being dependent on the geographic source of the material. The latter could be due to inhalation of dust particles polluting the atmosphere; this has been pointed out by others (Tipton and Cook, 1963). Geographic variations in V concentrations in human tissues were investigated by Allaway et al (1968) who found that over 90% of whole blood samples from men in 19 cities of the United States contained less than 1 μg V/100 ml. The V content of human blood is entirely in the plasma and is not located in the erythrocytes.

Sources of V in the food vary, but fish appear to contain concentrations higher than most foods (Table 8-1) (Söremark, 1967). The earlier analyses of V levels in foods and in man (Schroeder et al, 1963) have subsequently been questioned as inaccurate (Bryne and Kosta, 1978), probably because of insufficient sensitivity of the analytic techniques used. Generally V concentrations in foods are very low although parsley appears to contain substantial amounts, and recent analyses by Bryne and Kosta (1978) showed spinach, buckwheat, and tea all to contain high V levels. An average daily intake is, therefore, of a very low order of a few tens of micrograms. Contamination of vegetables with soil particles may greatly vary the intake; young children may also ingest increased levels of V from play with soils or pica.

Several possible physiologic roles for V have been suggested. Inhibition of the biosynthesis of cholesterol has been ascribed to V (Curran and Burch, 1967). Söremark (1964) remarked that the longevity of Scandinavians might be explained by the high consumption of ocean fish and sea salt, which have high concentrations of V.

Studies using radiovanadium by Söremark (1964) showed that the distribution was highest in the bones and teeth of the various tissues he tested. A possible role for V in dental disease was suggested as early as

Table 8-1
Vanadium Concentrations[a] in Animal Tissues and Vegetables

Animal Tissues		Vegetables	
Calf liver	0.51	Lettuce	0.58
Pork	$< 10^{-4}$	Apples	0.86×10^{-2}
Sardines	0.24×10^{-3}	Potatoes	0.64×10^{-2}
Fresh milk	0.24×10^{-3}	Carrots	$< 10^{-4}$
Skim milk	0.50×10^{-3}	Peas	$< 10^{-4}$

[a]Concentration of V in ppm dry wt.
Data from Söremark, 1967.

1949 by Rygh who noted that stores of Ca salts in bones and teeth are affected by an interaction of V and Sr. Animals fed diets deficient in these elements showed increasing susceptibility to dental decay.

CARIES

Man

An inverse relationship between V concentrations in water supplies and dental caries was reported by Tank and Storvick (1960) in a study of children in Wyoming, USA. The exact location of these study sites within the state was not given, other than that 15 study areas were chosen because of the geological distribution of Se and the incidence of selenosis in livestock. The areas were noted as having widespread V deposits.

In these studies water supplies were analyzed and V found in all specimens. Only minute amounts were found which ranged from 0.03 to 0.22 mg/l. It is of note that these concentrations were considerably lower than those used in animal experiments described below. In all areas a decrease of dental caries in deciduous teeth was found related to an increase of V in drinking water. Within areas these differences were not statistically significant, but the reduction in dental caries by V was significant over the combined areas. The study is clouded by the presence of Se and F, necessitating the subdivision of the subject into nonseleniferous and seleniferous; high and low F; V concentrations of 0.03 to 0.06, 0.07 to 0.09, and 0.10 to 0.22 ppm. Consequently there are study cells with very low numbers of subjects or missing subjects. The inferences drawn from the research are, therefore, not well substantiated.

Another study of V and dental caries in man was reported by Sandor and Denes (1972) on 6- to 14-year-old children in towns and villages in Borsod, an area of Hungary, where caries levels were related to drinking waters of varying hardness. Low frequencies of caries were found where V concentrations were high, and were unrelated to the F water concentrations. This low level of caries was also found to be related to increased levels of Ca and Mg, as well as total water hardness. High levels of V in soils has been associated with low caries prevalence in a study in South America (Rothman et al, 1972).

In only one study has the relationship of V in enamel and the prevalence of dental caries been reported. In this study no difference in V concentration was found between tooth donors with a DMFT of 0 compared with donors with mean DMFT of 10+ (Curzon et al, 1974). Further studies on trace elements in general in human enamel and caries using over 350 samples (Curzon and Crocker, 1978) also did not show any relationship of V to caries.

Table 8-2
Summary of Previous Reports of the Effects of Vanadium on Caries in Animals

Author	Year	Animal	Vanadium		Effect on Caries
			Salt	*Conc.*	
Rygh	1949	Rat	NG	NG	Improved calcification
Geyer	1953	Hamster	V_2O_5	0.08 mg diet	Reduced
Hein and Wisotsky	1955	Hamster	V_2O_5	10 mg/l water	Increased
Winiker	1957	Hamster	V_2O_5	0.01–0.02 μg/g	Reduced
Muhler	1957	Rat	V_2O_2	10, 20, 40 μg/ml	No effect
Kruger	1958	Rat	VCl_4	0.025, 0.005 mg inj.	Reduced
Pappalardo	1959	Rat	$NaVO_3$	35 mg total dose	Reduced
Tempestini and Pappalardo	1960	Rat	$NaVO_3$	0.5 mg of 1% soln. inj. with and without NaF	Reduced caries more than fluoride
Büttner	1963	Rat	NH_4VO_3	10 mg/l water	No effect
Muhleman and Köenig	1964	Rat	V_2O_3	2.1 mg/l water	No effect
Bowen	1972	Monkey	V_2O_5	2.0 mg/l water	Increased
Tamura	1976	Rat	$NaVO_3$	3.6, 36 mg/kg diet	No effect

NG = Not Given

Animals

Rygh (1949) showed that the addition of V, together with Sr, to a purified diet promoted mineralization of bones and teeth. A high degree of protection against dental caries has been reported in hamsters with V (Geyer, 1953; Winiker, 1957) and similar results were found in rats (Kruger, 1958; Pappalardo, 1959; Tempestini and Pappalardo, 1960). Conversely in other studies on caries in animals, no beneficial effect was obtained or caries was actually increased.

The caries rate in monkeys was found by Bowen (1972) to be increased if low levels of V and Se were added to drinking water. It appeared that V promoted the development of caries regardless of whether the element was administered before or after tooth eruption. The significant differences in the V concentration in the plaque from V-fed monkeys compared with the controls is of interest in consideration of any possible role of V in dental caries.

MECHANISMS OF ACTION

Incorporation in Human Enamel

Drea (1936) and Lowater and Murray (1937) concluded that the concentration of V in human enamel was below their analytic limit of detection for that time. A number of studies have been published since concerning V in enamel (Table 8-3). Many of these researchers found V concentration well below the limits of detection of the analytic methods used.

In one recent study, besides analyzing enamel for V, samples were grouped by geographic origin to see if any geographic variation occurred (Curzon et al, 1974). No such differences were found. Either no such variation exists, or relevant examples were not chosen.

Table 8-3
Reported Concentrations of Vanadium in Human Enamel

			Concentration of V μg/g	
Author	year	Method of Analysis	*Mean ± SE*	*Range*
Drea	1936	Spectrographic	not detected	
Lowater and Murray	1937	Spectrographic	not detected	
Söremark	1964	Activation analysis	10^{-5}	
Hardwick and Martin	1967	Mass spectrography	1	
Nixon et al	1967	Activation analysis	0.01	
Losee et al	1974	Optical emission	0.06	
Losee et al	1974	Mass spectrography	0.017 ± 0.00	0.01 – 0.03
Curzon et al	1974	Mass spectrography	0.054 ± 0.00	0.01 – 2.7
Curzon and Crocker	1978	Mass spectrography	0.02 ± 0.00	0.00 – 0.2

The reports of V concentrations in human enamel have been questioned by Bryne and Kosta (1978) on the grounds of poor analytic techniques. In addition, these authors consider the cutting, drilling, and burning operations involved in enamel preparation as suspect because of contamination. This depends, however, on how the cutting and drilling operations are carried out, since they can be done with a minimum amount of contact between steel and enamel. Fracturing of the enamel, after freezing, also includes errors due to contamination with dentin. Whether or not V concentrations in enamel are true, all published reports give very low concentrations.

The incorporation of V into enamel does not directly appear to be related to dental caries (Curzon et al, 1974). However, early work by Manley and Bibby (1949) showed that V_2O_5 was active in reducing acid solubility of enamel.

Effects on Oral Bacteria

A recent study by Beighton and McDougall (1981) studied the effects of V, at concentrations found in drinking waters, on fissure plaque bacteria in rats. They found that V markedly stimulated the growth of *A. viscosus* (strain WVU 626) at 0.38 to 3.2 μg/ml with a maximum stimulation at 1.6 μg V/ml. It was suggested that these findings might explain the effects of V on dental caries by reducing the percentage of *S. mutans* and increasing the percentage of actinomycetes.

In a study on acid production by streptococci and actinomycetes, V exerted major inhibitory effects when tested at concentrations to be found in the outer 20 μ of human enamel (Gallagher and Cutress, 1977). The effect was not seen for all types of bacteria tested but occurred in cultures of *S. mutans* and *A. viscosus*.

The evidence for an effect of V on bacteria is based on the two studies discussed above. V may have an effect on oral bacteria, and this may be the mechanism by which the element affects caries.

SUMMARY

Although more than 12 animal experiments have been completed on V and dental caries, the relationship between the two remains unclear. There are only two reports of a cariostatic effect of V in man, and in both cases the presence of either Se or water hardness confused the results. It is not clear at this time whether V has any effect. Further work should concentrate on more carefully planned studies in man and additional animal experiments using V concentrations more closely related to those known to occur naturally. Interactions between V and fluoride should also be considered.

REFERENCES

Allaway WH, Kubota J, Losee FL, et al: Selenium, molybdenum and vanadium in human blood. *Arch Environ Health* 1968;16:342–348.

Beighton D, McDougall WA: The influence of certain added waterborne trace elements on the percentage bacterial composition of tooth fissure plaque from conventional Sprague-Dawley rats. *Arch Oral Biol* 1981;26:419–425.

Bertrand D: The biochemistry of vanadium. *Bull Am Mus Nat Hist* 1950;94: 407–409.

Bowen WH: The effects of selenium and vanadium on caries activity in monkeys (*M. irus*) *J Irish Dent Assoc* 1972;8:83–89.

Bryne AR, Kosta L: Vanadium in foods and in human body fluids and tissues. *Sci Total Environ* 1978;10:17–30.

Büttner W: Action of trace elements on the metabolism of fluoride. *J Dent Res* 1963;42:453–460.

Curran GL, Burch RE: Biological and health effects of vanadium. *Trace Subst Environ Health I.* Columbia, University of Missouri, 1967, pp 96–105.

Curzon MEJ, Losee FL, Brown R, et al: Vanadium in whole enamel and its relationship to dental caries. *Arch Oral Biol* 1974;19:1161–1165.

Curzon MEJ, Crocker DC: Relationships of trace elements in human tooth enamel to dental caries. *Arch Oral Biol* 1978;23:647–653.

Drea WF: Spectrum analysis of dental tissues for "trace" elements. *J Dent Res* 1936;15:403–406.

Gallagher IHC, Cutress TW: The effect of trace elements on the growth and fermentation by oral streptococci and actinomyces. *Arch Oral Biol* 1977;22: 555–562.

Geyer GF: Vanadium, a caries inhibiting trace element in the Syrian hamster. *J Dent Res* 1953;32:590–595.

Hardwick JL, Martin CJ: A pilot study using mass spectrometry for the estimation of the trace element content of dental tissues. *Helv Odont Acta* 1967;11: 62–70.

Hein JW, Wisotsky S: The effect of 10 ppm vanadium drinking solution in dental caries in male and female Syrian hamsters. *J Dent Res* 1955;34:756.

Kruger BJ: The effect of "trace elements" in experimental dental caries in the albino rat. II. A study of aluminum, boron, fluorine, iodine and vanadium. *Aust Dent J* 1958;58:298–302.

Losee FL, Curzon MEJ, Little MF: Trace element concentrations in human enamel. *Arch Oral Biol* 1974a;19:467–471.

Losee FL, Cutress TW, Brown R: Natural elements of the periodic table in human dental enamel. *Caries Res* 1974b;8:123–134.

Lowater F, Murray MM: Chemical composition of teeth: Spectrographic analysis. *Biochem J* 1937;31:837–843.

Manley RS, Bibby BG: Substances capable of decreasing the acid solubility of tooth enamel. *J Dent Res* 1949;28:160–168.

Muhleman HR, König KG: The effect of some trace elements on experimental fissure caries and on growth in Osborne-Mendel Rats. *Helv Odont Acta* 1964;8:79–81.

Muhler JC: The effect of vanadium pentoxide, fluorides and ten compounds on dental caries experience in rats. *J Dent Res* 1957;36:787–794.

Nixon GS, Livingston HD, Smith H: *Trace Elements in Human Tooth Enamel. Nuclear Activation Techniques in the Life Sciences.* Vienna, International Atomic Energy Agency, 1967; pp 455–462.

Pappalardo G: Effects of vanadium on experimental caries. *Rev Ital Stom* 1959;7:966–970.

Rothman KJ, Glass RL, Espinal F, et al: Dental caries and soil content of trace metals in two Colombian villages. *J Dent Res* 1972;51:1686.

Rygh O: Recherches sur les ologio-elements: De l'importance du thallium et du vanadium, du silicon et du fluor. *Bull Soc Chim Biol* 1949;31:1403–1407.

Sandor T, Denes I: Caries es az ironiz nyomelemci. *Or Hetil* 1972;113:1062–1064.

Schroeder HA, Balassa JJ, Tipton IH: Abnormal trace metals in man—Vanadium. *J Chron Dis* 1963;16:1047–1071.

Schwarz K, Milne DB: Growth effects of vanadium in rats. *Science* 1971;174: 426–428.

Söremark R, Andersson N: Uptake and release of vanadium from intact human enamel following $V_2^{48}O_5$ application in vitro. *Acta Odont Scand* 1962;20:81–93.

Söremark R: Vanadium in some biological specimens. *J Nutr* 1967;92:183–190.

Söremark R: Studies on the concentration of vanadium in some biological specimens. 3^e Collogue International de Biologie de Saclay. CEA, 1964, pp 223–238.

Talvite NA, Wagner WD: Studies in vanadium toxicology. *Arch Indust Hyg* 1954;9:414–420.

Tamura S: Effect of vanadium on the incidence of experimental dental caries in rats. *Shaka Gakoho* 1976;76:825–828.

Tank G, Storvick CA: Effect of naturally occurring selenium and vanadium on dental caries. *J Dent Res* 1960;39:473–488.

Tempestini O, Pappalardo G: Combined effect of action of vanadium and fluoride on experimental caries. *Panminerva Med* 1960;2:334–348.

Tipton IH, Cook MJ: Trace elements in human tissue—II. Adult subjects for the United States. *Health Phys* 1963;2:103–109.

Winker M: Weitere Versuche zur Kanisbe kamp fung mit Vanadin verbindungen. *Odont Revy* 1957;8:196–201.

30 Zn

65.38

8-18-2

9 Zinc

S. Hsieh
R. N. Al-Hayali
J. M. Navia

Next to iron, zinc (Zn) is the most abundant trace element in humans and animals. A 70 kg adult man is estimated to contain 1.4 to 2.3 g of Zn (Underwood, 1977), which is about half the amount of Fe, 10 to 15 times that of Cu, and 100 times that of Mn. Despite its abundance, Zn was not established as an essential nutrient for animals until 1934, when researchers in Wisconsin showed that Zn prevented poor growth and alopecia in rats (Todd et al, 1934). Interestingly, the Wisconsin group discovered early that the less abundant trace elements, eg, Cu in 1928 (Hart et al, 1928) and Mn in 1931 (Kemmerer et al, 1931), were essential to life. This reflects the difficulty in preparing a Zn-deficient but otherwise adequate diet and in maintaining a Zn-free environment. Because of its ubiquitous presence, Zn deficiency in humans was not suspected for decades. In 1961, Prasad and his associates reported that their Iranian patients had anemia, hepatosplenomegaly, dwarfism, and hypogonadism (Prasad et al, 1961). Fe deficiency was diagnosed, and Zn deficiency was speculated on. Further studies (Prasad et al, 1963) in Egypt established that Zn deficiency was present in humans. Since then, Zn deficiency has been shown to occur in Denver school children (Hambidge et al, 1972) and in patients with various diseases (Sandstead et al, 1976), and interest in Zn in nutrition has surged.

Metabolism of Zinc

In animals Zn occurs widely throughout the body with highest concentrations found in the choroid and iris of eye tissues, up to 14,600 ppm Zn (dry basis) (Weitzel and Fretzdorff, 1953). The prostate gland also contains unusually high concentrations of Zn (Table 9-1). The function of such localized Zn is unknown, but it may not be a coincidence that both tissues are metabolically very active. In bones and muscle Zn is relatively high and represents a substantial amount of total body burden. Liver, as well as kidney, contains higher Zn levels than most other soft tissues except those mentioned above, although levels in soft tissues do not necessarily reflect dietary Zn status (Kirchgessner et al, 1976).

Human plasma Zn levels range from 85 to 159 μg/100 ml (Parr and Taylor, 1964; Butt et al, 1964), with the majority of reported values being 100 $\pm$ 10 μg/100 ml. Most data show that plasma Zn levels are the same in men and women, but Bjorksten et al (1978) found that men's were higher than women's while McKenzie (1979) reported the opposite.

Table 9-1
Zinc Concentrations in Tissues*

Tissue	Human[a]	Rat[b]	Monkey[c]
Bone	66	141	—
Brain	13	18	—
Heart	27	21	22
Kidney	55	37	29
Liver	55	35	51
Lung	15	22	19
Muscle	54	53	24
Prostate	102	223	—
Spleen	19	24	21
Testes	17	22	17
Enamel	231[d]	129[f]	—
Dentin	173[e]	249[f]	—

*mg/g wet tissue, except those for enamel and dentin where dry weight was used.
[a]Tipton and Cook, 1963.
[b]Macapinlac et al, 1966.
[c]Macapinlac et al, 1967, 1968.
[d]Losee et al, 1974.
[e]Retief et al, 1971.
[f]Huxley and Leaver, 1966.

Table 9-2
Normal Zinc Concentrations in Human Blood

Whole blood	550–730 μg/100 ml blood[a]
Plasma	104 ± 14 μg/100 ml plasma[b]
Erythrocytes	1,250 μg/100 ml erythrocytes[c] 1.6 μg/10⁹ cells[d]
Leukocytes	14 μg/10⁹ cells[e]
Platelets	0.2–0.45 μg/10⁹ cells[f]

[a]Dennes et al, 1961, Auerbach, 1965.
[b]Prasad et al, 1961.
[c]Prasad et al, 1963.
[d]Auerbach, 1965.
[e]Dennes et al, 1961.
[f]Foley et al, 1968.

Red blood cells contain approximately tenfold more Zn than does plasma (Table 9-2), with most of the Zn present as carbonic anhydrase (Hove et al, 1940), and the remainder associated with other proteins or enzymes such as superoxide dismutase (McCoy and Fridovich, 1969). Concentrations of Zn in leukocytes are even higher than those in red blood cells (Table 9-2), but little is known about their nature except that most of the Zn is protein bound (Vallee et al, 1954).

Like humans, most animals have a normal plasma Zn level of approximately 100 μg/100 ml (Underwood, 1977), and numerous studies have shown that plasma Zn decreased when animals were fed a Zn-deficient diet. In rats, plasma Zn fell almost 40% one day after receiving a Zn-deficient diet (Wilkins et al, 1972). Thereafter, the fall was slower and in five days the level stabilized at 50% of the control value. Plasma Zn levels show circadian rhythm. Walker and Kelleher (1978) found that mean plasma Zn for 12 rats was 21.3 ± 0.5 μmol/L and 15.8 ± 0.6 μmol/L for 0900 hours and 1400 hours samples, respectively. When rats were fasted for 12 hours and blood was collected at 0900 hours, the mean plasma level was 18.8 ± 0.4 μmol/L, as compared with the nonfasting value of 21.3 ± 0.5 μmol. Thus rats' plasma Zn concentrations vary depending on the time of collection and the state of food intake. This is also true in humans (Walker et al, 1978). The mechanism for the diurnal variation is not clear.

Absorption Dietary Zn is mainly absorbed in the small intestine; in vivo studies show that the duodenum is the site of maximal Zn absorption (Schwarz and Kirchgessner, 1975). The ileum has been shown to absorb Zn most rapidly in in vitro studies (Antonson et al, 1979). The absorption of Zn is usually in the range of 20% to 30%, but it could be as high as 60% (Ritchey et al, 1979). The degree of absorption is important when considering the amount available from the diet.

Many dietary factors can interfere with Zn absorption. High Cu levels reduced Zn absorption (Evans et al, 1974) and vice versa (Van

Campen and Scaife, 1967). Similarly, divalent ions like Ca, Fe, Cd, and Cr can act as antagonists of Zn, presumably competing with it for binding sites at the absorption level (Underwood, 1977). High Ca accentuated Zn deficiency in pigs fed a corn and peanut meal diet (Tucker and Salmon, 1955), and reduced Zn absorption in animals fed diets containing phytate (Morris and Ellis, 1980). Thus, Ca alone depressed Zn absorption but it also aggregated the phytate-Zn interaction (Morris and Ellis, 1980). However, Ca-Zn interaction has not been observed in humans when meat was the protein source (Spencer et al, 1979). Pécoud et al (1975) found that peak serum Zn levels of subjects taking an oral dose of 50 mg Zn and 480 mg phosphate did not differ from the level in those taking the same dose plus 500 mg Ca, suggesting that there was no Ca-Zn interaction. Dietary fiber can bind Zn and reduce its bioavailability, resulting in negative Zn balance (Reinhold et al, 1976). But not all forms of dietary fiber were effective. In humans, fecal Zn loss was significantly increased by the addition of 14.2 g of hemicellulose to a basal diet but was not affected by the same amount of cellulose or pectin (Drews et al, 1979). In contrast, Davies et al (1977) reported that phytate, not fiber, in bran was the major dominant factor of Zn availability. These findings are of interest with the current concern to increase fiber intake in man.

Intermediate metabolism After being absorbed Zn is transported to the liver, which plays an important role in Zn metabolism. Zn in liver cytosol was found to be associated with high molecular weight proteins, and a metallothionein-like protein, thought to be a detoxifying protein since it binds Zn as well as other heavy metals such as Cd (Kagi and Vallee, 1960).

Excretion Whether it is ingested or injected Zn is mainly excreted in the feces. Spencer et al (1979) observed that with a Zn intake of 13.0 g/day, only 0.6 mg appeared in the urine whereas 11 to 12 g was in feces.

Zinc deficiency Naturally occurring Zn deficiency in humans was first speculated on by Prasad and his associates in their Iranian patients (Prasad et al, 1961). Further studies carried out in Egypt demonstrated that Zn deficiency was present in young Egyptians (Prasad et al, 1963). Dwarfism and hypogonadism were conspicuous features; Zn concentrations in plasma, red cells, and hair were low, symptoms reminiscent of Zn deficiency in rats. There was no report of any dental symptoms. Both the Iranians and Egyptians under investigation had anemia, which responded to Fe treatment. The beneficial effect of Zn on other Iranian patients was also observed in different studies (Halsted et al, 1972); dietary habit was the cause of deficiency. Their diet consisted principally of bread, so prepared that it was high in phytate and fiber; consequently Zn became unavailable (Reinhold et al, 1976).

In 1972, Hambidge et al found that a number of children living in the Denver area were marginally deficient in Zn. The signs included low hair Zn levels, subnormal growth, poor appetite, and poor taste acuity. Supplementation of 0.4 to 0.8 mg Zn/kg of body weight improved the condition (Hambidge et al, 1972). Furthermore, it was found that hair Zn levels of children of low-income families, who were originally studied for poor growth, were significantly lower than those of middle-income children of the same age. Addition of Zn to the formula to 5.8 mg/1 resulted in increased tissue Zn concentrations and in increased growth rate in males, but not females (Walravens and Hambidge, 1977).

Zinc requirements Recent metabolic studies showed that in healthy adults, Zn balance was reached with an intake of 12.5 mg/day (Spencer et al, 1976). This probably reflects a minimum requirement since dermal loss was not considered. The turnover rate of body Zn was calculated from radioisotope studies at 6 mg/day (Engel et al, 1966). Thus, according to Recommended Dietary Allowances (RDA) established for adults living in the United States, 15 mg Zn/day are required with an additional 5 mg recommended during pregnancy and 10 mg during lactation.

The Zn requirement of preadolescent children was estimated to be 6.2 mg/day (Engel et al, 1966). For girls the mean daily loss of Zn through sweat was 1.43 mg, and a minimum daily intake of 7 mg was required (Ritchey et al, 1979). An allowance of 10 mg is recommended (National Research Council, 1980) for preadolescent children. Meanwhile the Zn requirement of infants still is unknown and is tentatively set at 3 mg/day.

Foods in the meat group have the highest Zn content (0.40 to 6.77 mg/100 g of wet material), followed by the bread and cereals group (0.30 to 2.54 mg/100 g), milk and milk products (0.36 to 0.49 mg/100 g), vegetables (0.12 to 0.60 mg/100 g), and fruits (0.02 to 0.26 mg/100 g), (Haeflein and Rasmussen, 1977). Individually, oysters, seafoods, muscle meats, and nuts are the richest Zn sources while the poorest sources are refined sugar, citrus fruits, nonleafy vegetables, and vegetable oils which contain less than 1 ppm on dry basis (Sandstead, 1973).

CARIES

Man

The evidence for a relationship of Zn intake to dental caries in man is very thin. Unlike studies on other trace elements such as Se, Sr, and Mo there have been no attempts to relate variations of Zn intake, via food or drinking water to caries. In the several studies discussed in Chapter 2 Zn did not figure in any of the caries-related results. This was so even though Zn was found in all water supplies. Only in the first

Papua-New Guinea report did Zn show a direct association with caries, and that was for its presence in foods (Barmes, 1969). The only other report for an effect of Zn in humans comes from Russia (Khrosh, 1966) who reported that a daily administration of 3 mg of $ZnSO_4$ for three months, twice a year for three years, gave a drop in increasing caries incidence. A small epidemiologic study by Curzon and Bibby (1970) indicated Zn (with Cu and Pb) was associated with increased caries.

Helle and Haavikko (1977a) attempted to correlate the prevalence of dental caries with the presence of nine macro- and microminerals in drinking water. They found that fluoride showed the strongest negative correlation ($r = -0.484$, $p < 0.001$), while Zn also had a negative correlation with caries ($r = -0.179$, $p < 0.01$).

Zn in water samples collected from 194 sources of public water supply in the United States was reported to range from 0.06 to 7.0 μg Zn/ml with a mean value of 1.33 μg Zn/ml (Taylor, 1963). Hadjimarkos (1967) believed that with the exception of fluoride, water supplies contribute only a small amount of trace elements to the daily diet compared with foodstuffs. Therefore, he suggested that in conducting epidemiologic studies of associations between intake of trace elements and dental caries, determination of elemental concentration in water alone does not provide a reliable indicator of the total amount ingested daily by individuals.

There is a significant difference in the concentration of fluoride and Sr in human enamel between high and low caries areas in the United States, but insignificant differences in concentration of other trace elements in enamel (Curzon, 1977, 1978). However, a similar study conducted in South Africa revealed that Zn concentration was significantly higher in the enamel in white population groups with higher caries incidence in comparison to black population groups with low caries incidence (Retief et al, 1978). An investigation conducted in Finland to correlate the macro- and micromineral composition of deciduous teeth from different geographic areas with dental caries indicated that Zn concentrations in dentin had a positive correlation with caries prevalence (Helle and Haavikko, 1977b). This finding is in contradiction to a previously mentioned observation by the same investigators (Helle and Haavikko, 1977a) and was explained by high fluoride in drinking water, which might alone be responsible for low caries prevalence and not necessarily be due to the presence of Zn and other trace elements.

The data from the limited number of human studies are therefore equivocal, and it is not clear whether Zn increases or decreases caries. As with other trace elements, such as fluoride and Sr, it is possible that Zn may do both, depending on the concentrations used. There is as yet no evidence, from human studies, to support this supposition. The bulk of the research on caries has involved the use of animals.

Animals

Although Navia (1970) classified Zn as an element with doubtful effect on dental caries, and supplementation of diets fed to rats with $ZnCl_2$ or $ZnSO_4$ was found to be ineffective in reducing dental caries (Navia et al, 1968; McClure, 1948), Hendershot and Forsaith (1959) found that when Zn was given to rats as Zn-versenate, it moderately reduced dental caries. More recently, it has been reported that 220 μg Zn/ml (as $ZnSO_4$) in drinking water of rats, infected with *S. mutans* at the time of tooth eruption, significantly reduced the buccal caries score (Bates and Navia, 1979); this reduction could also be shown with a topical application of 500 μg Zn/ml. Animal experiments have shown that supplementation with 250 ppm Zn, 10 ppm Mo, and 4 ppm Cr enhanced the cariostatic action of carbamyl phosphate, and completely suppressed dental decay in rats fed a caries-promoting diet (Steinman and Leonora, 1975).

Early studies of Zn deficiency and dental caries indicated that rats fed a low Zn, caries-promoting diet exhibited a lower dental caries incidence than did controls (Sortino and Palazzo, 1971a, 1971b). More recent investigations indicate that there is a significantly greater severity of caries in enamel and dentin of pups nursed by Zn-deficient rats for 20 or 21 days and then challenged with microbial and cariogenic diet for five weeks than in pups nursed by Zn-adequate pair-fed and ad libitum rats (Brown et al, 1979) (Table 9-3). Furthermore, analysis of Zn concentrations in maxillary molars, after feeding with a caries-promoting diet for five weeks, shows that there is a significant difference between experimental groups and controls. The mean Zn concentration of rat molars was 124 $\pm$ 14 ppm, 157 $\pm$ 16 ppm, and 150 $\pm$ 11 ppm ($\bar{x} \pm$ SD) for the Zn-deficient, pair-fed and ad libitum groups, respectively. However, not all groups showed significant differences in Zn concentration in plasma nor in femurs after the experimental period. Carious lesions in rats suckled by Zn-deficient dams showed greater penetration into the dentin than controls, and there were no significant differences in caries scores between the ad libitum and the pair-fed groups (Brown et al, 1979).

Fang et al (1980) investigated the effect of graded doses of Zn on development and mineralization of bone and teeth and their effect on dental caries. Rats (21 days old) were fed a diet containing < 1, 12, 36 or 108 ppm Zn for four weeks, with the last three groups being pair fed. All animals were then sacrificed and samples used for analysis and determination of caries scores. It was found that Zn-deficient rats had significantly smaller tibia and jaws than Zn pair-fed controls.

Tibia, maxillary tooth, and incisor Zn concentrations, based on dry weight, were significantly lower than those of pair-fed controls. The

Table 9-3
Summary of Studies Reporting the Effects of Zinc on Caries in the Rat

Author (year)	Zn Used		Vehicle	Age of Rats (days)	Effect on Caries[a]	Comments
	Salt	*ppm*				
Hendershot et al, 1960	$ZnSO_4$	85, 170 340, 680	Diet	21–161	Increase	
Navia et al, 1968	$ZnCl_2$	50, 2000	Diet	14–74	N.S.	Rats fed control (50 ppm) or supplemented diet (2000 ppm).
Sortino and Palazzo, 1971	NA	1.5 mg/ 100 gr diet	Diet	NA	Decrease	↓ 20.8%
Steinman and Leonora, 1975	$ZnSO_4$	250, 500, 1000	Diet	21	Cariostatic	Suppressed cariogenic effect of carbamyl phosphate.
Brown et al, 1979	$ZnCO_3$	0.9, 89	Diet	1–36	+ 39% B	Pups nursed by – Zn[a] pair fed then challenged + Zn.
Bates and Navia, 1979	A. $ZnSO_4$	25, 50 100, 200	Water	16–44	– 15% B	Only 200 ppm effective
	B. $ZnSO_4$	500, 1000	Swabbing	22–37	– 21% B	Molars swabbed twice
Fang et al, 1980	NA	< 1, 12 36, 108	Diet	21–49	+ 167% B	
Al-Hayali et al, 1981	$ZnSO_4$	2, 50	Diet	Gestation and lactation	+ 38% B – 60% S	Dams fed – Zn (2 ppm) or + Zn (50 ppm) pair fed and ad libitum contr. Caries diet 3 weeks postweaning

NA = Not available
[a] – Zn or + Zn denotes whether a Zn deficient or adequate diet was used.
Caries result is shown as + 1% or – %, indicates an increase or decrease. B = buccal lesions, S = sulcal lesions.

values of each group reflected the graded concentration of dietary Zn, being the lowest of the Zn-deficient group and highest for rats fed 108 ppm Zn. Furthermore, a greater incidence of enamel lesions was observed in Zn-deficient rats than in pair-fed Zn controls (Table 9-3).

Recently, the effect of pre- and postnatal Zn deficiency on dental caries has been studied (Al-Hayali et al, 1981). Fourteen-day pregnant rats were randomly divided into five groups and fed various dietary Zn levels (Table 9-4). Pups were randomized within their group and nursed by their respective dams. At 18 days of age, pups were weaned, fed a cariogenic diet, and challenged with *S. mutans* 6715. At 40 days of age, rats were sacrificed and caries scores determined.

Table 9-4
Dietary Regimens of Zinc Fed to Rat Dams During Gestation and Lactation

Group	Number of Dams	Dietary Zn (ppm)	
		Gestation[a]	*Lactation*[b]
A	7	2	2
B	7	4	2
C	5	4	4
D[c]	5	50	50
E[d]	4	50	50

[a]last week only
[b]for 18 days
[c]pair-fed to group A
[d]ad libitum

All caries scores of groups A and B were significantly higher than the ad libitum control, ie, group E. Group A also showed significantly higher buccal and sulcal caries scores than its pair-fed control (group D). Compared with groups D and E, moderate Zn deficiency during the gestation and lactation periods (group C) increased buccal but not sulcal caries scores. Thus, severe Zn deficiency imposed during the gestational and lactation periods significantly increased dental caries.

In their study of Zn and dental caries, both Brown et al (1979) and Al-Hayali et al (1981) followed a similar protocol, ie, pups were first nursed by Zn-deficient (Zn) dams pair fed or ad libitum, followed by a challenge of a caries-promoting diet, which was provided ad libitum. This eliminated the difference in dental caries due to difference in the pattern of food intake. On the other hand, Fang et al (1980) took a different approach, ie, rats were fed either -Zn or three different Zn levels as pair-fed controls, followed by sacrifice of all rats and determination of caries scores. Obviously, the difference in caries incidence may also be affected by the pair-feeding process.

The most recent work on Zn and dental caries in animals therefore has shown increases in caries in animals that are Zn deficient. Several studies show that there is no longer any doubt of this function of Zn. The converse, effecting a reduction in caries with Zn supplementation, has not yet been fully investigated; further work to determine optimum Zn levels required for cariostasis is indicated.

MECHANISM OF ACTION

Zinc Distribution in Teeth

Early reports indicated that human enamel contained 233 ± 27 ppm of Zn dry weight (Cruickshank, 1940). Enamel from patients with tuberculosis has been reported to have higher Zn concentration than that of healthy individuals (Cruickshank, 1940), and this finding is more likely to be associated with differences in the amount of Zn ingested than with the specific disease (Brudevold et al, 1963). Concentrations of Zn in surface layers of enamel in human teeth range from 430 to 2100 ppm (Brudevold et al, 1963), with the highest concentration in the outer surface, so that the distribution pattern of Zn is similar to that exhibited by fluoride and Pb. Dentin exhibits a similar gradient in Zn concentration, but the highest concentration occurs in areas adjacent to the pulp (Brudevold et al, 1963). Lower values for enamel were reported by Nixon et al (1967) using neutron activation analysis (58 to 2000 ppm), but the Zn distribution pattern was in agreement with previous observations (Brudevold et al, 1963). In enamel, the major deposition of Zn takes place before eruption and, in contrast to fluoride, posteruptive deposition appears to be irregular (Brudevold et al, 1963). Deposition of Zn in bone and teeth is a relatively slow process, but once incorporated it remains bound for a relatively long time and exchanged slowly (Orten, 1966).

The mean Zn concentration for both deciduous and permanent teeth of children living in Oslo was reported as 130 ppm with a range of 91 to 180 ppm Zn (Attramadal and Jonsen, 1976). A later report (Lappalainen and Knuuttila, 1979), concerning five different geologic areas in Finland, indicated that the mean Zn concentration of these samples was higher (182 ± 37.1 ppm Zn). Söremark and Samsahl (1961, 1962) reported the mean Zn concentration of 15 tooth samples from children 12 to 16 years of age was 276 ± 106 ppm for enamel, and 199 ± 78 ppm for dentin. Meanwhile, using neutron activation analysis, Retief et al (1971) found that enamel of adult teeth contained 263 ppm of Zn and dentin, 173 ppm. There appears to be no significant differences in tooth Zn concentration with respect to sex, or to mandibular versus maxillary teeth (Söremark and Samsahl, 1961, 1962).

Geographic location, geologic variability, food intake, water supply, and racial factors may influence the amount of Zn deposited in the teeth. Thus, Curzon et al (1975) showed that enamel Zn concentrations of New Zealand children were lower than those reported for children in the United States, with identical sampling and analytic techniques (Losee et al, 1974).

It has been shown that the mean Zn concentration in teeth from people living in four different localities in Japan (Annaka, Tokyo, Okitsue, and Hachijo Island) was 198 $\pm$ 69 ppm on a dry weight basis, and the mean values of Zn in teeth obtained from Hachijo Island were significantly higher than those for all other areas (Kaneko et al, 1974). Retief et al (1978) indicated that enamel Zn concentration in a white population was significantly higher than in a black population of South Africa.

While human enamel contains higher Zn than dentin, available animal data seem to show the opposite. When rats were fed a control diet containing 47 ppm Zn, the concentrations found in dentin and enamel were 249 $\pm$ 132 ppm and 129 $\pm$ 23.5 ppm Zn, respectively (Huxley and Leaver, 1966). The large standard deviation of the dentin values, however, suggests that the difference may not be statistically significant. Deficiency of Ca and excessive dietary Zn intake significantly increased the concentration of Zn in bone and dentin. A similar trend was observed in enamel, but the small number of individual analyses made these results less conclusive. Brudevold et al (1963) showed that Zn was incorporated into synthetic hydroxyapatite and that it competes with Ca for position on the surface of the hydroxyapatite crystal.

Hove et al (1938) reported a low Zn concentration in teeth from Zn-deficient rats. However, Bergman (1970) could not detect significant differences in Zn concentration of incisors between Zn-deficient and control rats. Recent studies have shown that Zn concentrations in bone and teeth from Zn-deficient rats are significantly lower than the values obtained from rats fed a Zn-supplemented control diet (Fang et al, 1980; Al-Hayali et al, 1981). Furthermore, jaw bones from Zn-deficient rats were significantly lighter in weight than those from Zn pair-fed controls (Fang et al, 1980).

Zinc and Oral Microflora and Plaque

Concentrations of Zn ranging from 8 to 32 ppm have been shown to have a definite antibacterial effect. In vitro tests, where *S. mutans* was grown in chemically defined media of known trace element composition, containing either 0.5, 5, or 16 ppm Zn (as $ZnSO_4$), showed depressed *S. mutans* growth, initial plaque formation, and inhibition of acid production at Zn:cell ratio of 2 and 4 μg Zn/mg cell wet weight (Bates and

Navia, 1979). It was suggested that Zn might inhibit the growth by interfering with cysteine metabolism.

Wildra (1964) reported a remarkably complete reversal of M-phase and overgrowth of Y-phase for *Candida albicans* by the Zn ion in a defined culture medium. This finding was substantiated later by Yamaguchi (1975) who suggested that Zn participates in morphogenesis of a wide range of micro-organisms. In contrast to these earlier reports, Bedell and Soll (1979) examined the effects of micromolar Zn concentration on growth and dimorphism of *C. albicans,* and found that Zn neither depressed growth nor suppressed the formation of the invasive mycelium under all conditions. The reason for such differences is unknown.

Using mass spectrometry, for semiquantitative estimation of the trace element contents of dental plaque, Hardwick and Martin (1967) reported that the Zn concentration in plaque ranged from 100 to 1000 ppm. Recently Schamschula et al (1977) found that the Zn content of dental plaque collected from three communities in New South Wales (Australia), namely Kataomba, Sydney, and Yars, was 108.8 ± 47.0, 93.8 ± 52.2, and 106.1 ± 37.1 ppm Zn respectively. These authors further reported that the mean Zn plaque concentrations were significantly higher for men (115.3 ± 51.8 ppm) than for women (87.0 ± 27.2 ppm). Comparable Zn concentrations were reported in primitive populations in New Guinea (Schamschula et al, 1977). It is not clear if the presence of such amounts of Zn in dental plaque will modify dental caries or periodontal disease.

Zinc and Oral Mucosa

Epidermal parakeratosis is one of the classical signs of Zn deficiency in animals (Follis et al, 1947), and occurs in the epithelium of cheek, tongue, and esophagus (Follis et al, 1947; Alvares and Meyer, 1968; Diamond et al, 1971; Chen, 1977; Joseph et al, 1981). Alvares and Meyer (1968) found that oral mucosa showed regional differences in response to mild Zn deficiency (1.7 ppm Zn in the diet), ie, parakeratosis consistently and dramatically occurred in the buccal mucosa, but to a lesser degree in the dorsal aspect and in the posterior third of the ventral aspect of tongue. The palatal mucosa were not affected unless the diet was severely deficient in Zn (Follis et al, 1947; Osmanski and Meyer, 1969).

The effect of Zn deficiency on the cellular and biochemical changes of the rat buccal mucosa has been extensively studied by Meyer and her associates. Histologically, the affected buccal mucosa showed a marked increase in the number of dividing cells in the deeper cellular layers, with enlarged nucleoli and increased cytoplasmic basophilia in the upper cellular layers (Alvares and Meyer, 1968; Osmanski and Meyer, 1969). In addition, there was increased retention of nuclei in the keratin layer with thickening of both keratin and cellular layers (Alvares and Meyer, 1968). Studies using tritiated thymidine showed that both the rate of cell division and that of cell migration from basal layer to surface were accelerated, indicating the hyperplastic and hyperproliferative nature of affected cells (Alvares and Meyer, 1974). It was suggested that this hyperplasia was related to the reduced nucleohistone production due to Zn deficiency (Chen, 1977). When successive histologic layers, ie, basal, lower spinous, upper spinous, granular, inner one-third of keratin, and outer two-thirds of keratin layers, were removed by microdissection for relative dry weight or density (g/cm^3) measurement, it was found that density of every layer of − Zn group was significantly higher than that of controls (Meyer and Alvares, 1974). Furthermore, the size of granular cells was also increased, indicating cellular hypertrophy. Quantitatively, buccal mucosa of − Zn rats showed the following increases over controls: length of capillaries, 2.67 × ; thickness of cellular layer, 1.47 ×; thickness of keratin layer, 2.82 × (Meyer et al, 1981). However, Zn contents (based on dry weight) of both areas were reduced to half that of the control (Gerson et al, 1981).

Biochemically, enzymes including acid phosphatase (Alvares et al, 1973; Rijhsinghani et al, 1975) and lactate dehydrogenase (Gerson and Meyer, 1977) were found to be increased in the buccal mucosa. Deficiency of Zn caused a more than twofold increase in lactate dehydrogenase activity of buccal mucosal origin but the same enzyme in palatal mucosa was not affected by Zn depletion (Gerson and Meyer, 1977). This is in agreement with the histologic findings previously described, that is, that parakeratosis due to Zn deficiency occurred in buccal but not in palatal mucosa (Alvares and Meyer, 1968).

Deficiency of Zn also affects cells underneath epithelium, but to a lesser degree. Thus, most cells beneath the hyperplastic buccal mucosa were found to be normal (Ashrafi et al, 1981), although their number was increased (Kravich et al, 1981).

Epithelium of tongues also shows characteristic changes similar to those observed in buccal mucosa, ie, thickening of epithelium and increase of cell numbers (Barney et al, 1967; Alvares and Meyer, 1968; Mann et al, 1974; Joseph et al, 1981). The dorsal surface of tongues seemed to be most affected by Zn deficiency and the filiform papillae were flattened. Similarly, esophageal lesions in Zn-deficient rats consisted of epithelial hyperplasia, hyperkeratosis, and parakeratosis (Diamond et al, 1971), which were reversible by Zn repletion. Partial reversal was evident by day six, and complete disappearance of the lesion was evident after 15 days of Zn therapy (Diamond et al, 1971).

Recently, the effect of Zn deficiency on rat gingiva and alveolar mucosa has been studied (Armitage et al, 1981). In Zn-deficient rats, epithelium of the free gingival crest showed parakeratinization and slight thickening but that of attached gingiva was normal. Parakeratosis, however, was found throughout the alveolar mucosa which also showed great thickening. There is, thus, a considerable body of evidence to show that Zn deficiency produces several signs visible in the mouth.

Zinc, Saliva and Salivary Glands

Concentrations of Zn in human mixed saliva vary greatly, ranging from 88,000 mg/ml to 135 mg/ml as reported in nine independent studies (Freeland-Graves et al, 1981). This variation may reflect the difference of saliva collection or analytical procedures. However, two recent studies show comparable results, 266 mg/ml (Greger and Sickles, 1979) vs 266 mg/ml (Freeland-Graves et al, 1981). When whole saliva was centrifuged, it was found that more than 70% of the whole saliva Zn was present in the sediment fraction (Baratieri et al, 1979; Freeland-Graves et al, 1981). In addition, Zn concentration in human whole saliva showed circadian variations, decreasing upon rising in the morning, followed by an increase in late morning and eventually decreasing again in the late afternoon and evening (Snowden and Freeland, 1978).

In their study on Zn and hypogeusia, Henkin et al (1975) found that parotid salivary Zn was a better indicator of human Zn status than plasma Zn level. This could be an important finding for future dental studies in humans. However, animal studies showed that Zn levels of mixed saliva were not decreased by severe Zn deficiency. In humans, when dietary Zn intake was slightly reduced (from 14.7 mg/day to 11.5 mg/day), Zn levels of the whole mixed saliva were unchanged, but those in the supernatant fraction, collected after centrifugation of whole saliva, were significantly decreased (Greger and Sickles, 1979). It was suggested that Zn levels in the supernatant, but not the whole saliva, might be a potential indicator of Zn status. Similarly, Freeland-Graves et al (1981) found that women fed a low Zn diet (3.2 mg/day) for 22 days showed no significant reduction in whole saliva Zn. Although Zn contents of the supernatant fraction were not analyzed, those in the sediment fraction were found to be significantly reduced, suggesting that Zn levels in salivary sediment may be a sensitive parameter of Zn status. Therefore, whole saliva Zn probably is not a good indicator of Zn status, but Zn contents in either salivary supernatant or sediment may deserve some consideration for this purpose, if precautions are taken to prevent contamination during collection.

The effect of Zn deficiency on salivary glands has received only limited attention. Recently, Gandor et al (1981) showed that activities of carbonic anhydrase and alkaline phosphatase in submandibular glands were reduced when rats were fed a Zn-deficient diet for four weeks.

However, Zn contents in the glands were unchanged. It was suggested that submandibular AP activity might be a useful indicator of rats' Zn status.

Zinc and Taste

Of considerable interest to dentists is the relationship of Zn to taste. Henkin et al (1967) established an association between trace elements and impairment of taste (hypogeusia). They reported that 23 out of 73 patients suffering from a variety of diseases treated with a chelating agent, D-penicillamine, manifested hypogeusia and developed Cu deficiency. On the other hand, 4 out of 100 patients with Wilson's disease who were treated with D-penicillamine manifested hypogeusia, which could be improved by Cu administration. Administration of Ni^{++} or Zn^{++} also improved taste acuity in patients with hypogeusia (Henkin and Bradley, 1970). Furthermore, mean Zn level in parotid saliva in 34 normal subjects was 51 ng/ml but that of 47 subjects with idiopathic hypogeusia was 10 ng/ml, which indicated a role of Zn in taste acuity.

Hambidge et al (1972, 1976) indicated that impaired taste acuity is a typical feature of Zn deficiency in children. Dietary supplementation of $ZnSO_4$ (0.1 to 4.0 mg/kg body weight) reduced the hypogeusia in these children. In a single-blind study involving 103 patients with hypogeusia, Henkin et al (1974) reported that taste acuity improved with Zn therapy. Laser microprobe studies (Henkin et al, 1974) revealed that Zn was found in the vallate papilla in association with Sr, Ba, and P. Furthermore, analysis of taste buds from patients with untreated idiopathic hypogeusia had no measurable Zn. In contrast to this, patients treated with 100 mg Zn/day recovered from hypogeusia and had measurable Zn in their taste buds.

The effect of Zn on smell and taste dysfunction was tested in a double-blind study involving 106 subjects by Henkin et al (1976) who found no significant differences between taste acuity at the beginning of treatment and after three or six months of treatment for any group. These contradicting reports arise because hypogeusia could be caused by several factors, only one being Zn deficiency. Therefore, treatment with Zn compounds of hypogeusia caused by poor oral hygiene will be as ineffective as that of a placebo. Supplementation of Zn in older women (41 to 78 years) significantly increased the Zn content of saliva but there was no significant difference in taste acuity (Buchanan et al, 1980). Daily Zn supplementation given to healthy, normal women for six days produced a transient rise in plasma Zn and may have improved the ability of this population to taste sweetness. Slight improvement in the detection threshold for NaCl and sucrose was reported in a double-blind study of aged, institutionalized groups after 95 days of Zn supplementation.

However, the recognition thresholds for NaCl and sucrose were unaffected in this population (Greger and Geissler, 1978).

Animal studies indicate that Zn-deficient rats develop a strong preference for NaCl (Hastings, 1980) as compared with pair-fed or ad libitum controls. In addition to increased preference for NaCl, Zn-deficient rats showed significantly higher preference for 3.0×10^{-2} sucrose, 1.3×10^{-6}M quinine sulfate, and 2.5×10^{-3}M hydrochloric acid as compared with controls (Catalanotto and Lacy, 1977). In an attempt to correlate physiologic alteration and pathologic changes due to Zn deficiency, Catalanotto and Nunda (1978) studied the histology of Zn-deficient rat tongues including the taste buds. They found that in the epithelial cell lining the papilla appeared in disarray and exhibited loss of basal cell polarity. The taste buds appeared smaller, and the individual cell detail composing the taste buds was obscured. Mann et al (1974) indicated that in Zn-deficient lambs, the outer taste pores were not patent, and the inner taste pores were small or vestigial. Taste cells were relatively few and in some instances appeared to be absent. Catalanotto (1978) suggested the depletion of Zn can lead to decreased taste acuity, but decreased taste acuity is not necessarily associated with depletion of Zn. Thus Zn appears to have an important function in relation to taste, perhaps an indirect one in relation to its role in the integrity of the oral epithelium and taste buds.

SUMMARY

As an essential nutrient, Zn plays important roles in many metabolic processes. The earliest and most consistent effect of Zn deficiency on animals is the reduction of food intake. To eliminate the secondary effect due to reduced food intake, most Zn-deficiency studies must include pair-fed animals as a control, ie, animals fed the same amount of diet as that consumed by Zn-deficient group. Unfortunately, pair-fed controls usually still grow better than experimental animals, and therefore, the effect contributed by inanition cannot be completely dismissed. Weight-paired controls can be used, but this is technically burdensome. Nevertheless, the available evidence shows a clear relationship of Zn deficiency to increased caries in rodents. The role of increased Zn intake, as a caries-preventing agent, still needs to be properly studied, but it is likely that Zn may have a role in the remineralization process.

Next to hypophagia, parakeratosis of skin is the most common finding in Zn-deficient animals. Many studies have found that parakeratosis also occurs in oral mucosa such as the epithelium of cheek, tongue, and esophagus. Among these tissues, buccal mucosa are most profoundly affected, having hypertrophic and hyperplastic effects.

The element also has a role in taste; Zn deficiency in humans and animals brings about impaired taste acuity. Meanwhile, our knowledge of the effect of Zn deficiency on other oral soft tissues such as periodontal tissues and salivary glands is limited.

REFERENCES

Al-Hayali R, Hsieh HS, Navia JM: Gestational and postnatal dietary zinc and dental caries. *J Dent Res* 1981;60:401.

Alvares OF, Meyer J: Regional differences in parakeratotic response to mild zinc deficiency. *Arch Dermatol* 1968;98:191–221.

Alvares OF, Meyer J, Gerson SJ: Activity and distribution of acid phosphatase in zinc-deficient parakeratotic rat buccal epithelium. *Scand J Dent Res* 1973; 81:481–488.

Alvares OF, Meyer J: Thymidine uptake and cell migration in the cheek epithelium of zinc-deficient rats. *J Oral Pathol* 1974;3:86–94.

Antonson DL, Barak AJ, Vanderhoof JA: Determination of the site of zinc absorption in rat small intestine. *J Nutr* 1979;109:142–147.

Armitage S, Stablein M, Meyer J: Effect of zinc deficiency on rat gingiva and alveolar mucosa. *J Dent Res* 1981;60:639. Abstract No. 1320.

Ashrafi SH, Stoncius L, Meyer J: Subepithelial mast cells in buccal mucosa of zinc-deficient rats. *J Dent Res* 1981;60:639. Abstract No. 1321.

Attramadal A, Jonsen J: The content of lead, cadmium, zinc and copper in deciduous and permanent human teeth. *Acta Odont Scand* 1976;34:127–131.

Auerbach S: Zinc content of plasma, blood and erythrocytes in normal subjects and in patients with Hodgkin's disease and various hematologic disorders. *J Lab Clin Med* 1965;65:628–637.

Baratieri A, Picarelli A, Piselli D: Zinc distribution in human saliva. *J Dent Res* 1979;58:540–541.

Barmes DE: Caries etiology in Sepik villages. Trace element, micronutrient and macronutrient content of soil and food. *Caries Res* 1969;3:44–59.

Barney GH, Macaplinlac MP, Pearson WN, et al: Parakeratosis of the tongue: A unique histopathologic lesion in zinc-deficient squirrel monkey. *J Nutr* 1967;93:511–517.

Bates D, Navia JM: Chemotherapeutic effect of zinc on *Streptococcus mutans* and rat dental caries. *Arch Oral Biol* 1979;24:799–805.

Bedell GW, Soll DR: Effects of low concentrations of zinc on the growth and dimorphism of Candida albicans: Evidence for zinc-resistant and sensitive pathways for mycelium formation. *Infect Immun* 1979;26:348–354.

Bergman B: The distribution of ^{65}Zn in the endochondral growth sites of the mandibular condyle and the proximal end of the tibia in young rats: An autoradiographic and gamma scintillation study. *Ondont Revy* 1970;21: 261–271.

Björksten R, Aromaa A, Knekt P, et al: Serum zinc concentrations in Finns. *Acta Med Scand* 1978;204:67–74.

Brown ED, Calhoun NR, Larson RH, et al: An effect of zinc deficiency on dental caries. *Life Sci* 1979;24:2093–2098.

Brudevold F, Steadman LT, Spinelli AA, et al: A study of zinc in human teeth. *Arch Oral Biol* 1963;8:135–144.

Buchanan D, Geders J, Freeland-Graves J: Response of zinc status parameters to zinc supplementation in older women. *Fed Proc* 1980;39:652. Abstract #2047.

Butt EM, Nusbaum RE, Gilmour TC, et al: Trace metal levels in human serum and blood. *Arch Environ Health* 1964;8:52–57.

Catalanotto FA: The trace metal zinc and taste. *Am J Clin Nutr* 1978;31:1098–1103.

Catalanotto FA, Lacy P: Effect of zinc-deficient diet upon fluid intake in the rat. *J Nutr* 1977;107:436–442.

Catalanotto FA, Nanda FA: The effect of administration of zinc-deficient diet on taste acuity and tongue epithelium. *J Oral Pathol* 1978;6:211.

Chen SY: Cytochemical and autoradiographic study of nuclei in zinc-deficient rat buccal epithelium. *J Dent Res* 1977;56:1546–1551.

Cruickshank DB: The natural occurrence of zinc in teeth: III. Variation in tuberculosis. *Br Dent J* 1940;68:257–271.

Curzon MEJ, Losee FL: Dental caries and trace element composition in whole human enamel: Eastern United States. *JADA* 1977;94:1146–1150.

Curzon MEJ, Losee FL: Dental caries and trace element composition of whole human enamel: Western United States. *JADA* 1978;96:819–822.

Curzon MEJ, Bibby BG: Effect of heavy metal on dental caries and tooth eruption. *J Dent Child* 1970;37:463–465.

Curzon MEJ, Losee FL, Macalister AS: Trace elements in the enamel of teeth from New Zealand and the U.S.A. *NZ Dent J* 1975;71:80–83.

Davies NT, Hristie V, Flett AA: Phytate rather than fibre in bran as the major determinant of zinc availability to rats. *Nutr Rept Int* 1977;15:207–214.

Dennes E, Tupper R, Wormall A: The zinc content of erythrocytes and leucocytes of blood from normal and leukaemia subjects. *Biochem J* 1961;78: 578– 587.

Diamond I, Swenerton H, Hurely LS: Testicular and esophageal lesions in zinc-deficient rats and their reversibility. *J Nutr* 1971;101:77–84.

Drews LM, Kies C, Fox HM: Effect of dietary fiber on copper, zinc and magnesium utilization by adolescent boys. *Am J Clin Nutr* 1979;32:1893–1897.

Engel RW, Miller RF, Price NO: Metabolic patterns in preadolescent children: XIII. Zinc balance, in Prasad AS (ed): *Zinc Metabolism.* Springfield, IL, Charles C Thomas, 1966, pp 326–338.

Evans GW, Grace GI, Hahn C: The effect of copper and cadmium on ^{65}Zn absorption in zinc-deficient and zinc-supplemented rats. *Bioinorg Chem* 1974;3:115–120.

Fang MM, Lei KY, Kilgore LT: Effect of zinc deficiency on dental caries in rats. *J Nutr* 1980;110:1032–1036.

Follis RN, Day HG, McCollum EV: Histological studies of tissues of rats fed a diet extremely low in zinc. *J Nutr* 1947;22:223–233.

Freeland-Graves JH, Hendrickson PJ, Ebangit ML, et al: Salivary zinc as an index of zinc status in women fed a low-zinc diet. *Am J Clin Nutr* 1981;34:312–321.

Gandor DW, Meyer J, Gerson SJ: Response of rat submandibular gland to increasing periods of zinc deficiency. *J Dent Res* 1981;60:355. Abstract No. 177.

Gerson SJ, Meyer J: Increased lactate dehydrogenase activity in buccal epithelium of zinc-deficient rats. *J Nutr* 1977;107:724–729.

Gerson SJ, Gandor D, Meyer J: Zinc concentration in oral epithelium of zinc deficient rats. *J Dent Res* 1981;60:639. Abstract No. 1319.

Greger JL, Geissler AH: Effect of zinc supplementation on taste acuity of the aged. *Am J Clin Nutr* 1978;31:633–637.

Greger JL, Sickles VS: Saliva zinc levels: Potential indicators of zinc status. *Am J Clin Nutr* 1979;32:1859–1866.

Hadjimarkos DM: Effect of trace element in drinking water on dental caries. *J Pediatr* 1967;70:697–699.

Haeflein KA, Rasmussen AI: Zinc content of selected foods. *J Am Dietet Assoc* 1977;70:610–616.

Halsted JA, Ronaghy HA, Abadi P, et al: Zinc deficiency in man: The Shriz Experiment. *Am J Med* 1972;53:277–284.

Hambidge KM, Walravens PA, Brown RM, et al: Zinc nutrition of preschool children in the Head Start Program. *Am J Clin Nutr* 1976;29:734–738.

Hambidge KM, Hambidge C, Jacobs M, et al: Low levels of zinc in hair, anorexia, poor growth, and hypogeusia in children. *Pediatr Res* 1972;6: 868–874.

Hambidge KM, Walravens PA, Neldner KH: The role of zinc in the pathogenesis and treatment of acrodermatitis enteropathica, in Brewer GJ, Prasad AS (eds): *Zinc Metabolism.* New York, Alan R Liss, 1977, pp 329–340.

Hardwick JL, Martin CJ: A pilot study using mass spectrometry for the estimation of the trace element content of dental tissue. *Helva Odont Acta* 1967; 2:62–70.

Hart EB, Steenbock H, Waddell J, Elvehjem CA: Iron in nutrition: VII. Copper as a supplement to iron for hemoglobin building in the rat. *J Biol Chem* 1928;77:797–812.

Hastings MM: Adrenocortical hormone response to sodium deprivation in the zinc-deficient rat. *Fed Proc* 1980;39:896. Abstract #3307.

Helle A, Haavikko K: The concentrations of nine macro- and microminerals in drinking water examined in relation to caries prevalence. *Proc Finn Dent Soc* 1977a;73:87–98.

Helle A, Haavikko K: Macro- and micromineral levels in deciduous teeth from different geographical areas correlated with caries prevalence. *Proc Finn Dent Soc* 1977b;73:87–98.

Hendershot LC, Forsaith J: Effects of metals on rat caries and enamel metal levels. *J Dent Res* 1959;38:669.

Henkin RI, Keiser HR, Jaffe IA, et al: Decreased taste sensitivity after D-penicillamine reversed by copper administration. *Lancet* 1967;2:1268–1271.

Henkin RI, Bradley DF: Hypogeusia corrected by Ni^{++} and Zn^{++}. *Life Sci* 1970; 9:701–709.

Henkin RI, Schechter PI, Raff MS, et al: Zinc and taste acuity: A clinical study including a laser microprobe analysis of the gustatory receptor area, in Pories WJ, Strain WH, Hsu JH, Woosley RL (eds): *Clinical Applications of Zinc Metabolism.* Springfield, IL, Charles C Thomas, 1974.

Henkin RI, Mueller CW, Wolf RO: Estimation of zinc concentration of parotid saliva by flameless atomic absorption spectrophotometry in normal subjects and in patients with idiopathic hypogeusia. *J Lab Clin Med* 1975;86:175–180.

Henkin RI, Schechter PI, Friedewald WT, et al: A double-blind study of the effects of zinc sulfate on taste and smell dysfunction. *Am J Med Sci* 1976; 272:285–299.

Hove E, Elvehjem CA, Hart EB: Further studies on zinc deficiency in rats. *Am J Physiol* 1938;124:750–758.

Hove E, Elvehjem CA, Hart EB: The relation of zinc to carbonic anhydrase. *J Biol Chem* 1940;136:425–434.

Huxley HG, Leaver AG: The effect of different levels of dietary zinc and calcium upon the zinc concentration of the rat femur and incisor. *Arch Oral Biol* 1966;11:1337–1344.

Joseph CE, Ashrafi SH, Waterhouse JP: Structural changes in rabbit oral epithelium caused by zinc deficiency. *J Nutr* 1981;111:53–57.

Kagi JHR, Vallee BL: Metallothionein: A cadmium and zinc containing protein from equine renal cortex. *J Biol Chem* 1960;235:3460–3465.

Kaneko Y, Inamori I, Nishimura IT: Zinc, lead, copper, and cadmium in human teeth from different geographical areas in Japan. *Bull Tokyo Dent Coll* 1974;15:233–243.

Kemmerer AR, Elvehjem CA, Hart EB: Studies on the relation of manganese to nutrition of the mouse. *J Biol Chem* 1931;92:623–630.

Khrosh TM: The use of zinc for the prevention of dental caries. *Stomatologiia* (Mosk) 1966;45:38–41.

Kirchgessner M, Roth HP, Weigand E: Biochemical changes in zinc deficiency, in Prasad AS (ed): *Trace Elements in Human Health and Disease*. New York, Academic Press, 1976.

Kravich ME, Meyer J, Waterhouse JP: Increased numbers of mast cells in the hyperplastic buccal mucosa of the zinc-deficient rat. *J Oral Pathol* 1981;10: 22–31.

Lappalainen R, Knuuttila K: The distribution and accumulation of Ca, Zn, Pb, Cu, Co, Ni, Mn and K in human teeth from five different geological areas of Finland. *Arch Oral Biol* 1979;24:363–368.

Losee FL, Cutress TW, Brown R: Trace elements in human dental enamel, in Hemphill DD (ed): *Trace Substance in Environmental Health,* vol 7. Columbia, MO, University of Missouri, 1974, pp 13–17.

Macapinlac MP, Barney TH, Pearson WN, et al: Production of zinc deficiency in the squirrel monkey. *J Nutr* 1967;93:499–510.

Macapinlac MP, Pearson WN, Barney TH, et al: Protein and nucleic acid metabolism in the testes of zinc deficient rats. *J Nutr* 1968;95:569–577.

Mann SO, Fell BF, Dalgarno AC: Observations of the bacterial flora and pathology of the tongue of sheep deficient in zinc. *Res Vet Sci* 1974;17:91–101.

Mathur A, Wallenius K, Abudlla M: Relation between zinc content in saliva and blood in healthy human adults. *Scand J Clin Lab Invest,* 1977;37:269–472.

McClure FJ: Observation on induced caries in rats. *J Dent Res* 1948;27:34–40.

McCoy JM, Fridovich I: Superoxide dismutase, and enzymic fraction for erythrocuprein (hemocuprein). *J Biol Chem* 1969;244:6049–6054.

McKenzie JM: Content of zinc in serum, urine, hair, and toenails of New Zealand adults. *Am J Clin Nutr* 1979;32:570–579.

Meyer J, Alvares OF: Dry weight and size of cells in the buccal epithelium of zinc-deficient rats: A quantitative study. *Arch Oral Biol* 1974;19:471–476.

Meyer J, Stohle MR, Stablein MJ: Correlation of changes in capillary supply and epithelial dimensions in the hyperplastic buccal mucosa of zinc-deficient rats. *J Oral Biol* 1981;10:49–59.

Morris ER, Ellis R: Effect of dietary phytate zinc molar ratio on growth and bone zinc response of rats fed semipurified diets. *J Nutr* 1980;110:1037–1045.

National Research Council: *Recommended Dietary Allowances,* ed 9. Washington, DC, National Academy of Sciences, 1980; pp 144–147.

Navia JM: Effect of minerals on dental caries, in Harris RS (ed): *Dietary Chemical vs Dental Caries, Advances in Chemistry* No. 94. Washington, DC, American Chemical Society, 1970.

Navia JM, Cagnone LD, Lopez H, et al: The effect of $MgCl_2$, $ZnCl_2$, $MnCl_2$ and

Na trimetaphosphate on dental caries when fed alone or in combinations. *J Dent Res* 1968;47:128. Abstract #361.

Nixon GS, Livingston HD, Smith H: Estimation of zinc in the human enamel by activation analysis. *Arch Oral Biol* 1967;12:411–416.

Orten JM: Biomedical aspects of zinc metabolism, in Prasad AS (ed): *Zinc Metabolism.* Springfield, IL, Charles C Thomas, 1966.

Osmanski CP, Meyer J: Ultrastructural changes in buccal and palatal mucosa of zinc-deficient rats. *J Invest Dermatol* 1969;53:14–28.

Parr RM, Taylor DM: The concentrations of cobalt, copper, iron and zinc in some normal human tissues as determined by neutron activation analysis. *Biochem J* 1964;91:424–431.

Pécoud A, Donzel P, Schelling JL: Effect of foodstuffs on the absorption of zinc sulfate. *Clin Pharmacol Ther* 1975;17:469–474.

Prasad AS, Halstead JA, Nadimi M: Syndrome of iron deficiency anemia, hepatosplenomegaly, hypogonadism, dwarfism and geophagia. *Am J Med* 1961; 31:532–546.

Prasad AS, Schulert AR, Miale A, et al: Zinc and iron deficiencies in male subjects with dwarfism but without ancyclostomiases, schistosomiasis or severe anemia. *Am J Clin Nutr* 1963;12:437–444.

Reinhold JG, Faradji B, Abadi P, et al: Decreased absorption of calcium, magnesium, zinc and phosphorus by humans due to increased fiber and phosphorus consumption as wheat bread. *J Nutr* 1976;106:493–503.

Retief DH, Cleaton-Jones PE, Turkstra J, et al: The quantitative analysis of sixteen elements in normal human enamel and dentin by neutron activation analysis and high resolution gamma spectrometry. *Arch Oral Biol* 1971;16: 1257–1267.

Retief DH, Turkstra J, Cleaton-Jones PE, et al: Mineral composition of enamel from population groups with high and low caries incidence. *J Dent Res* 1978; 57:150. Abstract #302.

Rijhsinghani K, Squier CA, Meyer J: The ultrastructural localization of acid phosphatase in zinc-deficient rat buccal epithelium. *Arch Oral Biol* 1975; 20:461–464.

Ritchey SJ, Korslund MI, Gilbert LM, et al: Zinc retention and losses of zinc in sweat by preadolescent girls. *Am J Clin Nutr* 1979;32:799–803.

Sandstead HH: Zinc nutrition in the United States. *Am J Clin Nutr* 1973;26: 1251–1260.

Sandstead HH, Vo-Khactu KP, Solomons N: Conditioned zinc deficiencies, in Prasad AS (ed): *Trace Elements in Human Health and Disease,* vol 1. New York, Academic Press, 1976, pp 33–49.

Schamschula RG, Agus H, Bunzel M, et al: The concentration of selected major and trace minerals in human dental plaque. *Arch Oral Biol* 1977;22:321–325.

Schwarz FJ, Kirchgessner M: Experimental studies on the absorption of zinc from different parts of the small intestine and various zinc compounds. *Nutr Metab* 1975;18:157–166.

Snowden J, Freeland JH: Circadian rhythms of zinc in saliva. *Fed Proc* 1978; 37:890, Abstract #3557.

Söremark R, Samsahl K: Gamma-ray spectrometric analysis of elements in normal human enamel. *Arch Oral Biol* 1961;6:275–283.

Söremark R, Samsahl K: Gamma-ray spectrometric analysis of elements in normal human dentin. *J Dent Res* 1962;41:603–606.

Sortino G, Palazzo H: Zinco e carie superimentale nel ratto. *Riv Ital Stomatol* 1971a;26:509–513.

Sortino G, Palazzo H: Azione della somministrazione di dieta priv di zinco nei ratti. *Riv Ital Stomatol* 1971b;26:587–592.

Spencer H, Osis D, Kramer L, Norris C: Intake, excretion, and retention of zinc in man, in Prasad AS (ed): *Trace Elements in Human Health and Disease,* vol 1. New York, Academic Press, 1976.

Spencer H, Asmussen CR, Holtzman RB, et al: Metabolic balances of cadmium, copper, manganese, and zinc in man. *Am J Clin Nutr* 1979;32:1867–1875.

Steinman RR, Leonora J: Effect of selected dietary additives on the incidence of dental caries in the rat. *J Dent Res* 1975;54:570–577.

Taylor FB: Significance of trace elements in public finished water supplies. *J Am Water Works Assoc* 1963;55:619.

Tipton IH, Cook MJ: Trace elements in human tissue: II. Adult subjects for the United States. *Health Phys* 1963;2:103–109.

Todd WR, Elvehjem CA, Hart EB: Zinc in the nutrition of the rat. *Am J Physiol* 1934;107:146–156.

Tucker HF, Salmon WD: Parakeratosis or zinc deficiency disease in the pig. *Proc Soc Exp Biol Med* 1955;88:613–616.

Underwood EJ: *Trace Elements in Human and Animal Nutrition,* ed 4, ch 8. New York, Academic Press, 1977.

Vallee BL, Hoch FL, Hughes WL Jr: Studies on metalloproteins. Soluble zinc-containing protein extracted from human leukocytes. *Arch Biochem Biophys* 1954;48:347–360.

Van Campen DR, Scaife PV: Zinc interference with copper absorption in rats. *J Nutr* 1967;91:473–476.

Walker BE, Hughes S, Simmons AV, et al: Plasma zinc after myocardial infarction. *Eur J Clin Invest* 1978;8:193–195.

Walker BE, Kelleher J: Plasma, whole blood and urine zinc in the assessment of zinc deficiency in the rat. *J Nutr* 1978;108:1702–1707.

Walravens PA, Hambidge KM: Nutritional zinc deficiency in infants and children, in Brewer GJ, Prasad AS (eds): *Zinc Metabolism: Current Aspects in Health and Disease.* New York, Alan R Liss, 1977, pp 61–72.

Weitzel G, Fretzdorff AM: Zink in den Augen von Säugetieren. *Z Physiol Chem* 1953;292:221–231.

Wildra A: Phosphate directed Y-M variation in *Candida albicans. Mycopathol Mycol Appl* 1964;23:197–202.

Wilkins PJ, Grey PC, Dreosti IE: Plasma zinc as indicator of zinc status in rats. *Br J Nutr* 1972;27:113–120.

Yamaguchi H: Control of dimorphism in *Candida albicans* by zinc: Effect on cell morphology and composition. *J Gen Microbiol* 1975;86:370–372.

$_{12}$ Mg
24.30
2-8-2

10 Magnesium

J.D.B. Featherstone

The second of the alkaline earth elements, magnesium (Mg), is the eighth most abundant element in the earth's crust, and is essential for life. Chlorophylls are Mg-centered porphyrins, and indicate the great importance the element plays in the plant and animal kingdoms. The element is never found uncombined, and because of its highly reactive nature, Mg salts are many and varied. Notable among these salts are:

Epsom Salts	- $MgSO_4 \cdot 7H_2O \cdot$	Brucite	- $Mg(OH)_2$
Magnesite	- $MgCO_3$	Periclase	- MgO
Spinel	- $MgAl_2O_4$	Dolomite	- $CaCO_3MgCO_3$

Other salts, such as the chloride, fluoride, and silicate, as well as various other carbonates and phosphates, all occur naturally. Many of these naturally occurring salts are mined, and Mg is primarily used commercially for flash photography, flares, pyrotechnics, and alloys in airplanes and space vehicles, and as an alloying agent.

Metabolism

As early as 1926 Mg was shown to be essential as a nutrient, and a number of experiments have shown the need for Mg for proper growth and development in rats. Recent years have seen greater research interest in the element, partially because deficiency states may develop in the course of various diseases or in states of stress. The mechanism of action of Mg in human tissues is little understood.

The effect of Mg deficiency on bodily functions in a number of animals has been reported, and the metabolic function of Mg is mainly as an activator of many enzyme systems in the body. Included among these are those using adenosine triphosphate (ATP) to provide energy to cells. Since ATP is necessary for the synthesis of many metabolites, such as nucleic acid and coenzymes, the importance of Mg is readily apparent.

The Mg requirement for the average adult is about 300 mg/day, although this can be as high as 400 mg/day for pregnant and lactating mothers. The average diet provides adequate quantities of Mg if there are plenty of cereals and vegetables.

Table 10-1
Magnesium Concentrations in Human Tissues

Tissue	Concentrations in ppm (μg/g) wet weight Mean ± SE
Brain	150 ± 3.4
Heart	180 ± 5.5
Muscle	189 ± 7.2
Kidney	129 ± 3.3
Lung	71 ± 2.2
Liver	172 ± 5.3
Gastrointestinal tract	125 ± 9.6
Enamel	16,700 ± 1200
Dentin	87,000 ± 300

Data from Tipton and Cooke, 1963; Schroeder et al, 1969; Losee et al, 1974; Retief et al, 1971.

If present as soluble salts, Mg is readily absorbed in the small intestine; this process does not appear to be affected by pH or carbonates. Thirty percent of a daily dose of Mg can be recovered in the urine, and when present in feces Mg represents unabsorbed salts. Tipton and Cooke (1963), in their classic studies on "reference man," reported that Mg showed the least variation from tissue to tissue of any of the elements studied. This is demonstrated in the concentrations reported for various tissues shown in Table 10-1.

In human enamel Mg appears to be essential for normal growth of this tissue. Developmental disturbances in enamel were demonstrated using Mg-deficient diets by Becks and Furuta in 1941, and recently Kiely and Domm (1977) showed that a diet deficient in Mg will bring about a marked decrease in the rate of eruption of maxillary incisors in rats.

Magnesium in Calcified Tissues

Being in group IIa of the periodic table, along with Ca, Mg is a normal constituent of dental enamel, dentin, and bone. Analyses of tooth tissues in many studies have consistently identified Mg at concentrations of between 0.06% to 0.80% in enamel, but most commonly less than 0.4%. (See also the chapters on trace elements in tooth tissues.) Dentin Mg concentrations are generally higher (up to 1.1%), and in bone Mg is found at about 0.6%.

In bone Mg seems to occur in two forms which have differing solubility. It has been suggested that the more soluble form of Mg is adsorbed onto the surfaces of bone crystals as Mg^{++} or $MgOH^{+}$, while the less soluble form replaces Ca within the crystal lattice. Whether this same action occurs in enamel and dentin is not clear.

Human Food and Water Sources

Foods which are rich in Mg fall into two main groups, nuts and cereals. Thus, expressed as mg/100 g of edible protein, the highest levels are found in cocoa (420), cashews (267), almonds (252), and brazil nuts (225). Cereals such as barley (171), whole wheat (165), oatmeal (145), corn (121), and brown rice (119) all contain good levels of Mg. In addition legumes such as lima beans (181) also contain quantities of the element. Many water supplies contain appreciable quantities of Mg which contribute to the hardness of water. In the United States water supplies contain Mg ranging from not detected to 120 ppm with a mean of 6.25 ppm (Dufor and Becker, 1962). Most water supplies, therefore, contain ample concentrations of Mg although the availability of the salts may vary.

Because there appears to be no recorded case of true Mg deficiency in man, it is assumed that the normal diet contributes all that is required. In a number of circumstances, such as severe renal disease, pregnancy toxemia, and chronic alcoholism, Mg deficiency may be induced in man. There is no evidence of any dental diseases being affected during these induced Mg-deficiency states.

CARIES

Man

A number of surveys have associated high levels of Mg in water supplies with low caries prevalence in man (Table 10-2). Most surveys found a negative association between Mg in water and caries prevalence; that is, areas of low caries prevalence were associated with diets or water supplies high in Mg. It should be pointed out, however, that in most cases the level of Ca present was also very high, and several authors commented upon the hardness of the water. By the nature of the chemistry and geology involved, high Mg will usually be present with high Ca (and here will sometimes be high Sr, Ba, and Fe as well). The question arises as to how much the high Ca, or the hardness of water per se, is producing the low caries and whether there is any effect from the Mg.

Table 10-2
Reports on the Effects of Environmental Magnesium on Dental Caries in Man

Author (year)	Country	Magnesium Source	Effect on Dental Caries
Ritchie, 1961	New Zealand	Diet, 50% $MgPO_4$	Protective in enamel
Losee et al, 1961	New Zealand	Water, high Ca + Mg	Low caries
Plevova, 1966	Czechoslovakia	Water, high Ca + Mg	Low caries
Rothman et al, 1972	Colombia	Soil, high Mg	Low caries
Glass, et al, 1973	Colombia	Water, high Mg	Low caries
Vejrosta et al, 1975	Czechoslovakia	Water, high Mg	Low caries
Helle and Haavikko, 1977	Finland	Water, Mg (+ Ca)	High caries

The use of Mg was noted as "protective to enamel" in a human study where dietary Ca appeared to be relatively low (Ritchie, 1961) and where the added Mg salt was 50% $Mg_3(PO_4)_2$. Here the possible anticaries effect of the phosphate perhaps masks any effect of the Mg, for there is ample evidence that phosphates alone have a marked cariostatic effect (Lilienthal, 1977). Therefore, no conclusions on the effect of Mg on caries can be drawn from this study.

Differences in caries prevalence between two Colombian villages have been reported in two papers (Rothman et al, 1972, Glass et al, 1973) where a negative association of Mg to caries was found. In these reports the Mg was present in high concentrations in both soil and water in the low caries village, although a concomitant presence of high Ca was again noted. Nevertheless, in these studies extensive trace element analyses were carried out, and the association of Mg to caries remained strong.

Recent research in Finland (Helle and Haavikko, 1977) considered the drinking water concentrations of nine macro- and microminerals, including Mg, that were related to caries prevalence. Among the findings was a strong positive correlation between Ca and Mg, and caries. These authors showed that there was a relationship between F and other elements such as Mg. In one study district an F concentration as high as 2.16 ppm not only seemed to have a caries-producing effect but also appeared to prevent the caries-positive effects of Mg and Na, even though these elements were present in high concentrations.

These epidemiologic surveys have generally indicated that Mg, when present in hard water, is associated with low caries. This finding would seem to contradict the theoretical implications of the incorporation of Mg in hydroxyapatite. Considering all the Mg studies on caries in man, we must conclude that it is difficult to separate any possible effect of Mg alone from the simultaneous presence of Ca and/or other alkaline earth trace elements. Thus, it would appear that good studies on the relationship of Mg to caries are still lacking; they can only be done by controlled experiments where the influence of Ca (as well as phosphate) is kept to a minimum. A few studies have been completed in animal models, which fulfill such a function.

Animals

Little work has been carried out to date on the effect of Mg on caries. Several animal studies (Table 10-3) have generally found either no effect of Mg or an increase in caries. This type of research presents difficulties because of the need to maintain animals on a diet commensurate with good health, but low in Mg so that known quantities of the element can be added and the effects studied. In addition, the presence of Ca must affect any action of the Mg, but all experimental animal diets are rich in Ca. In the animal studies cited, no data were given on the Ca content of the diets used, so that a definite conclusion cannot be drawn from the evidence.

The conclusions of McClure (1948) were that Mg had no effect upon caries. Careful study of the caries scores given in his paper, however,

Table 10-3
Reports on the Effects of Magnesium on Experimental Caries in Animals

Author (Year)	Animal	Magnesium Salt Used*			Other* Treatment	Effect on Caries
		Salt	*Conc.*	*Vehicle*		
McClure, 1948	R	$MgCl_2$	500	W	none	Increased
		$MgCl_2$	500		NaF – 10	Reduced F effect
Wisotsky and Hein, 1958	H	$MgSO_4$	24.3	W	none	Increased caries
McClure and McCann, 1960	R	$MgCO_3$	1.6%	D	none	Increased caries
		$Mg_3(PO_4)_2$	2.3%		$CaCO_3$	Increased caries
		$Mg_3(PO_4)_2$	2.3%		none	No effect
		$MgCO_3$	2.3%		none	Increased caries
Rosen and Penter, 1972	R	$MgCl_2$	45	W	none	No effect
		$MgCl_2$	45		NaF – 45	No change on F
		$MgCl_2$	2000		none	Slight reduction
		$MgCl_2$	2000		NaF – 45	No change on F
Luoma et al, 1975	R	$MgSO_4$	20	D	$MgCO_3$	Increased caries
		$MgSO_4$	20		$MgCO_3$ + NaF – 10 + PO_4	Greater than F alone
Luoma et al, 1977	R	MgO	250	D	NaF – 15 + PO_4 1%	Increased caries
		MgO	460		NaF – 15 + PO_4 1.5%	Increased caries
		MgO	870		NaF – 15 + PO_4 2.0%	Reduced caries
		MgO	970		NaF – 15 + PO_4 2.0%	Reduced caries

Animal used: H = Hamster; R = Rat.
Vehicle used for Mg administration: W = water, D = Diet.
*Concentrations used in ppm (mg/l) in water or as % in diet.

shows that the Mg, as $MgCl_2$ (500 ppm), caused an increase in caries, although not one that was statistically significant. When the same salt and concentrations were used in combination with NaF (10 ppm), the Mg appeared to offset the cariostatic effect of the F; McClure, however, did not draw attention to this finding. In a second series of rat experiments, McClure and McCann (1960) used $MgCO_3$ either alone or with $CaCO_3$, and then $Mg_3(PO_4)_2$ with $CaCO_3$ added to the diet. Caries scores for these rats, compared with control groups, were all considerably increased on these dietary regimens. In a final group fed $Mg_3(PO_4)_2$ alone as the mineral supplement, there was no apparent effect on caries.

Wisotsky and Hein (1958) used $MgSO_4$ at a concentration of 24.3 ppm, based upon milliequivalents. With the $MgSO_4$ given via the water supplies to hamsters living on a cariogenic diet, an increase in caries was noted. Recently Luoma et al (1975) used $MgSO_4$, also in the diet, and reported an increase in caries. Rosen and Penter (1972) again considered the action on caries of Mg alone and in combination with F. The Mg salt used was not defined in the abstract (the only report of this study) but was the chloride (Rosen, personal communication, 1980). These experiments showed no effect of Mg on caries at 45 ppm, a slight reduction at 2000 ppm, and no change in the F effect when used in combination with F. No details are available as to the diet composition.

Luoma et al have completed further series of experiments (1977) on the question of Mg and caries in animals. Their results are complicated since they used NaF and various phosphates in addition to the Mg. The previous papers did not report details of the Ca content of the diets, and the same holds true for these more recent studies. The complex nature of the study designs used by Luoma makes interpretation very difficult. The anticaries effect of the phosphate is clear from these experiments. This was also apparently offset, to a modest degree, by the presence of Mg.

In conclusion, the animal studies, like the human epidemiologic surveys, are inconsistent. It is not possible at this time to say whether the evidence shows that Mg promotes, retards, or is in any way associated with dental caries. Theoretically, evidence points toward cariogenic action, but good experimental work is still needed for clarification.

MECHANISM OF ACTION

There is considerable debate, and confusion, as to any action of Mg on dental tissue. Experimental evidence in animal and in vitro experiments points to an increase of caries by Mg, while epidemiologic studies suggest the opposite. There is no definite evidence at this time as to whether Mg is incorporated in, or on, the enamel crystals, or whether Mg affects enamel solubility, tooth morphology, plaque bacteria, mineralization, or remineralization. Although Mg occurs in enamel as one of the group of constituents second only to Ca and P, much has yet to be learned about its importance.

Incorporation in Whole and Surface Enamel

In enamel Mg is incorporated in substantial quantities. The concentration increases from the surface toward the dentino-enamel junction. Besic et al (1969), using electron probe analysis, reported an almost linear rise from about 0.1% wt/wt near the surface to about 0.3% wt/wt near the junction, followed in the dentin by 3 to 4 times this concentration. Cutress (1972) reported uniform values in teeth from several geographic and ethnic groups, of 0.12% to 0.19% Mg in three successive surface etchings of about 30 μm thick. Values in whole enamel ranged from 0.05% to 0.30% with a mean of 0.17% in a comprehensive study by Losee et al (1974). Surface enamel of acid-resistant and acid-susceptible teeth was analyzed to about 10 nm by ion probe (Besic et al, 1975), and Mg means were reported as 0.032% and 0.026% respectively. A neat selected area analysis of Mg in thin sections of enamel was recently reported by Robinson et al (1981). This study produced contours of Mg concentration and confirmed that concentrations increase from the surface toward the interior, generally in the range of 0.2% to 0.4% wt/wt. The increase was rarely smooth, and occasional pockets of high Mg concentration were found, for example, near the molar fissure and near dentinal cornua. In each case high Mg concentrations correlated well with the presence of low density enamel; the authors suggest that some of the Mg is associated with enamel protein. The pattern of Mg distribution shown by these workers is somewhat similar to that of carbonate (Weatherell et al, 1968).

There is, therefore, no doubt that Mg is an intrinsic component of enamel. It is obvious, however, that precise sampling and precise analytical techniques must be used to provide new insight into the role of Mg in enamel, and surface enamel must be clearly differentiated from whole enamel. This is particularly relevant to the question of whether Mg is part of the crystal lattice and what function it has in caries production.

Magnesium in Carious Enamel

Bowes and Murray (1936) and Okerse (1943) considered that enamel from high and low caries areas showed no differences in Mg content. Wei and Koulourides (1972) stated that the distribution of Mg, Na, and Cl in carious enamel, as assessed by electron microprobe analysis, did not appear to be significantly different from that in sound enamel. Conversely, other authors have reported low concentrations of Mg in carious enamel (Nordback and Johansen, 1962; Helle and Haavikko, 1977), and Hals and Halse (1975) found lower Mg levels in carious lesions of cavity walls. It is obvious from the discussion above that great care must be taken in comparing Mg concentrations because of the natural several-fold increase from the surface to the interior. Thus interpretation of these studies mentioned is difficult.

Hallsworth et al (1972, 1973) used precise selected area analysis of natural carious lesions and clearly demonstrated that Mg was selectively lost from the translucent zone, dark zone, and body of early carious lesions. A more dramatic selective loss of carbonate-containing mineral was also demonstrated. At first sight it appears that Mg itself formed a point in the enamel where initial dissolution could take place. However, it may be that carbonate is the major determinant and that Mg is simply associated with it. Mg in the surface layers of the lesions studied by Hallsworth et al (1972) was present at similar low concentrations to those found in adjacent, unattacked, sound surface enamel.

Effect on Mineralization, Dissolution, and Caries Mechanism

Since the ionic radius of Mg is smaller (0.65 Å) than that of Ca (0.99 Å), its incorporation in the apatite lattice is not self-evident. Duff (1973) suggested that the divalent cations of ionic radius greater than 0.8 Å were necessary for ready formation of solid solutions in apatite. Several workers have studied the precipitation of calcium phosphates in the presence of Mg. Duff (1971) showed that Mg inhibited the transformation of brushite ($CaHPO_4.2H_2O$) to fluorapatite. Increases in the Mg:Ca ratio in solution promote the formation of stable amorphous calcium phosphate in both synthetic (Trautz and Zapanta, 1962; LeGeros et al, 1972) and also in biological systems (LeGeros et al, 1973, 1975). Furthermore, in some systems the presence of Mg can suppress the growth of octacalcium phosphate and favor the growth of whitlockite (LeGeros and Morales, 1973). It seems that the Mg:Ca ratio is at least one of the determining factors of the type of calcium phosphate to be formed. Although these phosphates are considered to be precursors to apatite formation, the relevance of these studies to Mg in enamel apatite is still not clear.

In other studies it has been shown that Mg inhibits hydroxyapatite formation (eg, Neuman and Mulryan, 1971; Tomazic et al, 1975; Eanes and Rattner, 1981). Further, Nancollas et al (1976) showed that in spontaneous precipitation of calcium phosphate in the presence of Mg, the

Mg ion is excluded from the precipitates formed. Other workers have produced Mg in the crystal form of Mg-whitlockite, a solid solution of β-$Ca_3(PO_4)_2$ with various amounts of Mg in the crystal structure (Dickens et al, 1974; Gopal et al, 1974).

These combined chemical studies would suggest that Mg cannot occur in enamel apatite crystals. However, most of the studies excluded carbonate and used relatively high Mg concentrations. Significantly, dental enamel crystals, although similar to hydroxyapatite, are a carbonated apatite with carbonate present at about 2% to 5% wt/wt incorporated in the crystal lattice. The carbonate chiefly substitutes for phosphate ions but a small amount replaces some of the hydroxyl ions. Several other divalent metals, such as Sr, Pb, and Zn, are believed to be incorporated in the enamel apatite crystal lattice, together with Na. It is possible that the carbonate-apatite structure can incorporate Mg in the lattice whereas pure hydroxyapatite will exclude it. Carbonate has been shown by several authors to disturb the apatite structure but a recent study (Featherstone and Nelson, 1980) indicated that this crystalline disturbance can be partially compensated for by the incorporation of some metal ions and F. Combinations of Zn and F were particularly effective. Significantly, in the same study, Mg actually appeared to improve the crystallinity of carbonated apatites formed in its presence.

In a subsequent study (Featherstone and Mayer, 1982) a range of carbonated apatites were precipitated in the presence of Na and Mg to give precipitates with Ca, PO_4, CO_3, Na, and Mg contents in the range found in dental enamel. X-ray diffraction and infrared spectroscopy identified the precipitates as carbonated apatites. The Mg uptake into, or onto, the crystals was 90% to 100% of that present in solution. Crystallographic data suggested that at least some of the Mg was incorporated in the apatite lattice. The results indicated some form of coupling between Mg and carbonate. Although these results are only preliminary and need to be substantiated by some extensive crystallographic work, the implications are that Mg can be incorporated as an intrinsic part of the enamel apatite crystal and need not exist as part of a separate phase.

In the early stages of enamel caries, Mg and more particularly carbonate are preferentially lost (Hallsworth et al, 1972, 1973; Coolidge and Jacobs, 1957; Featherstone et al, 1978). It has been suggested that Mg plays a part in this process in one of three ways: (1) a preferential dissolution of areas of the crystal surface which are rich in Mg; (2) a preferential dissolution of apatite crystallites containing Mg throughout their structure; or (3) selective dissolution of a separate calcium phosphate phase, rich in Mg, that might exist in enamel. Such a separate phase has not been detected by X-ray diffraction or any other technique, but it is possible that crystal size is too small (amorphous) to permit detection at the low concentrations that would be present. Further investigation is necessary using high resolution techniques which are now becoming available. Robinson et al (1981) in their Mg contour study of enamel sections demonstrated similar concentration profiles to carbonate and suggested a link between the two. This, together with the chemical studies of Featherstone and Mayer (1982), indicates that the Mg may be incorporated with carbonate in the apatite structure, and hence when carbonate is preferentially lost during caries formation, or acid dissolution, Mg is lost also. In depth investigation of this phenomenon may be very useful.

Dissolution studies using enamel, or synthetic apatites, do not as yet define the role of Mg. Thiradilak and Feagin (1978) showed that treatment of enamel surfaces with $MgCl_2$ in both the absence and presence of NaF increased the amount of surface minerals subsequently dissolved into the acid. This suggests that Mg inhibits remineralization. Brudevold et al (1962), using synthetic hydroxyapatite found that the presence of Mg in solution did not increase the solubility of the hydroxyapatite. Conversely Higuchi and coworkers (Dedhiya et al, 1974) reported an effect of Mg on the inhibition of hydroxyapatite dissolution under partial saturation conditions. In these experiments Mg^{2+} had modest retarding effects on dissolution rate, and a large synergistic effect was found for Mg^{2+} plus F. A preliminary study by Featherstone et al (1982b) has measured acid dissolution rates for a range of carbonated apatites with Mg in the crystals. The dissolution rate was a function of the carbonate content, was independent of Mg, and was not increased by Mg. More comprehensive studies must be made with enamel and synthetic carbonated apatites before any conclusions can be drawn.

It seems therefore that (1) Mg inhibits hydroxyapatite and fluorapatite formation, (2) Mg can readily be incorporated into carbonated apatites, (3) Mg readily forms other Ca-Mg phosphates, and (4) Mg is preferentially lost during the early stages of enamel caries formation, but this may be primarily due to preferential carbonate loss. There may be two opposite effects of Mg: one is to interact with carbonate during enamel formation, and the other is to inhibit remineralization. These would have opposing effects on caries susceptibility.

Presence and Effects of Magnesium in Plaque and Saliva

The occurrence of Mg in plaque and saliva has been reviewed in Chapters 4 and 5. The wide variation of Mg concentrations in saliva (0.1 to 0.7 mg%) varies inversely with flow rate and also with individuals. On the other hand, reported concentrations of Mg in plaque are remarkably similar among studies and people (0.15% to 0.22% dry weight). These figures translate to about 10 to 20 mmol/l (200 to 400 ppm) in total plaque. Plaque fluid (most relevant in terms of solution in contact with the enamel surface), has about 4 mmol/l Mg (100 ppm). Some of this will be bound and not available but it seems that the Mg is present in solution in contact with enamel at about half the concentrations of Ca. This may be important for remineralization and caries prevention, or lack of caries prevention. Studies of available Mg^{2+} in plaque fluid in relation to caries would be valuable, as would thorough in vitro studies of the effect of Mg on remineralization.

Interactions with Other Trace Elements

The interaction of Mg with carbonate (not a trace element) has been discussed above. This relationship is intimately linked with Na and the presence of other trace metals. Since Mg^{2+} is a small ion compared with Ca^{2+} it might be expected that Mg^{2+} and Sr^{2+} (radius 1.13 Å) might conveniently fit together in the apatite lattice. This type of association has yet to be studied, and must also include carbonate and Na. Studies on the structural, chemical, and dissolution properties of synthetic apatites with this composition would be of value. Preliminary work has been done by Featherstone and Nelson (1980). Theoretical justification for the importance and existence of these compounds has been provided by Driessens (1973) and Driessens et al (1978) who have taken numerous published data for apatites and enamel and demonstrated chemical relationships among Ca, PO_4 CO_3, Na, Mg and F, and made predictions about other ions such as Pb^{2+} and Sr^{2+}. This model deserves extension by systematic experimental study, including an investigation of possible links between Mg and F. This approach would clarify the conflicting report of McCann and Bullock (1957) that Mg and F in enamel are related versus that of Brudevold et al (1965) who stated that the interrelationships of F, carbonate, and Mg, seen in bone, are not found in enamel. With the sampling techniques, analytic techniques and physicochemical methods now available the interaction of Mg with other enamel components can be investigated. It is possible, for example, that the apparently reactive and strained central core of the enamel crystallites might contain a mineral structure similar to that of Huntite, a mineral like apatite which contains a considerable amount of Mg and carbonate. The techniques of Raman spectroscopy (Casciani et al, 1979), high resolution lattice imaging (Featherstone et al, 1981), selected area diffraction, proton probe, SIMS, and other similar methods should be used to resolve the importance or lack of importance of Mg in enamel.

SUMMARY

The evidence relating Mg to dental caries is equivocal. Of the in vivo studies, epidemiologic evidence points to an inverse relationship of Mg to caries in man, while animal studies show the opposite. Of the in vitro studies the evidence is that Mg can enter and affect the properties of the enamel apatite crystallite, but whether this results in any resistance or susceptibility to dissolution remains unclear. A further factor requiring investigation is the interaction of Mg with other ions, such as Ca^{2+}, PO_4^{3-}, F^-, Sr^{2+}, CO_3^{2-}, and Ba^{2+}, both by incorporation in the crystallite, and at the plaque-enamel interface. The presence of these other ions together with Mg forms a complex system which must in some way be related to dental caries.

REFERENCES

Becks H, Furuta WJ: Effects of magnesium deficient diets on oral and dental tissues: II. Changes in the enamel structure. *JADA* 1941;26:1083–1088.

Besic FC, Knowles CR, Wiemann MR, et al: Electron probe microanalysis of noncarious enamel and dentin and calcified tissues in mottled teeth. *J Dent Res* 1969;48:131–139.

Besic FC, Bayard M, Wiemann MR, et al: Composition and structure of dental enamel: Elemental composition and crystalline structure of dental enamel as they relate to its solubility. *JADA* 1975;91:594–601.

Bowes MM, Murray JH: Enamel and dental analyses. *Br Dent J* 1936;61:473–476.

Brudevold F, Amdur B, Rasmussen S: Magnesium in human teeth. Proc 40th meeting IADR, N. Amer. Div., Abstract 33, 1962.

Brudevold F, Gron P, McCann H: Physico-chemical aspects of the enamel-saliva system. *Adv Fluorine Res* 1965;3:63–78.

Casciani FS, Etz ES, Newburg DE, et al: Raman microscope studies of mineralizing tissues. *Scanning Electron Microscopy,* vol 2, 1979.

Coolidge JB, Jacobs MH: Enamel carbonate in caries. *J Dent Res* 1957;36:765–768.

Cutress TW: The inorganic composition and solubility of dental enamel from several specified population groups. *Arch Oral Biol* 1972;17:93–109.

Dedhiya MG, Young F, Hefferen JJ, et al: The inhibition of hydroxyapatite dissolution by Sr^{++} and Mg^{++} under partial solutions containing F′. IADR. Abstract # 204, *J Dent Res* 1974;53:105.

Dickens B, Schroeder LW, Brown WE: Crystallographic studies of the role of Mg as a stabilizing impurity in β-$Ca_3(PO_4)_2$. *J Solid State Chem* 1974;10:232–248.

Driessens FCM: Fluoride incorporation and apatite solubility. *Caries Res* 1973;1: 297–314.

Driessens FCM, vanDijk JWE, Borggreven JMP: Biological calcium phosphates and their role in the physiology of bone and dental tissues. I. Composition and solubility of calcium phosphates. *Calcif Tiss Res* 1978;26:127–137.

Duff EJ: The inhibition of brushite to fluoroapatite transformation by Mg^{2+} and HPO_4^{2-}. *Chem Ind* 1971;16:1191–1193.

Duff EJ: Interactions of divalent cations with apatites. *Colloq Int CNRS* 1973; 230:420–421.

Dufor CN, Becker E: Public water supplies of the 100 largest cities in the U.S.A. U.S. Geol. Surv. Water Supply Paper 1812. Washington, DC, Govt. Printing Office, 1962.

Eanes ED, Rattner SJ: The effect of magnesium on apatite formation in seeded supersaturated solutions at pH 7.4. *J Dent Res* 1981;60:1719–1723.

Featherstone JDB, Duncan FJ, Cutress TW: Crystallographic changes in human tooth enamel during in vitro caries simulation. *Arch Oral Biol* 1978;23: 405–413.

Featherstone JDB, Nelson DGA: The effect of F, Zn, Sr, Mg and Fe on the crystal-structural disorder in synthetic carbonated apatites. *Aust J Chem* 1980;33:2363–2368.

Featherstone JDB, Nelson DGA, McLean JD: An electron microscope study of modifications of defect regions in dental enamel and synthetic apatites. *Caries Res* 1981;15:278–288.

Featherstone JDB, Mayer I: Magnesium in carbonated apatites related to dental enamel. Abstracted Fifth International Workshop on Calcified Tissue; Israel, 1982.

Featherstone JDB, Mayer I, Shields CP, et al: Acid reactivity of carbonated-apatites with incorporated magnesium. Abstracted, ORCA Congress, Annapolis, 1982b.

Glass RL, Rothman KJ, Espinal F, et al: The prevalence of human dental caries and water borne trace metals. *Arch Oral Biol* 1973;18:1099–1104.

Gopal R, Calvo C, Ito J, et al: Crystal structure of synthetic Mg-Whitlockite, $Ca_{18}Mg_2H_2(PO_4)_{14}$. *Can J Chem* 1974;52:1155–1164.

Hals E, Halse A: Electron probe microanalysis of secondary carious lesions associated with silver amalgam fillings. *Acta Odont Scand* 1975;33:149–160.

Hallsworth AS, Robinson C, Weatherell JA: Mineral and magnesium distribution within the approximal carious lesion of dental enamel. *Caries Res* 1972; 6:156–168.

Hallsworth AS, Weatherell JA, Robinson C: Loss of carbonate during the first stages of enamel caries. *Caries Res* 1973;7:345–348.

Helle A, Haavikko K: The concentration of nine macro -and micro-minerals in drinking water examined in relation to caries prevalence. *Proc Finn Dent Soc* 1977;73:76–86.

Kiely ML, Domm LU: The effect of a magnesium deficient diet and cortisone on the growth of the rat incisor. *J Dent Res* 1977;56:1577–1585.

LeGeros RZ, Shirra WP, Miravite MA, et al: Amorphous calcium phosphates synthetic and biological. *Colloq Int CNRS* 1972;230:105–115.

LeGeros RZ, Morales P: Renal stone crystals grown in gel systems. *Invest Urology* 1973;11:12–20.

LeGeros RZ, Contiguglia SR, Alfrey AC: Pathological calcifications associated with uremia. *Calc Tiss Res* 1973;13:173–175.

Lilienthal B: Phosphates and dental caries. *Monographs in Oral Science* #6. Basel, Switzerland, Karger, 1977.

Losee FL, Cadell PB, Davies GN: Caries, enamel defects and soil: Owaka-Cheviot Districts. *New Zealand Dent J* 1961;57:135–143.

Losee FL, Cutress TW, Brown R: Natural elements of the periodic table in human dental enamel. *Caries Res* 1974;8:123–134.

Luoma H, Nuuja T, Nummikoski P: Changes in dental caries and calculus development in rats through addition of magnesium, orthophosphate and fluoride to high-sucrose diets. *Arch Oral Biol* 1975;20:227–230.

Luoma H, Nuuja T: Caries reductions in rats by phosphate, magnesium and fluoride additions to diet. *Caries Res* 1977;11:100–108.

McCann HG, Bullock FA: The effect of fluoride ingestion on the composition and solubility of mineralized tissues of the rat. *J Dent Res* 1957;36:391–398.

McClure FJ: Observations on induced caries in rats. *J Dent Res* 1948;27:34–40.

McClure FJ, McCann H: Dental caries and composition of bones and teeth of white rats. Effect of dietary mineral supplements. *Arch Oral Biol* 1960;2: 151–161.

Nancollas G, Wefel JS: Seeded growth of calcium phosphates: Effect of different calcium phosphate seed material. *J Dent Res* 1976;55:617–624.

Neuman WF, Mulryan BJ: Synthetic hydroxyapatite crystals: Magnesium incorporation. *Calcif Tiss Res* 1971;7:133–138.

Nordbark LG, Johansen E: The chemistry of carious lesions I. Calcium, magnesium and carbonate. IADR Abst #141. *J Dent Res* 1962;51:39.

Okerse T: Chemical composition of enamel and dentin in high and low caries areas of South Africa, *J Dent Res* 1943;22:441–446.

Plevova V: Stav chrupe deti v zavislosti na pitne vode bohate na soli horciku a vapniku. *Cesk Stomat* 1966;66:214–217.

Retief DH, Cleaton-Jones PE, Turkstra J, et al: The quantitative analysis of sixteen elements in normal human enamel and dentine by neutron activation analysis and high resolution gamma-spectrometry. *Arch Oral Biol* 1971;16: 1257–1267.

Ritchie DB: Surface enamel magnesium and its possible relation to incidence of caries. *Nature* 1961;190:458–459.

Robinson C, Weatherell JA, Hallsworth AS: Distribution of magnesium in mature human enamel. *Caries Res* 1981;15:70–77.

Rosen S, Penter S: Effect of fluoride and magnesium on streptococci and dental caries in rats. IADR Abst. #400. *J Dent Res* 1972;51:148.

Rothman K, Glass RL, Espinal F, et al: Dental caries and soil content of trace metals in two Colombian villages. *J Dent Res* 1972;51:1686.

Schroeder HA, Nason AP, Tipton IH: Essential metals in man—Magnesium. *J Chron Dis* 1969;21:815–841.

Thiradilak S, Feagin F: Effects of magnesium and fluoride on acid resistance of mineralized enamel. *Ala J Med Sci* 1978;15:144–148.

Tipton IH, Cooke MJ: Trace elements in human tissues. II. Adult subjects in U.S.A. *Health Phy* 1963;9:103–145.

Tomazic B, Tomson M, Nancollas GH: Growth of calcium phosphates on hydroxyapatite crystals: The effect of magnesium. *Arch Oral Biol* 1975;20:803–808.

Trautz OR, Zapanta RR: Effect of magnesium on various calcium phosphates. IADR Abst. #101. *J Dent Res* 1962;41:28.

Vejrosta Z, Sindelka Z, Feiler M, et al: A study in the decaying of teeth in children drinking water with a high percentage magnesium content. *Cesk Stomat* 1975;5:346–354.

Wei SH, Koulourides T: Electron microprobe and microhardness studies of enamel remineralization. *J Dent Res* 1972;51:648–651.

Weatherell JA, Robinson C, Hiller CR: Distribution of carbonate in the sections of dental enamel. *Caries Res* 1968;2:1–9.

Wisotsky J, Hein JW: Effects of drinking solutions containing metallic ions above and below hydrogen in the electromotive series on dental caries in Syrian Hamsters. *JADA* 1958;57:796–800.

25 **Mn**
54.94
8-13-2

11 Manganese

D. Beighton

Manganese (Mn), as an essential trace element, has been rather thoroughly studied, for its effects not only on man but also on domestic and agricultural animals. However, the role of Mn in the caries process is not known, although there are indications that it may be deleterious and increase caries. One of the most abundant elements, and a transition element, Mn is found in association with many other metals such as Cd and Zn. In bacterial and animal metabolism Mn has been shown to have interactions with these and other trace elements.

In human tissues Mn is widely distributed but it does not occur to a great degree in any one tissue (Table 11-1). The element is known to be a cofactor for a number of enzyme systems and appears to be essential for cholesterol biogenesis, lipid metabolism, reproductive function, bone growth, prothrombin formation, and carbohydrate metabolism, among many functions in animals (Underwood, 1977). In enzyme systems, the number of metalloenzymes requiring Mn is small (Silver and Jasper, 1979), but many are activated by the metal. Abnormalities of function

Table 11-1
Manganese Concentrations[a] in Human Tissues

Tissue	Mean ± SE	Tissue	Mean ± SE
Brain	0.2 ± 0.03	Heart	0.23 ±
Muscle	0.04 ± 0.0007	Kidney	1.3 ± 0.5
Lung	0.2 ± 0.03	Liver	0.5 ± 0.8
Lymph nodes	1.1 ± 0.6	Teeth—enamel	0.59 ± 0.02
		—dentin	0.63 ± 0.02

[a]Concentrations in μg/g (ppm)
Data from Underwood, 1977; Curzon and Crocker, 1978; Retief et al, 1971.

can therefore arise as a result of Mn deficiency. In animals as diverse as rats, pigs, poultry, sheep, and cattle Mn deficiency has been known to occur with the use of certain types of diet (Underwood, 1977). Weight loss, transient dermatitis, nausea, slow growth of hair, and hypocholesterolemia resulted from the failure to add Mn to a purified diet administered to a human volunteer undergoing a study of vitamin K deficiency in a metabolic unit (Doisy, 1973). Toxic reactions to excess Mn have been noted in exposed miners (Borg and Cotzias, 1958; Cotzias, 1958).

Metabolism and Nutrition

The Mn content of foods is highly variable, related to their origin and to food handling and processing. This is particularly noted in cereal grains where whole wheat originally contained 31 ppm while after processing the germ contained 160 ppm and the resulting white flour 5 ppm (Schaible et al, 1938). In descending order of Mn content, concentrations ranged from 23 to 0.2 ppm for nuts, whole cereals, dried fruits, roots, stalks, fruits, nonleafy vegetables, animal tissues, poultry, fish, and seafoods (Underwood, 1977).

Daily intake is variable since it depends on dietary and cultural practices. In tea-drinking countries a considerable proportion of daily Mn intake may be derived from this beverage; dry tea contains 350 to 900 ppm (Wenlock et al, 1979). The presence of Mn may also be related to the availability of fluoride in tea (Spiers et al, 1981). The total daily intake of Mn is highly variable depending on the composition of the diet; an estimate of average daily intake for British adults is 4.6 mg Mn/day (within the range found in numerous other studies) (Wenlock et al, 1979). The minimum daily requirement for Mn is not known, but intakes as low as 0.71 mg/day lead to a negative Mn balance (De, 1949).

After absorption through the small intestine (only a small percentage is absorbed there), Mn is used throughout the body in enzyme systems. Excretion is almost totally via the intestinal wall, an efficient homeostatic mechanism which regulates the level of Mn in the body. The absorption of Mn can be affected by excess Ca; also Fe and Co compete with Mn for binding sites in the absorption process. Addition of Fe to the diet of rats has been shown to inhibit Mn absorption, and conversely a high Mn intake reduces absorption of Fe (Underwood, 1977). Microorganisms possess highly specific Mn transport systems that enable organisms to concentrate Mn even in the presence of much higher levels of Ca and Mg (Silver and Jasper, 1979).

Skeletal abnormalities due to Mn deficiency have been described in a number of species, although not in man. They are likely to arise primarily from the effect of its deficiency on the synthesis of mucopolysaccharides rather than from impairment of calcification. Chondroitin sulfate is the mucopolysaccharide most severely affected, and this could interfere with structural components of cartilage and result in skeletal defects. No evidence of any relationship to tooth formation is known.

CARIES

Man

The effects of Mn on dental caries in man are not well understood. A small amount of information indicates that increased intake of Mn is associated with higher caries levels, but epidemiologic evidence for this comes mainly from the Colombia (South America) study reported by Glass et al (1973). In this report there were significantly higher concentrations of Mn, as well as Cu and Fe, in the drinking water of people living in Don Matias, where the incidence of caries is high. This finding was significant regardless of the statistical test used. In these findings from South America the association of Mn with high caries confirmed the earlier statistical analysis of Adkins and Losee (1970) using published epidemiologic and water analysis data. The findings of Glass et al (1973) are also in some agreement with those of Ludwig (1963) in a dental caries study in New Zealand. In these studies there was always a positive correlation of Mn in the drinking water to presence of caries.

The evidence for a significant positive relationship between Mn in saliva and caries is confused. Green (1970) could find no statistical relationship between Mn in saliva and caries; only Cu could be related to such an increase. Analysis of plaque and saliva samples collected from 75 children in each of the low and high caries areas in the South American study (Glass and Rothman, 1975) revealed that high levels of Mn in both saliva and plaque were positively related to an increase in caries.

Two other studies involving Mn in human caries in clinical trials should be mentioned. Gerdin and Torell (1969) used $MnCl_2$, at 100 ppm, with NaF and KF in mouthwashes, and later Koch (1972) used $KMnF_3$ in a toothpaste over three years. In the mouthwash study the 0.2% KF solution with Mn ions reduced the increment of caries in children more than did a 0.2% of NaF solution although the difference was not statistically significant. In the third year of the toothpaste trial, a significantly lower increment of caries was found for the $KMnF_3$ compared with the F. In both these studies the reduction in caries was not great and only statistically significant in one year of one trial. It is also not clear what effect Mn had as a metal ion or whether there was any significant uptake of Mn by the enamel.

Of all the studies on trace elements in human enamel only two have shown that increased concentrations of Mn in the enamel are significantly related to increased caries. In a South American study Glass and Rothman (1975) showed that the Mn concentration in the enamel of a high-caries group was greater than that of a low-caries group. Curzon and Losee (1978) showed that increased concentrations of Mn in enamel from the western United States were significantly related to increased caries. Further studies are therefore needed to consider the relationship to enamel concentrations (in both whole and surface enamel) and dental decay of Mn in food and drinking water.

Animals

A report by Berthold (1959) showed that subcutaneous injection of potassium permanganate at 0.10 to 0.40 mg per animal, per week, in Syrian hamsters reduced caries incidence by 15%. Berthold speculated that the procedure increased the resistance of the hard tooth structure to caries.

The acetate of Mn was shown to inhibit caries in Wistar rats but, in contrast with the human study (Glass and Rothman, 1975), there was no correlation between caries activity and the Mn level in enamel (Hendershot et al, 1960). Similarly, a subcutaneous injection of a potassium manganese solution resulted in a significantly lower rate of caries, again in Syrian hamsters (Munch and Winiker, 1959).

In a recent study (Beighton, 1982) germ-free WAGG rats were monoinfected with *S. mutans* (strain Ingbritt), fed a caries-promoting diet, and given drinking water supplemented with 0, 0.1, and 0.5 mM Mn (as $MnSo_4.6H_2O$) for 21 days. Both the Mn-treated groups had mean caries scores significantly greater than did the control group.

MECHANISMS OF ACTION

The relationship of Mn to dental caries has been so little studied that any mechanism of action remains purely speculative. Increased caries levels have been associated (but not consistently) with increased concentrations of Mn in saliva, plaque, and enamel. The investigations outlined above indicate that Mn may influence the incidence of caries either by its presence in dental plaque or by its incorporation in enamel. The concentration of Mn in saliva may only serve as an indicator of the total dietary Mn intake. The presence of Mn in increased concentrations could exert changes in the dental plaque, since numerous studies have shown it to be essential for bacterial metabolism. However, the incorporation of Mn in enamel and its possible effects on enamel solubility have hardly been studied. Limited evidence available suggests that if Mn does increase caries activity it is likely to do so as a result of its effects on bacterial metabolism.

Effects on Oral Bacteria

Stephan (1949) found that Mn inhibited the growth of oral bacteria, and Hein (1955) reported that Mn, at concentrations of up to 0.73 mmol, was without effect on the growth of oral microorganisms. Stamar et al (1964) found that Mn was the only cation tested that stimulated the growth of lactobacilli, but Hendershot et al (1960) noted that the growth of lactobacilli and streptococci, isolated from rats, was inhibited by Mn. These studies suggest that the bacteria might respond differently to different Mn concentrations. Gallagher and Cutress (1977) reported that 0.91 mmol Mn did not inhibit growth or acid production by oral bacteria but that 9.1 mmol Mn was inhibitory for all bacteria examined. Beighton and McDougall (1981) found that Mn (up to 0.91 mmol) had no effect on the growth of *S. mutans* FA-1, *Actinomyces viscosus* WVU 626, or *Neisseria flava* NCTC 4590, and on *Actinobacillus sp.* isolated from rats, nor did 0.5 mmol Mn influence the growth of seven *S. mutans* serotype *c* strains (Beighton, 1980). The administration of drinking water supplemented with 0.91 mmol Mn to rats did not significantly alter the percentage bacterial composition of the dental plaque (Beighton and McDougall, 1981).

These previous studies were largely concerned with the growth of bacteria. However, bacteria are unlikely to be cariogenic unless they have the ability to utilize carbohydrates and to produce acid. Glucose utilization was stimulated by Mn and in strains of *S. mutans, S. mitior, S. milleri,* and *S. sanguis* (Beighton, 1980). Acid and iodophilic polysaccharide (IPS) production from glucose by seven *S. mutans* serotype *c* strains was also stimulated by the addition of Mn but the growth of these organisms was unaffected. Beighton (1981, 1982) recently showed that Mn also stimulated acid and IPS production from sucrose and fructose by *S. mutans* strain Ingbritt. It is of interest that 0.05 mmol Mn antagonized the inhibitor effect of 0.53 mmol NaF on the glucose metabolism of all the or al streptococcal species examined.

These results indicate that Mn may owe its caries-promoting effect to its ability to stimulate carbohydrate metabolism by oral streptococci. This might explain the positive association found between increased levels of Mn in dental plaque and dental caries in the South American study of Glass and Rothman (1975).

Effects of Enamel

From analysis of human enamel samples there is no consistent evidence of any correlation between levels of Mn in enamel and solubility (susceptibility or resistance) or caries prevalence or incidence. In the studies of Curzon and Losee (1978) Mn was higher in enamel samples of teeth from the high caries areas of western United States, and Glass and Rothman (1975) found a similar association in their South American study. Berthold (1959) speculated that Mn might change the resistance to caries of the hard tooth structure, and there is evidence to show an increased incorporation of Mn in enamel of animals fed Mn-supplemented diets (Hendershot et al, 1960).

If there are changes in the physical chemistry of enamel caused by Mn incorporation, this should be seen in tests of the resistance of Mn-apatites to acid dissolution. Munch and Winiker (1959) tested areas of human enamel by submersion in potassium-manganese solution for 24 hours before exposure to organic acids (malic, tartaric, and citric acids) for various time periods. They found that in the majority of instances Mn rendered enamel less susceptible to the effects of acid. Early studies by Manley and Bibby (1949), using various metal ions with fluoride, showed a reduction of enamel solubility after treatment with MnF. Overall, therefore, the evidence for an effect of Mn on enamel is meager, and the work that has been done has not been designed to allow for effects of the anion.

SUMMARY

The epidemiologic evidence that Mn influences the incidence of dental caries is sparse; however, available evidence suggests that caries-promoting activity may be mediated by stimulating the carbohydrate metabolism of plaque streptococci. There is a need to clarify the relationships between levels of Mn in plaque and enamel and dental caries and to determine the effect of Mn incorporation in enamel on its dissolution properties.

REFERENCES

Adkins BL, Losee FL: A study of the co-variation of dental caries prevalence and multiple trace element content of water supplies. *NY State Dent J* 1970;36: 618–622.

Beighton D: Manganese antagonizes the inhibitory effect of fluoride on the glucose metabolism of *Streptococcus mutans. Microbio* 1980;28:149–156.

Beighton D: The influence of manganese on carbohydrate metabolism and caries induction by *Streptococcus mutans* strain Ingbritt. *Caries Res* 1982;16: 189–192.

Beighton D, McDougall, WA: The influence of certain added water-borne trace elements on the percentage bacterial composition of fissure plaque from conventional Sprague-Dawley rats. *Arch Oral Biol* 1981;26:419–425.

Berthold H: Caries-reducing effects of manganese. *Zahnartzl Praxis* 1959;10:238–241.

Borg DC, Cotzias GC: Incorporation of manganese into erythrocytes as evidence for a manganese porphyrin in man. *Nature* 1958;182:1677–1678.

Cotzias, GC: Manganese in health and disease. *Physiol Rev* 1958;38:503–532.

Curzon MEJ, Losee FL: Trace element composition of whole human enamel and dental caries. Part II. Western USA. *JADA* 1978;96:819–822.

Curzon MEJ, Crocker DC: Relationships of trace elements in human tooth enamel to dental caries. *Arch Oral Biol* 1978;23:647–653.

De HN: Copper and manganese metabolism with typical Indian dietaries and assessment of their requirement for the Indian adult. *Indian J Med Res* 1949; 37:301–309.

Doisy EA: In proceedings of the University of Missouri's 6th annual conference on Trace Substances in Environmental Health. Columbia, MO, University of Missouri Press, 1973, pp 193–196.

Gallagher IHC, Cutress TW: The effects of trace elements on the growth and fermentation by oral streptococci and actinomyces. *Arch Oral Biol* 1977;22:555–562.

Gerdin PO, Torell P: Mouth rinses with potassium fluoride solutions containing manganese. *Caries Res* 1969;3:99–107.

Glass RL, Rothman KJ: Dental caries prevalence and levels of manganese. *J Dent Res* 1975;54A:L84.

Glass RL, Rothman KJ, Espinal F, et al: The prevalence of human dental caries and water-borne trace elements. *Arch Oral Biol* 1973;18:1099–1104.

Green I: Copper and manganese in saliva of children. *J Dent Res* 1970;49:776–782.

Hein JW: Effects of various agents on experimental caries. A resume, in Sognnes RF (ed): *Advances in Experimental Caries Research.* Washington, DC, American Association for the Advancement of Science, pp 197–222.

Hendershot LC, Mansell RE, Forsaith J: The effect of Zn, Ni, and Mn on rat dental caries and dental enamel metal levels, in *Metal Binding in Medicine.* New York, Lippincott, 1960.

Koch G: Comparison and estimation of the effect on caries of daily supervised brushing with a dentifrice containing sodium fluoride and a dentifrice containing potassium fluoride and manganese fluoride. *Odont Revy* 1972; 23:341–354.

Ludwig TG: Recent marine soils and resistance to dental decay. *Aust Dent J* 1963;8:109–112.

Manley RS, Bibby BG: Substances capable of decreasing the acid solubility of tooth enamel. *J Dent Res* 1949;28:160–166.

Munch J, Winiker M: Effects of manganese on caries susceptibility. *Zahnartzl Prax* 1959;10:173–174.

Retief DH, Cleaton-Jones PE: The quantitative analysis of Sr, Au, Br, Mn and Na in normal human enamel and dentine by neutron activation and high resolution gamma spectrometry. *J Dent Assoc S Africa* 1971;26:63–69.

Schaible PJ, Bandemer SL, Davison JA: *Manganese Content of Grains and Agricultural Products.* Mich. Agric. Exp. Stn. Tech. Bull. #159. 1938.

Silver S, Jasper P: Manganese transport in microorganisms, in Weinberg ED (ed): *Microorganisms and Minerals.* New York, Marcel Dekker Inc, 1979, pp 105–149.

Spiers RL, Murphy M, Das AK: Some fluoride correlates in tea infusions. *J Dent Res* 1981;60B:1158.

Stamar JR, Albury MN, Pederson CS: Substitution of manganese for tomato juice in the cultivation of lactic acid bacteria. *Appl Microbiol* 1964;12:165–168.

Stephan RM: In vitro studies of the effects of some chemical substances on the growth of oral microorganisms and their ability to dissolve tooth substances. *J Dent Res* 1949;28:652–653.

Underwood EJ: *Trace Elements in Human and Animal Nutrition,* ed 4. New York, Academic Press, 1977.

Wenlock RW, Buss DH, Dixon EJ: Trace nutrients. 2. Manganese in British food. *Br J Nutr* 1979;41:253–261.

16 **S**

32.06

2-8-6

12 Sulfur

B. E. Johnson
T. R. Shearer

Sulfur (S), atomic number 16, is the ninth most common element in the human body, comprising 0.05% of total atoms and 0.25% of total body weight (Anderson, 1977; Lehninger, 1977). In group IV of the periodic table S is found with tellurium and selenium. The concentrations of S in various entities and human tissues are listed in Table 12-1. The highest concentrations of S in human tissues are found in hair and nails, the lowest in the teeth. In enamel and dentin S is present primarily in the amino acids cystine and methionine (Table 12-2).

The trace element S is most often found in the amino acids methionine, cysteine, and cystine, and also in peptides and proteins such as glutathione, insulin, keratin, and numerous enzymes. It also serves many important biochemical functions in nonproteinaceous forms such as chondroitin sulfate, sulfatides, taurocholic acid, thiamine, and sulfonamides (Figure 12-1). Inorganic forms of S of biologic interest include sulfate ($SO_4^=$), sulfite ($SO_3^=$), thiosulfate ($S_2O_3^=$), and hydrogen sulfide (H_2S).

Table 12-1
Sulfur Concentrations

Location	Sulfur (ppm)
Earth's crust	5200
Salt water	885
Fresh water	4
Hair	38,000
Heart	9500
Liver	8400
Muscle	6800
Brain	6700
Kidney	6600
Skin	3200
Bone	2600
Blood	2040
Plasma	1220
Enamel	281
Human body (average)	5400

Data from Bowen, 1966; Schroeder, 1973; Losee et al, 1974.

Table 12-2
Amino Acid Composition of Human Enamel Protein

Fraction	Total[a]	Unerupted Third Molar[a]
Aspartic acid	31	96
Threonine	38	51
Serine	63	72
Glutamic acid	142	130
Proline	251	64
Glycine	65	108
Alanine	20	79
Half-cystine	<4	3
Valine	40	60
Methionine	42	19
Isoleucine	33	43
Leucine	91	96
Tyrosine	53	17
Phenylalanine	23	37
Lysine	18	53
Histidine	65	25
Arginine	23	49
Tryptophan	—	—

[a]Residues per 1000 total residues.
From Eastoe, 1979.

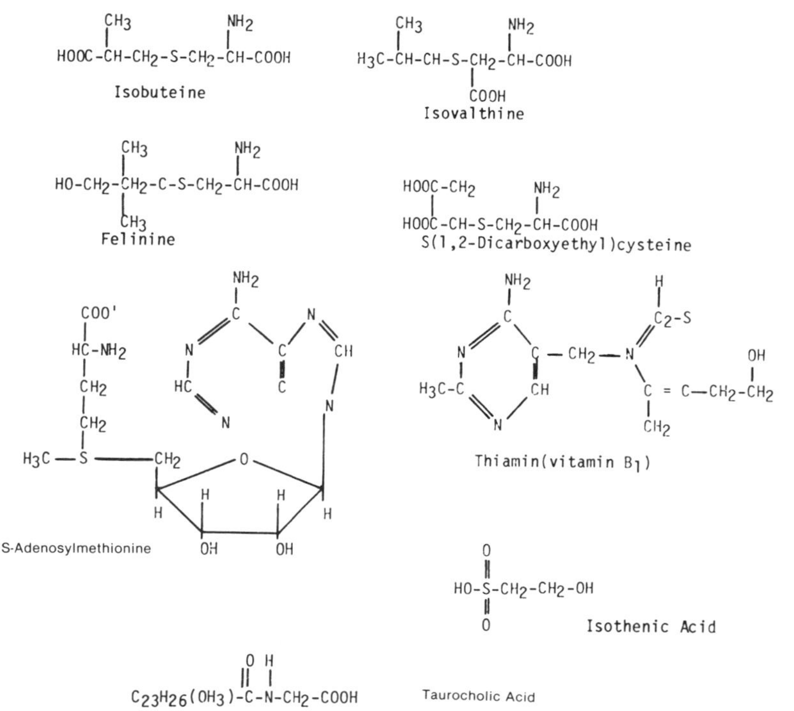

HOOC-CH(CH_3)-CH_2-S-CH_2-CH(NH_2)-COOH
Isobuteine

H_3C-CH(CH_3)-CH(COOH)-S-CH_2-CH(NH_2)-COOH
Isovalthine

HO-CH_2-CH_2-C(CH_3)(CH_3)-S-CH_2-CH(NH_2)-COOH
Felinine

HOOC-CH_2-CH(COOH)-S-CH_2-CH(NH_2)-COOH
S(1,2-Dicarboxyethyl)cysteine

S-Adenosylmethionine

Thiamin(vitamin B_1)

HO-S($=O$)($=O$)-CH_2-CH_2-OH
Isothenic Acid

$C_{23}H_{26}(OH_3)$-C(=O)-N(H)-CH_2-COOH
Taurocholic Acid

Figure 12-1 Some non-proteinaceous sulfur compounds.

METABOLISM AND NUTRITION

The importance of S in the nutrition and metabolism of mammals cannot be overstated. The S atom contributes to numerous electrochemical, catalytic, and structural functions in mammalian physiology. Because of these numerous roles, S has great potential for exerting important influences on the health of the oral cavity.

Metabolism and Biologic Roles of Sulfur

Mammals are able to utilize the dietary S ingested as the S-amino acids, methionine and cystine, in protein foods. The incorporation of S into amino acids occurs in the microbial and vegetative synthesis of S-amino acids. Methionine can be converted to cysteine and cystine in mammals (Figure 12-2). Therefore, mammalians require a regular dietary intake of at least methionine to fulfill their S requirements.

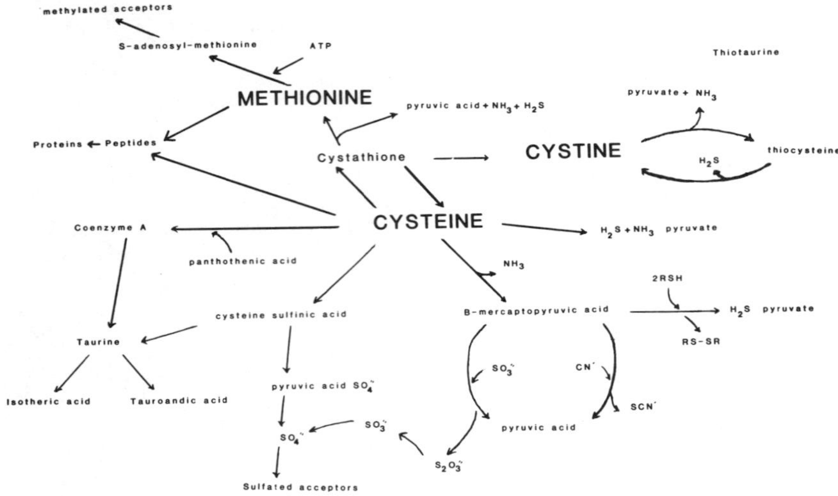

Figure 12-2 Major metabolic pathways of sulfur.

Figure 12-2 shows some of the major catabolic fates of methionine and cysteine in mammals (Bowen, 1966). Three of the major routes involve the enzymatic conversion to H_2S, NH_3, and pyruvate; they are as follows: (1) direct enzymatic conversion from cysteine, (2) cleavage of the intermediate cystathionine, and (3) cleavage of cystine-yielding 5-thiocystine, pyruvate, NH_3, and eventually formation of H_2S. Another major pathway of cysteine catabolism includes the initial oxidation of the organically bound S, forming cysteine sulfinic acid and eventually yielding pyruvic acid and sulfite. Initial deamination of cysteine yields β-mercaptopyruvic acid in which the organically bound S can be reduced or oxidized. Hydrogen sulfide is expelled through the respiratory system. Likewise, most of the inorganic sulfate ion, SO_4, is subsequently secreted in the urine.

Also shown in Figure 12-2 are the major anabolic pathways of S. The sulfate ion can be utilized for the synthesis of 3′ phosphoadenosine 5′ phosphosulfate which is used in diverse esterification reactions such as the formation of the sulfatides or the excretion derivatives of unwanted metabolites. The sulfur amino acids are, of course, utilized in the synthesis of peptides and proteins. In these macromolecules the amino sulfhydryls perform countless structural and catalytic duties. Cysteine is also an important precursor in the synthesis of coenzyme A, an important cofactor in fatty acid metabolism. Cysteine sulfinic acid is utilized for the synthesis of taurine which participates in the formation of taurocholic acid, a bile acid, and isothenic acid, a nervous tissue metabolite of unknown significance.

In the sulfhydryl form S is responsible for many important structural and catalytic effects. For example, cystine residues in the α-keratins are a good demonstration of interchain cross-bridging. The α-keratins are responsible for the structural qualities of hair, hooves, and quills. Disulfide bridging is also responsible for maintaining the permanent physical conformation of many regulatory and enzymatic proteins such as in insulin, chymotrypsin, and the immunoglobulins. Many enzymes require reduced glutathione, a small sulfhydryl peptide, or cysteine as essential cofactors. Protein sulfhydryl groups may also bind heavy metal ions such as Ca^{2+} or Mg^{2+}, affecting a protein's intended activity, eg, myosin. Some sulfhydryls simply mediate the attachment or positioning of coenzymes and substrates to an enzyme.

Sulfhydryls also play an active role in biochemical reactions. They may act as hydrogen donors in reduction reactions and are also thought to participate in electron transfer in oxidative enzymes.

Organic S also provides a source of high energy bonding, and coenzyme A and its high energy acetylthioester derivatives are essential to the oxidation of fatty acids. S-adenosyl-methionine serves as a methyl donor in many biosynthetic pathways. The free energy required for its many methylation reactions is stored in its thiomethyl bond.

Thus sulfur is involved in a multitude of biochemical functions in the overall metabolism of mammalian cells. It remains to be determined which of these functions are significant to the integrity of the oral cavity, and if sulfur influences the ultimate caries susceptibility of enamel.

Excretion

Most S is excreted from the body through the urine, and the average daily output is reported to be between 0.7 and 1.4 g, although in some cases it may reach 2.0 g daily. Approximately 75% of the expended S is in the form of inorganic sulfate ions. Another 5% is excreted in the ethereal form, and the remaining 20% as organic sulfhydryls or thiomethyl derivatives (Anderson, 1977).

Smaller amounts of other S compounds can be found by urine analysis, although their metabolic origin is unknown. These include isobuteine, isovalthine, felinine, and S-(1,2 dicarboxyethyl cysteine).

Sulfur Deficiency and Metabolic Disorders

There are several diseases caused by genetic disorders in the metabolism of sulfur. Cystathionuria causes increased urinary excretion of cystathionine, convulsed states, and mental retardation. Homocystinuria is characterized by increased urinary excretion of homocystine, convulsive states, and mental retardation. The hereditary absence of microsomal sulfite oxidase results in detectable SO_2 expiration, mental retardation, and liver disease in sulfite oxidase deficiency. In metachromatic leukodystrophy there is an unusual accumulation of sulfatides in the white matter of the brain, and to a lesser degree in the gray matter. Clinical examination of the oral tissue from such patients would provide important information on the role and metabolism of S compounds in the oral cavity and their relation to oral health.

Food Sources and Dietary Requirements

The trace element S is ingested primarily as the amino acids methionine, cystine, and cysteine in food containing protein. Dietary inorganic S, such as sulfates and sulfites, is absorbed and transported as sulfate esters. Most absorbed sulfates and sulfites are believed to be excreted almost immediately. Smaller amounts of S are also ingested as sulfolipids, sulfatides, and glycoproteins such as mucin, ovomucoid, and chondroitin sulfate.

Adults require approximately 10 mg of S-containing amino acids per kilogram body weight per day. A diet containing adequate protein will provide an adequate amount of S depending on the quality of the protein (Schroeder, 1973). Meats, milk, and nuts are good sources of S amino acids, while legumes are frequently deficient in methionine and should be complemented with other protein sources.

CARIES

Man

No direct evidence exists that shows a correlation between dietary amino acids and dental caries in humans (Nizel, 1970), although protein deficiencies in developing children are believed to cause several dental defects. Protein deficiency in Nigerian children is thought to have caused delayed eruption and hypoplasia (Enwonwu, 1969). Crowded and rotated teeth have also been reported in children suffering from protein malnutrition (Trowell et al, 1954). Although these studies underscore the importance of protein in proper oral development, to date no studies have explicitly explored specific deficiencies of sulfur amino acids with respect to dental development or caries in man.

One study attempted to correlate the incidence of caries with the S concentration of teeth (Curzon and Losee, 1978). No significant differences in S concentration between geographic regions of high caries incidence and low caries incidence were found (Table 12-3). However, the same study reported a positive relationship between caries occurrence and Cu content. These results are consistent with the finding in two other reports that showed a positive relation between Cu content and human dental caries (Barmes, 1969; Ludwig et al, 1970). The significance of these findings is that sulfate has been reported to affect the status of Cu in numerous mammalian animals (Underwood, 1977). It should be noted by the potential investigator that Mo is also involved in the apparent Cu-sulfate balance of mammals.

Table 12-3
Human Caries Incidence in Relation to Copper and Sulfur

Caries Incidence	Caries Score DMFT	Enamel Copper (ppm)	Enamel Sulfur
High	> 7	0.71 ± 0.20	not different between groups
Low	< 4	0.17 ± 0.04	

Data from Curzon and Losee, 1978.

Laboratory Animals

Caries studies with laboratory animals receiving inorganic S compounds have produced some significant results (Jordan et al, 1960; Jordan et al, 1962). The study shown in Table 12-4 compared several different inorganic S compounds and caries occurrence in rats. The mean caries scores for those rats fed sulfate diets do not vary significantly from those of the control group. However, the caries scores for those rats fed sulfite compounds were significantly lower than those of the control rats. This investigation indicates that, whereas sulfates seem to have no activity against experimental caries, sulfites appear to be significantly cariostatic.

Originally the mode of action of sulfites against caries was believed to be a growth inhibition of oral microorganisms produced by an apparent interference in glucose metabolism. Glucose metabolism was suppressed by metabisulfite in a *Streptococcus* from the oral cavity of the rat (Jordan et al, 1960). Believing that this suppression might be due to the carbonyl-binding abilities of metabisulfite, investigators studied other carbonyl-binding agents (Jordan et al, 1961b). The findings revealed that the cariostatic effectiveness of these agents generally paralleled their ability to inhibit oral microflora growth, but was not necessarily related to their inhibitory ability on glucose metabolism. Another hypothesis was that the caries-inhibitory nature of sulfites might be due to the antioxidant

Table 12-4
Inorganic Sulfur Compounds in Relation to Caries Incidence

Sulfur Compound	Concentration (g/kg diet)	Daily Intake (g)	Caries Score
Control	0	11.4	10.4 ± 1.9
Na Sulfite	5.0	11.0	1.3 ± .33
Na Metabisulfite	3.8	11.1	1.8 ± .50
Na Hydrosulfite	3.5	10.4	2.2 ± .49
Na Bisulfite	4.2	10.9	2.9 ± 1.04
Na Bisulfate	5.5	10.8	8.2 ± .96
Na Sulfate	5.7	10.6	10.0 ± 2.3

Data from Jordan et al, 1962.

properties of metabisulfite. Further studies with other known antioxidants revealed no significant correlation between antioxidant activity and caries inhibition (Jordan et al, 1961a). Though the exact mode of action has not yet been discovered, the overall effect of sulfite treatment in the reduction of caries is due to the lower production of acid from several oral microorganisms.

SUMMARY

Sulfur is an essential element for the proper function and maintenance of animal tissues. Its biologic activity is expressed primarily through the S amino acids and their conjugate peptides and proteins, though S is also actively involved in the biologic activity of other nonproteinaceous, organic compounds. Through these compounds, the S atom contributes to numerous electrochemical, catalytic, and structural functions. Mammals require a regular dietary intake of at least one S amino acid to fulfill their S requirement. Meats, milk, and nuts are good sources of the 10 mg/kg body weight S amino acids that adults require per day. Sulfur is expelled from the body mainly through the urinary tract as $SO_4^=$ or sulfated excretory metabolites, but also as sulfhydryls and thiomethyl derivatives.

There are several genetic disorders in S metabolism which generally cause metabolic imbalance and mental disorder. The implications of these metabolic sulfur imbalances on the integrity of the oral cavity are thus far unexplored. To date no evidence has been found to link human caries directly with any S compounds. In experimental animals, while sulfites were cariostatic, sulfates were found to have no effect on experimental caries. The overall effect of sulfites was to lower acid production in the oral microflora.

REFERENCES

Anderson CE: Minerals, in: *Nutritional Support of Medical Practice.* Hagerstown, Harper & Row, 1977.

Barmes DE: Caries etiology in Sepik villages—Trace element, micronutrient and macronutrient content of soil and food. *Caries Res* 1969;3:44.

Bowen HJM: *Trace Elements in Biochemistry.* London, Academic Press, 1966.

Curzon MEJ, Losee FL: Dental caries and trace element composition of whole enamel: Western United States. *JADA* 1978;96:879.

Eastoe JE: Enamel protein chemistry—Past, present, and future, in Nylen MU, Termine JD (eds): *Tooth Enamel. III. Its Development, Structure, and Composition.* (Publication No. 58B, p 753). Bethesda, National Institutes of Health, 1979.

Enwonwu CO: Prevalence of enamel hypoplasia in well-fed and malnourished Nigerians. General Meeting of International Association for Dental Research: 51, 1969, abstract.

Jordan HV, Bowler AE, Berger ND: Testing of antioxidants against experimental caries in rats. *J Dent Res* 1961a;40:878.

Jordan HV, Fitzgerald RJ, Berger ND: Carbonyl-binding compounds as inhibitors of experimental caries in rats. *J Dent Res* 1961b;40:199.

Jordan HV, Fitzgerald RJ, Berger ND: A comparison of the caries-inhibitory potential of sodium metabisulfite and related compounds. *J Dent Res* 1962;41:61.

Jordan HV, Fitzgerald RJ, Bowler AE: Inhibition of experimental caries by sodium metabisulfite and its effect on the growth and metabolism of selected bacteria. *J Dent Res* 1960;39:116.

Lehninger AL: *Biochemistry,* ed 2. New York, Worth Publishers, 1977.

Losee FL, Cutress TW, Brown R: Natural elements of the periodic table in human dental enamel. *Caries Res* 1974;8:123.

Ludwig TG, Adkins BL, Losee FL: Relationship of concentrations of eleven elements in public water supplies to caries prevalence in American schoolchildren. *Aust Dent J* 1970;15:126.

Nizel AE: Amino acids, proteins, and dental caries, in Harris RS (ed): *Dietary Chemicals vs Dental Caries* (Advances in Chemistry Series, No. 94). Washington, American Chemical Society, 1970.

Schroeder HA: *The Trace Elements and Man.* Old Greenwich, Devin-Adair Company, 1973.

Trowell HC, Davies JNP, Deard RFA: *Kwashiokor.* London, E. Arnold and Co, 1954.

Underwood EJ: *Trace Elements in Human and Animal Nutrition,* ed 4. London, Academic Press, 1977.

29 Cu
63.54
8-18-1

13 Copper

M. E. J. Curzon

A malleable ductile metal, copper (Cu) is found in its elemental form and also as a component of many minerals. It has been mined and widely used from earliest times. As an essential trace element its presence in animal tissues was demonstrated more than 150 years ago and its biologic role was proven 100 years ago. A measure of the biologic interest in Cu is that Underwood (1977) devotes nearly 50 pages of his authoritative textbook to the element. Although Cu has been studied for its toxicity and deficiency in man and animals, its role in dentistry has been sadly neglected. This is surprising because of its ubiquitous role in nature and its influence on so many biologic systems.

Metabolism and Nutrition

The human body usually contains about 80 mg of Cu although the distribution in various tissues varies according to age, Cu status, and dietary intake (Underwood, 1977). There appears to be a variation in tissue Cu concentration, such as in kidney, liver, lung, and spleen (Schroeder et al, 1966). The concentrations to be found in human tissues vary between 14.7 ppm in brain to 0.7 ppm in muscle. Dietary intakes will markedly affect the levels of Cu to be found in organs such as the liver. Levels are very high in the young child and drop with age (Underwood, 1977). Not only does the dietary intake of Cu vary because of different concentrations of Cu in food but also because of interactions of the element with Mo, Zn, Cd, Fe, and Ca. A marked Zn-Cu antagonism is known, as well as that of Mo-Cu (Underwood, 1977).

Several other trace elements such as Zn and Cd depress Cu absorption in the gut and so will reduce plasma Cu levels. Under different circumstances Mo can also reduce Cu levels, and if the Mo levels become excessive (see Chapter 6), striking changes in Cu blood concentrations occur. These interactions of trace elements with Cu as well as the complex metabolism of Cu are thoroughly reviewed by Underwood (1977).

CARIES

Man

The role of Cu in dental disease is not clear. Navia (1970) listed Cu as a mildly cariostatic trace element. This review was based on only a few reports, most of which concerned animal studies. In the last ten years, however, little has changed about knowledge of the influence of Cu on dental disease. One of the problems in research on the influence of Cu in man is the difficulty of separating the effects of Cu from those of other elements, such as Zn, Pb, Cd or Fe, which often occur together in water or food. Most reports associating Cu with caries have concerned soils or water supplies where other heavy metals have also been present.

A good example of the inability to separate Cu from other metals is the report of Anderson et al (1976). Caries prevalence was found to be increased in association with heavy metals in soils of the Bere Peninsula in the west of England, an area where Pb and Cu mining had been prominent in the past. With mapping techniques, DMFT scores were related to the geographic distribution of high metal mineralization, and high caries was significantly related to soil concentrations of Pb and Cu. The soils in question were described as being particularly rich in Cu and As as well as in Sn, Ag, Pb, W, and Mn. The chief conclusion from this study was that the high caries association was with Pb, but from data provided an equally strong case could have been made for Cu. Because analytical results for human samples, such as saliva, enamel, plaque, or urine were not reported in this study, definite conclusions as to which metal(s) could be involved were not made. Anderson et al (1976) commented that the lack of analysis left important questions unanswered, but to date no further information has been forthcoming.

In a statistical analysis of caries and trace elements in drinking water supplies, Adkins and Losee (1970) reported a number of significant relationships (Table 13-1). Using caries prevalence data for states in the United States (Dunning, 1953) and water analysis data (Dufor and Becker, 1962) various statistical analyses were completed. Negative associations were found for Ba, B, Li, Mo, and Sr, but it was concluded that the total concentration of these five elements was the important point. Of greater interest was the positive correlation of Cu and Mn to high caries. In particular it was found that the concentration of these two elements was, to a large extent, independent of the levels of other elements. Because of this independent pattern of occurrence Adkins and Losee suggested that a direct experimental attack on the problem would be promising, but this has not been done so far.

Table 13-1
Average Concentration of Elements and Statistical Significance of the Difference Between High and Low Caries States in the United States

	Mean Concentration (μg/1)[a]			
Trace Element	*Low Caries States*	*High Caries States*	*t*	*p*
Ba	102.1	25.0	4.3	< 0.01
B	113.8	16.2	6.3	< 0.01
Li	18.8	0.4	2.9	< 0.05
Mo	3.8	0.4	4.0	< 0.01
Sr	409.0	43.0	4.4	< 0.01
Cu	6.3	18.2	2.2	< 0.05
Mn	7.7	37.1	2.8	< 0.01

[a]μg/l is parts per billion as given in original report.
After Adkins and Losee, 1970.

A similar approach to that of Adkins and Losee was used by Ludwig et al (1970). Whereas in the former study preexisting data were used, in the second the authors conducted caries examinations and water analyses themselves in 19 cities in the United States (Table 13-2). The results showed product-movement correlations were highly significant for Cu (r = 0.80) and Pb (r = 0.72). As the correlation signs were positive, this indicated high Cu associated with high caries. These two studies together provide some evidence that high Cu intake in man may increase caries.

A small study in upstate New York reported high caries prevalence and delayed eruption of permanent teeth related to high concentrations

Table 13-2
Relationship of Caries Prevalence to Copper and Fluoride Concentrations in Water Supplies for 19 United States Towns

Town	Mean DMFT[a]	Mean Concentration ppm	
		F	*Cu*
Franklin (NH)	11.4	0.10	2.200
Milford (NH)	8.9	0.10	0.570
St. Johnsbury (VT)	8.6	0.10	0.700
Montpelier (VT)	8.2	0.15	0.075
Canastota (NY)	8.7	0.15	1.500
Geneva (NY)	8.1	0.15	0.030
Littleton (NH)	7.7	0.15	0.140
Kissimmee (FL)	7.7	0.15	0.004
Lima (NY)	7.3	0.10	0.060
Avon (NY)	7.5	0.10	0.015
Eufaula (AL)	7.3	0.20	0.013
Ashboro (NC)	6.6	0.10	0.093
Caledonia (NY)	7.1	0.013	0.013
Americus (GA)	6.3	0.10	0.002
Dillon (SC)	5.9	0.10	0.017
Bradenton (FL)	5.9	0.15	0.004
Lake City (SC)	5.8	0.10	0.002
Manning (SC)	5.4	0.10	0.002
Douglas (GA)	5.7	0.30	0.002

[a]For 14-year-old school children.
After Ludwig et al, 1970.

of heavy metals in the soil (Curzon and Bibby, 1970). As noted above, however, the occurrence together of Pb, Cu, and Zn prevented any conclusion as to which element was involved. Barmes (1969) has also reported Cu associated with caries. In the first Papua-New Guinea study high caries was associated with increased soil concentrations of Cu, as well as Pb. Recently Schamschula et al (1978) noted a significant correlation of Cu in plaque to caries in New Guinea tribesmen. Green (1970) reported significantly higher Cu concentrations in the saliva of children with a high rate of caries.

In contrast to all the above findings Rothman et al (1972) showed that Cu soil concentrations were significantly higher in the low caries town of Heliconia (Colombia, South America). Similarly, in the same general study, Glass et al (1973) showed the same relationship with Cu in the water supplies. However, many other elements such as Mg, Ca, Cr, Ni, and Sr were also negatively related to caries. The difficulty of separating the effects of individual trace elements was again apparent.

Many studies have analyzed enamel for trace elements (see Chapter 3) but very few have attempted to correlate caries with trace element concentrations. Pre-carious white and brown spots had a variable but high level of Zn and Cu, higher in younger teeth than older teeth (Little and Steadman, 1966). Using teeth from high and low caries areas of western United States Curzon and Losee (1978) found a significant positive association of Cu to caries prevalence. This appears to be the only study showing such an association.

The main evidence for Cu as a cariogenic element in man is therefore based upon analysis of water supplies. This is an interesting association because of the wide use of Cu piping for domestic water supplies. Schroeder et al (1966) noted that the daily increment of Cu ingested from soft water may amount to up to 20% of dietary intake. Soft water, with a low pH, readily corrodes Cu pipes and is a constant problem in certain areas of the United States such as New England. In this area surface waters are the predominant source of drinking water and are low in pH. It may be coincidental but these areas are also those previously reported as having high caries (Dunning, 1953, Ludwig et al, 1970). Evidence available for humans indicates that Cu is associated with an increase in caries. However, it has not been possible to reproduce this finding in animals.

Animals

Of the animal studies reported three have shown no effect of Cu and four have shown a reduction in decay (Table 13-3). There have, however, been many differences between the methods used so that interpretation is difficult. Three different salts have been used — $CuSO_4$, $Cu(NO_3)_2$, and Na-Cu-chlorophyllin. Reductions and no effects have been found in giving Cu either via the drinking water or by injection.

In the experiment reported by Gianpiccolo et al (1968) a reduction in caries was found with a 0.05% solution of $CuSO_4$. However, in this study the rats fed the Cu did not gain as much weight as did their controls. The authors commented on the toxic effect of the Cu but did not draw any conclusion that rats with poor weight gains were not eating as much and therefore would show a lower caries rate because of a lack of challenge. In this instance there is serious doubt as to whether the Cu, of itself, was having an effect on caries. This is a common problem in animal experiments and is discussed very thoroughly in Navia's (1979) book on the use of animals in dental research.

The conclusion that can be drawn from these animal studies is that as yet too little is known about the effects of Cu on animal caries. With existing studies being evenly distributed between Cu as cariostatic or inert more work is needed to resolve these issues. There may be an effect of Cu that varies according to the salt used, as in the case for Mo and Sr, as well as a variation due to the method of administration. Generally

Table 13-3
Summary of Reports on the Effect of Copper Salts on Dental Caries in Rodents

		Cu Used			
Author (year)	Animal	*Salt*	*Concentration*	Vehicle	Effect on Caries
McClure, 1948	Rat	$CuSO_4$	500 ppm	food	No effect
			250 ppm	food and water	
Shaw, 1950	Rat	Na-Cu-Ch[a]	0.2% Solution	water	No effect
Hein and Shafer, 1951	Hamster	Na-Cu-Ch	93.2% (1:500) Solution	water	Reduction
Hein, 1953	Hamster	$CuSO_4$	10, 25, 50 ppm	water	Reduction of initiation of decay
Kruger, 1958 I	Rat	$Cu(NO_2)_2$	0.005, 0.02 mg/day	injection	Caries reduction nearly significant
III	Rat	$Cu(NO_3)_2$	20, 40, 60 μg/day	injection	No effect
Gianpiccolo et al, 1968	Rat	$CuSO_4$	0.05%	water	Reduction but toxic effect

[a]Na-Cu-chlorophyllin

trace element feeding studies are carried out using esophageal intubation followed by the use of drinking water with the trace element. This has not been done for Cu.

Although there are indications that Cu may affect caries through its effects on oral bacteria, in the studies listed above no attention was given to any possible changes in oral microflora, another area where further work is needed.

MECHANISM OF ACTION

Copper in Enamel

Enamel takes up Cu very readily; a concentration as high as 1860 ppm has been reported in a tooth where a carious lesion was treated with Cu cement (De Renzis et al, 1969). In early carious lesions high levels of Cu were reported (Little and Steadman, 1966) indicating that altered enamel is very receptive to the Cu ion. As noted earlier, teeth from high caries areas (areas of soft water) were found to have significantly high Cu concentrations in enamel (Curzon and Losee, 1978). The Cu concentration of developing teeth is reported as higher than that in mature teeth (Coles et al, 1973), no doubt reflecting an association of Cu with the organic fraction of enamel. In Chapter 3 it was shown that Cu concentrations generally are relatively low but show fairly wide variation. This is no doubt due to differences in Cu intake derived from water and food, which are further influenced by geographic location.

Plaque and Bacterial Effects of Copper

Since Cu is essential to the metabolism of most organisms, it is surprising that more attention has not been devoted to the effects of Cu on plaque and oral bacteria. There is no doubt an optimal concentration of Cu is required for the growth of oral bacteria. Correspondingly, as with nearly all essential trace elements, there must also be a toxic concentration.

Hardwick and Martin (1967) reported a semiquantitative Cu concentration of 100 to 1000 ppm in plaque. In the Papua-New Guinea study, Schamschula et al (1978) found Cu concentrations of 9.84 ppm. In the statistical analysis of the trace element data a significant correlation of Cu to caries was noted. It has been shown that Cu ions may inhibit in vivo plaque formation (Skjörland et al, 1978) which may be by inhibition of enzyme systems in carbohydrate metabolism.

Driezen et al (1952) found that 2 ppm Cu was sufficient to reduce acid production in human saliva. In a series of experiments on the in vivo effect of Cu^{++} and Zn^{++} on the acidogenicity of dental plaque Afseth et al (1980) showed a significant reduction in acid production in plaque challenged with glucose solutions containing 0.25 mmol $CuSO_4$. A 0.1 mmol solution of $CuSO_4$ had no significant effect. Similarly Opperman and Rolla (1980) reported 5 mmol solutions of Cu inhibited acidogenicity, but this concentration was much higher than that of Afseth. Gallagher and Cutress (1977) reported a temporary inhibition of acid production with 50 ppm Cu as $CuSO_4$. It seems, therefore, that Cu inhibits acid production at most of the concentrations tested.

If Cu does have an inhibitory effect on oral bacterial carbohydrate metabolism and thus depresses acidogenicity of plaque, at what concentration is this most effective? Equally important is the apparent conflict between the epidemiologic data, albeit limited, indicating Cu as cariogenic and the bacteriologic studies which imply Cu is cariostatic through its effect on acidogenicity. The answer may be that there is a curvilinear relationship whereby low Cu concentrations have one effect and high ones have another. In addition, the incorporation of Cu in enamel may produce one effect and Cu in plaque, another. Obviously our knowledge of the effects of Cu on dental caries is still too limited to draw adequate conclusions, but it appears that Cu affects bacteria and may have a role in controlling cariogenic bacteria such as *S. mutans.*

Copper in Relation to Saliva, Taste, and Periodontal Disease

It would not be surprising if Cu had an effect on the supporting structures of the teeth and the oral mucous membranes. As with other aspects of Cu in relation to oral health, virtually nothing is known.

In human saliva Cu is found in low concentrations with some difference according to the type of saliva taken and method of analysis. Thus, Arwill et al (1967) found a concentration of 0.28 ppm for fresh stimulated whole saliva. This concentration was quite low and certainly lower than those found in enamel (see Chapter 3). The enamel surface is capable of accumulating and retaining high concentrations of elements such as Cu (De Renzis et al, 1969).

Various other relationships of Cu to oral tissues have been reported. Applications of cupric ions to the surface of rat tongues inhibited the sweet taste response. It was suggested that the cupric ion competed with sugar molecules for the same receptor site (Yamamato and Kawamura, 1971). Toxic effects of Cu in rats exposed to the metal in a Cu smelting works were reported as cyanosis of the oral mucous membranes and submucous hemorrhages (Velikov and Zlateva, 1969).

In bone, orofacial skeletal lesions have been observed in Cu-deficient (-Cu) swine. Changes in the alveolar bone were noted with decreased osteogenesis and increased osteoclastic activity. The result of this was an increase in bone resorption (Furstman and Rothman, 1972). Bergman and Söremark (1968) noted only moderate concentrations of Cu (1.48 to 4.49 ppm) in mandibular discs of various animals.

The effect of -Cu on alveolar bone resorption promotes the question of whether Cu has any relationship to periodontal disease. A limited study by Miglani et al (1969) showed no difference in blood serum Cu concentrations in relation to periodontal status in contrast to the finding of Chawala et al (1965). In both of these studies, however, the information given as to the extent of the periodontal disease was limited only to its presence or absence. On this basis it is impossible to evaluate these reports properly. The report on the effects of placing rats in a Cu smelter also recorded an increase in periodontal disease and resorption of bone (Velikov and Zlateva, 1969). Recently, Beighton et al (1979) determined Cu concentrations in 100 human supragingival plaque samples and related these to gingival index and plaque scores. Results showed a significant negative correlation between plaque Cu levels and gingival index, as well as mean pocket depth. The influence of Cu (as well as Zn) on periodontal disease is worthy of further investigation.

SUMMARY

Despite extensive research on Cu in relation to human metabolism and enzyme systems our knowledge of Cu and dental disease is negligible. Human studies, largely from inference, point to Cu as a caries-promoting element. Animal studies have been equivocal. Plaque and oral bacterial metabolism studies would indicate a Cu effect to reduce acidogenicity. There is an indication that Cu may have a role in influencing periodontal disease and alveolar bone resorption, but more work is needed in this field. Therefore, Cu should be included among a small group of trace elements that merit more dental research.

REFERENCES

Adkins BL, Losee FL: A study of the covariation of dental caries prevalence and multiple trace element content of water supplies. *NY State Dent J* 1970;36: 618–622.

Afseth J, Oppemann RV, Rolla G: The in vivo effect of glucose solutions containing Cu^{++} and Zn^{++} on the acidogenicity of dental plaque. *Acta Odont Scand* 1980;38:229–233.

Anderson RS, Davies BE, James PMC: Dental caries prevalence in a heavy metal contamination area of the West of England. *Br Dent J* 1976;141:311–314.

Arwill T, Myrberg N, Söremark R: The concentration of Cl, Na, Br, Cu, Sr and Mn in human mixed saliva. *Odont Revy* (Malmo) 1967;18:1–186.

Barmes DE: Caries etiology in Sepik villages—Trace element micronutrient and macronutrient content of soil and food. *Caries Res* 1969;3:44–59.

Beighton D, Fry T, Higgins T, et al: Associations between heavy metals in dental plaque and periodontal disease. *J Dent Res* 1979;58:British IADR Abst #29.

Bergman B, Söremark R: Analysis of the concentration of zinc, copper and manganese in the mandibular disc. *Acta Odont Scand* 1968;26:103–110.

Chawala TN, Kumar S, Mathur MN: Blood changes in periodontal disease. *J Indian Dent Assoc* 1965;37:224–227.

Coles S, Fletcher RP, Stack MV: Lead, zinc, cadmium and copper levels in mature and immature teeth. *J Dent Res* 1973;52:926.

Curzon MEJ, Bibby BG: The effect of heavy metals on dental caries and tooth eruption. *J Dent Child* 1970;37:463–465.

Curzon MEJ, Losee FL: Dental caries and trace element composition of whole human enamel: Western United States. *JADA* 1978;96:819–822.

DeRenzis FA, Aleo JJ, Baker WH: Copper localization on the root surface of a tooth. *J Dent Res* 1969;48:970.

Driezen SA, Spies HA, Spies TD: The copper and cobalt levels in human saliva and dental caries activity. *J Dent Res* 1952;31:137–142.

Dufor CN, Becker E: *Public Water Supplies of the 100 Largest Cities in the United States*. U.S. Geol. Surv. Water Supply Paper #1812. Washington, D.C., Government Printing Office, 1962.

Dunning JM: The influence of latitude and distance from the sea coast on dental disease. *J Dent Res* 1953;32:811–829.

Furstman L, Rothman R: The effect of copper deficiency on the mandibular joint and alveolar bone of pigs. *J Oral Path* 1972;1:249–255.

Gallagher IHC, Cutress TW: The effect of trace elements on the growth and formation by oral streptococci and actinomyces. *Arch Oral Biol* 1977;22: 555–562.

Gianpiccolo P, Mingani N, Tripi F, et al: Effects of copper on the experimental caries in the albino rat. *Ann Stomatol (Rome)* 1968;17:219–224.

Glass RL, Rothman KJ, Espinal F, et al: The prevalence of human dental caries and water borne trace metals. *Arch Oral Biol* 1973;18:1099–1104.

Green I: Copper and manganese in saliva of children. *J Dent Res* 1970;49:776–782.

Hardwick JL, Martin CJ: A pilot study using mass spectrometry for the estimation of the trace element content of dental tissues. *Helv Odont Acta* 1967;11: 62–70.

Hein JW, Shafer WG: Further studies on the inhibition of experimental caries by sodium copper chlorophyllin. *J Dent Res* 1951;30:510.

Hein JW: Effect of copper sulfate on initiation and progression of dental caries in the Syrian hamster. *J Dent Res* 1953;32:654.

Kruger BJ: The effect of trace elements on experimental dental caries in the albino rat. 1. A study of boron, copper, fluorine, manganese and molybdenum. *Aust Dent J* 1958;3:236–247.

Kruger BJ: The effect of trace elements on experimental caries in the albino rat. III. A study of boron and copper. *Aust Dent J* 1958;3:374–377.

Little MF, Steadman LT: Chemical and physical properties of altered and sound enamel. IV. Trace element composition. *Arch Oral Biol* 1966;11:273–278.

Ludwig TG, Adkins BL, Losee FL: Relationship of concentrations of eleven elements in public water supplies to caries prevalence in American school children. *Aust Dent J* 1970;15:126–132.

McClure FJ: Observations on reduced caries in rats. V. Results of various modifications of food and drinking water. *J Dent Res* 1948;27:34–40.

Miglani JC, Rajasekar A, Shyamala S, et al: Blood studies in periodontal disease. II. Serum iron and copper values. *J Indian Dent Assoc* 1969;41: 189–193.

Navia JM: Effect of minerals on dental caries, in Gould RF (ed): *Dietary Chemicals vs Dental Caries.* Advances in Chemistry Series. American Chemical Society Publications, 1970.

Navia JM: *Animal Models in Dental Research.* Birmingham, University of Alabama Press, 1979.

Opperman RV, Rolla G: Effect of some polyvalent cations on the acidogenicity of dental plaque in vivo. *Caries Res* 1980;14:422–427.

Rothman KJ, Glass RL, Epinal F, et al: Dental caries soil content of trace metals in two Colombian villages. *J Dent Res* 1972;51:1686.

Schamschula RG, Adkins BL, Barmes DE, et al: WHO study of dental caries etiology in Papua-New Guinea. WHO Publication #40, Geneva, 1978.

Schroeder HA, Nason AP, Tipton IH, et al: Essential trace elements in man: Copper. *J Chron Dis* 1966;19:1007–1034.

Shaw JH: Ineffectiveness of sodium copper chlorophyllin in prevention of experimental dental caries. *NY State Dent J* 1950;16:503–505.

Skjörland K, Gjermo P, Rolla G: Effect of some polyvalent cations on plaque formation in vivo. *Scand J Dent Res* 1978;86:103–107.

Underwood EJ: *Trace Elements in Human and Animal Nutrition,* ed 4. New York, Academic Press, 1977.

Velikov B, Zlateva M: Clinical and morphological studies of the changes in the oral cavity. *Stomatologica (Sofia)* 1969;51:41–50.

Yamamato T, Kawamura Y: Inhibitory effect of cupric and zinc ions on sweet taste response in the rat. *J Osaka Univ Dent Sch* 1971;11:99–104.

As, Cr, Co,

I, Fe, Ni,

K, Na, Si.

14 Other Essential Trace Elements

M. E. J. Curzon

ARSENIC

Normally considered a poison, arsenic (As) has for centuries been a valuable therapeutic agent. In earlier times As compounds were used in dentistry for root canal treatments as a filling material. Its effect was to mummify the infected tissues of the remnants of the pulp canal. However, because As did not remain in the root canal chamber but leaked out into surrounding bone, necrosis of the bone was often an outcome. The result was that As was dropped from dental use when other materials were developed.

Studies over long periods of time have indicated that As may be an essential trace element (Schroeder and Balassa, 1966), hence its inclusion in this chapter. The bad reputation of As as a poison was suggested by Schroeder and Balassa as due to the traditional use of the trivalent form. This reputation is undeserved in the case of the naturally occurring pentavalent form that is largely nontoxic. Although early reports associated

As with carcinogenicity, recent reports (Frost, 1967) have claimed that the element is quite free from such an effect and that there are wide differences in toxicity between the different chemical forms.

There is little evidence for any effect of As on dental disease. Navia (1970) did not mention As in his extensive review on the effects of trace elements on dental caries. In studies on the trace element composition of enamel As did not rate as important or as related to caries. Indeed, in most studies on enamel composition As was not even reported as present. Nixon and Smith (1960) specifically evaluated the As content of human enamel as a mean concentration of 0.06 ppm with a range of 0.031 to 0.145 ppm inorganic As. These are very low concentrations, and it would be surprising if As, at this level, would so affect the properties of enamel for significant change in solubility, hardness, or demineralization. Most trace elements associated with such changes are usually found at much higher levels.

A report in 1972 by Shintani et al gave information on dental disease in children who had been poisoned with As. Various measurements of dental disease in these children were compared with a control group of similar age, sex, and background. The subjects consisted of 29 infants poisoned in 1955 with As-tainted milk. Children from another district in Japan with a stable population and who were born between January 1954 and December 1955 were used as controls. This group consisted of 26 raised on As-free dried milk and 48 on their mothers' milk. Examinations were carried out for caries, as DMFT rate, and presence and severity of periodontal disease.

The As-tainted group had a higher prevalence of caries and abnormal dental enamel. The latter was not clearly defined and thus it is difficult to know whether there were actual enamel defects, leading to an increased risk of caries, or only enamel hypoplasia with mottling. The latter might or might not lead to greater susceptibility to caries. Thus, we do not know whether the caries increase was a direct one caused by the increased incorporation of As, or due to an increase in enamel defects.

There were fewer children with periodontal disease in the As-tainted group, but the difference was not statistically significant. However, Shintani et al reported that the periodontal disease rate, as a severity index, was higher in the As-tainted group. Presumably the interpretation can be made that in the As-tainted group the children were no more likely to have periodontal disease than the controls, but if they did get the disease, it would be more severe.

Shaw (1973) has published what appears to be the only study on the effects of As on dental disease in animals. A cariogenic diet was used for all rats, but in the test groups this was supplemented with 5, 10, and 20 ppm As as sodium arsenate. The base diet contained 0.7 ppm As. Results showed no influence of As on caries activity. There was, however, a tendency for the As-supplemented rats to show small increases in the severity of periodontal disease. Shaw noted that his negative findings were of some interest, because of the close chemical relationship of As to P, and the possible influence in replacing phosphate in the enamel crystallite. However, since the work of Shaw no other researchers have followed his suggestion for further work.

The above studies appear to be the total of our knowledge on As in relation to dental disease. Despite extensive literature on the effects of As on the health of man and animals, particularly for agricultural animals, the study of its relation to dental disease has been neglected.

CHROMIUM

Chromium (Cr) is one of the newer essential trace elements with a recommended dietary intake for an adult suggested as 0.05 to 0.2 mg/day (Mertz, 1981). The daily requirement is very low and in man appears to be generally provided in an average diet. The element has been determined as essential for the maintenance of normal glucose tolerance in rats (Schwartz and Mertz, 1959). It is suggested that Cr acts in the first step of sugar metabolism. Further studies in vitro and in vivo demonstrated that general insulin deficiency can be prevented and cured by Cr administration (Mertz, 1981). Deficiency of Cr in man has been implicated as a risk factor in cardiovascular disease (Schroeder et al, 1970, 1971). Whether Cr has any role in dental caries related to sugar metabolism remains unknown.

In dentistry, studies on Cr have been complicated by widespread use of steels for orthodontic and prosthetic appliances. All the metals used in these devices contain some Cr, which will be slowly released from the metals by action of the oral fluids. The released Cr ions are taken up by the enamel; analysis of teeth from mouths where such appliances have been used shows higher levels of Cr. Heavy contamination of enamel by Cr from orthodontic brackets has been shown to occur (Ceen and Gwinnett, 1980). As a result, the metal Cr is usually found in most samples, and in the study of Curzon and Crocker (1978) the mean Cr concentration was 0.44 ppm.

In various attempts to relate trace elements to dental caries Cr has never figured prominently, and in most cases has not even been mentioned. In man the only evidence for a Cr-to-caries relationship comes from the Colombia (South America) study previously referenced. Analysis of the trace element content of soils showed Cr to be significantly higher in the low caries village (Rothman et al, 1972). However, this finding was included among several other similar relationships for eight trace elements. Navia (1970) does not list Cr among the elements he discussed.

Within the last 20 years there have been two other reports concerning Cr and dental decay. Stawinski (1977) noted an increase in caries in workers exposed to Cr compounds in industry, but without sufficient details to interpret the findings, or means to know whether other factors might have been involved. In an earlier report Moreira (1972) used Cr anhydride, with acetic acid, to prevent caries; this work does not appear to have attracted any further interest or replication. There have been no studies on the effects of Cr on dental caries in experimental animals.

Therefore, although there is a slight indication that Cr, or Cr salts, may inhibit dental decay the evidence to prove a causal relationship does not exist. The important role of Cr in animal and human metabolism, particularly its suggested role in sugar metabolism, is of interest. It could be that low levels of Cr in the oral environment might affect oral bacteria and their growth, acid production, and polysaccharide production.

COBALT

The role of cobalt (Co) in animal metabolism was first investigated as a result of research in Australia on a debilitating disease of cattle and sheep known as "coast disease." It was shown that a deficiency of Co was associated with this disease, and the disease could be corrected with nutritional supplements (Underwood, 1977). Later work showed that pernicious anemia was related to a deficiency of vitamin B_{12}, also related to Co. Thus animals, including man, require the element Co as an integral part of this vitamin. It has also been suggested that man may require Co in addition for the B_{12} complex (Schroeder et al, 1967). Other research has demonstrated that excessive levels of Co may interfere with the calcification process (Goldenberg and Sobel, 1952).

Conclusive evidence of a dietary deficiency of Co in man, apart from pernicious anemia, has not been demonstrated even in geographic areas where coast disease is known to occur in animals. Similarly no studies have ever been attempted to relate Co to dental decay in Co-adequate, or Co-deficient, geographic areas. The question of Co in animal metabolism has been extensively reviewed by Underwood (1977).

In dental disease there is little evidence of a role for Co. Navia (1970) listed Co as a trace element of doubtful caries effect. Driezen et al (1952) analyzed human saliva for the presence of Co, as well as Cu (see Chapter 4), but could find no relationship of saliva Co concentrations to dental caries. Otherwise there have been no other human studies on Co.

None of the many enamel composition studies, although identifying Co in some samples, could find any relationship of Co in enamel to caries. It is surprising that despite the well-documented geographic distribution of Co (Kubota, 1964), so few studies have been attempted on such an essential trace element.

In animal experiments, Hendershot and Forsaith (1958) fed Co to rats as the Co-EDTA salt. There was higher caries in the Co-EDTA-treated group than in the controls. However, this study did not attempt to separate the effect of EDTA alone.

The possible role of Co in mineralization has already been referred to. Other research (Bird and Thomas, 1963) has shown that Co is unique in inhibiting apatite crystal formation at concentrations that would also produce prevention of mineralization of rachitic cartilage matrix. Nevertheless, from reviewing all available evidence on Co in enamel we find that the element has not been related to dental caries by incorporation in enamel.

IODINE

The relationship of iodine (I) to endemic goiter is well known, and I salts have also been used as anti-bacterial agents for many years. The action of I as an essential trace element, and as a useful agent in medicine, has a long history. Research on the dental aspects of I initially concerned its antibacterial activity, where salts such as iodoform were widely investigated for use in oral surgery and endodontics. Studies on any other possible dental effects of I on periodontal disease and dental caries have, however, been rather limited.

Of interest here is the early report of a possible association of I to dental caries by Sir Charles Hercus (1925). He suggested that there was an association between soil type, dental caries, and the prevalence of endemic goiter in New Zealand. This did not, however, promote any further interest in the subject. Although the geographic incidence of goiter is well documented, and therefore areas of I adequacy and deficiency are identified, no attempt has been made to study the epidemiology of dental caries and I.

There have been several studies to test the anticaries effect of I in animals (Table 14-1). When used, either by injection or as a dietary supplement, I reduced dental decay in nearly every experiment reported. In addition, the most recent report of Caufield (1981) showed that there was a significant additive effect of I when used with fluoride. Of particular interest was the finding that I reduced the buccal caries scores whereas fluoride, as NaF, reduced the sulcal scores. When given together, I and fluoride gave a greater caries reduction than did fluoride alone. This supports the early report of a similar effect by Dale and Keyes (1945).

As a possible topical preventive agent I shows some promise. Caufield et al (1981), using a 2% solution of I_2KI, produced a reduction in caries in rats. This caries reduction was significant, and most pronounced with buccal caries scores; it confirmed a previous report of a similar finding by the same group of workers (Caufield, 1981). It would

Table 14-1
Summary of Reports on the Effect of Iodine on Dental Caries in Experimental Animals

Author (year)	Animal	Iodine used		Vehicle	Effect on Caries
		Salt	*Conc.*		
McClure and Arnold, 1941	Rat	a. NaIac b. HIac	200 ppm 200 ppm	diet water	reduction in caries by both regimens
Powell and Dale, 1943	Rat	a. NaIac b. HIac	400 ppm 200 ppm	injection water	reduction in caries by both regimens
Dale and Keyes, 1945	Hamsters	HIac	200 ppm	diet	reduction in caries
Dale and Powell, 1949	Rat	a. HIac b. HIac	200 ppm 20 ppm	diet water	reduction in caries greater with I in food than in water
Lundquist, 1950	Rat	HIac	0.002%	food	reduction in caries
Kruger, 1958	Rat	HIac	0.005 mg 0.002 mg	injection	no effect
Caufield et al, 1981	Rat	I_2KI	2.0%	topical	reduced buccal caries
Caufield, 1981	Rat	a. I_2KI b. I_2KI + NaF	1.0% w/v 1.0% w/v 0.22%	water water	reduced caries additive reduction in caries

HIac = iodoacetic acid.
I_2KI = iodine potassium iodide.
NaIac = sodium iodoacetate.

therefore appear that I has a significant effect on smooth surface caries and would be a potential anticaries topical agent when used in conjunction with fluoride. However, in a previous study showing a reduction in caries with I in rats, Lundquist (1950) cautioned that the safety of I agents, such as iodoacetic acid, was not sure enough to permit its practical application as a prophylactic agent against caries.

Recently other work by Caufield (1981) indicated an antibacterial effect of I in addition to its cariostatic action. A combination of I_2KI and NaF exerted a depressing effect on the oral flora of the rats used to test the two agents for dental caries prevention. Previously Caufield and Gibbons (1979) had showed that I alone, when used topically, could suppress *S. mutans* in humans. This has also been shown to occur in animals where I_2KI affected oral bacteria in rats.

Despite the earlier caution of Lundquist (1950), I continues to be tested and evaluated as a cariostatic agent. Maltz-Turkienicz et al (1980) reported an effect of I on in vitro plaque formation. Both chlorhexidine and I showed greater antimicrobial effects on in vitro-grown plaque of *S. mutans* than on similar plaque grown with *S. sanguis*. Other authors have also reported such an effect of inorganic I (Tanzer et al, 1977).

Obviously much further work on the mechanism of action of I, and its use as a cariostatic agent, is warranted, in particular work concentrating on the antimicrobial aspects of I in relation to periodontal disease as well as dental caries.

IRON

All analyses of enamel have reported the presence of iron (Fe), and all tooth samples contain the element. This is hardly surprising considering its ubiquitness in the environment and its considerable importance to human metabolism. It is usually not considered to be a normal constituent of the apatite crystal (Navia, 1970) but merely a common contaminant.

The literature on Fe metabolism is extensive; the reader is referred to the excellent chapter by Underwood (1977). Of possible interest to dental research is the well-known interaction of Fe with other trace elements. This interest is not caused by any strong indication of an effect of Fe on dental caries alone, but because these interactions occur with elements such as Co, Cu, Zn, and Mn, for which there is some evidence of a cariogenic or cariostatic effect. However, little attention has been paid to this area of dental research.

McClure (1948) observed that ferric chloride in drinking water reduced caries in rats, but that ferric citrate did not. Emilson and Krasse (1972) obtained reductions in caries in the hamster using ferric chloride or ferrous sulfate. Other authors (Hendershot and Forsaith, 1958) have reported no effect of Fe salts on caries in animals. The later study used Fe-EDTA as the main experimental salt. Wynn et al (1958) also found no effect with Fe salts on animal caries.

Incorporated into whole enamel at concentrations of about 30 ppm (Curzon, 1977), there are considerable variations depending on different sampling techniques or the geographic origin of the teeth used. Since most drinking water samples contain appreciable levels of Fe, there is a continued posteruptive uptake of Fe by enamel. There is no evidence that this uptake has any effect on caries, although Torrell (1955) suggested that ferric solutions may give layers of ferric precipitates in surface enamel which might increase caries. However, Besic et al (1975) reported that increased concentrations of Fe in enamel were related to decreased solubility.

It is not known to what degree Fe affects enamel composition either before or after tooth development. In autoradiographic studies by Bawden et al (1977), using the developing rat molar, ^{59}Fe was localized in the ameloblast layer. However, uptake in the rat molar was also shown to be limited by the metabolic activity of the enamel organ cells. The incorporation of Fe into the enamel apatite was felt by Navia (1970) to be only a contaminant. Nevertheless, the levels of Fe in enamel are consistently high, and the presence of Fe might affect lattice parameters.

Part of the difficulty in interpreting different animal findings on Fe and dental caries may be due to the use of different Fe salts. The distinction between ferric and ferrous may be important, just as the form of the element used for experimental purposes has been shown to be important for other elements such as Mo.

In the study by Emilson and Krasse (1972) caries was produced by inoculating the rats with *S. mutans,* which produce extracellular dextrans from sucrose. These authors postulated a possible mechanism of action of Fe salts on caries, whereby an inhibiting effect upon the enzyme dextranase was brought about by the element. Recently Beighton and McDougall (1981) demonstrated that rats receiving $FeCl_3$ in drinking water at 500 ppm had significantly lower levels of *S. mutans* in their plaque. Whether Fe has any effect on caries by action in enamel or by changes in bacterial metabolism remains to be demonstrated.

The effect, therefore, of Fe on caries remains equivocal. Limited evidence suggests that Fe affects caries, but whether it is cariogenic or cariostatic remains to be further investigated, as does any role in other dental diseases, such as periodontal disease.

NICKEL

Widely distributed in the biosphere, nickel (Ni) has been suggested as an essential trace element (Schroeder et al, 1961), although the evidence is not completely accepted (Underwood, 1977). The element acts as an activator for a number of enzyme systems, such as arginase, carboxylase, and trypsin, which would indicate that the element is essential to animal metabolism even if at very low concentration.

There have not been any human epidemiologic studies specifically implicating Ni as either cariogenic or cariostatic. However, in the Colombian study of Rothman et al (1972) an analysis of soil samples showed Ni to be significantly higher in the low caries town of Heliconia. Thus, median soil concentrations of Ni were 70 and 0 ppm where the caries prevalence scores were 5.5 and 13.9 DMFT respectively. However, this same study also showed similar relationships for Ag, Cr, Cu, and V. In studies of this nature it is difficult to differentiate the true effects that can be ascribed to a single trace element. Where several metals are involved, as in this case, there is no means of knowing which metal is producing an effect and which metal is a "fellow traveler."

In Navia's (1970) review of trace elements and dental caries, he listed Ni as of doubtful effect. The evidence cited at that time was largely from the study of Forbes and Smith (1952) which showed that Ni salts exerted a marked inhibiting action on acid production in saliva.

In what appears to be the only animal study on the subject, Hendershot et al (1960) recorded that Ni salts had various effects on caries. The chloride increased caries whereas the acetate had no effect. In the same study a Ni-EDTA complex inhibited caries in male but not female rats.

Schroeder et al (1974) noted that Ni interacts with a number of other trace elements, such as Cu, Zn, and Mn, by promoting the excretion of these other elements. It remains to be shown whether these interactions of Ni have any effect on the cariostatic or cariogenic properties of these other trace elements.

In summary, data on the effect of Ni, like that on so many other trace elements, are very small. This trace element has not attracted much attention, but along with other metals such as Co, Cu, and Zn it may be shown to have a minor effect on dental decay.

POTASSIUM AND SODIUM

Many authorities would not consider either potassium (K) or sodium (Na) as trace elements. Underwood (1977) did not discuss them in his book but listed them in the introduction as major essential elements. These two elements were also not considered by Navia (1970). In enamel Na occurs at levels that are usually recorded as percentages rather than in ppm, while K is found at concentrations on the borderline between ppm and percent. Nevertheless, we mention these two elements for the sake of completeness.

Potassium

The presence of K in enamel is usually at a concentration of between 100 to 2000 ppm. In an analysis of over 450 teeth the mean concentration of K was found to be 961 ppm (Curzon, 1977). There was indication of a geographic variation in K concentrations in whole enamel. In the United States the mean K levels in enamel from teeth from various states were: New York, 1270; Oregon, 373; Montana, 1733; California, 185; South Carolina, 840; Ohio, 1372; and Florida, 842 ppm. In this same study K was found to be significantly higher ($p = < 0.2$) in teeth from low caries individuals (< 4 DMFT) than from high caries individuals (> 7 DMFT). However, when the results for all analyses were subjected to multiple regression, K did not show any significant relationship to dental caries, nor were there any significant interactions with other trace elements (Curzon and Crocker, 1978).

In the Papua-New Guinea study a significant relationship was found between soil K levels and arch widths. This was a negative association indicating that high soil levels of K were related to decreased arch widths (Schamschula et al, 1972). This finding has not been noted by any other researcher but is of some interest.

Apart from the early work on the possible use of fluoride salts of K as topical preventive agents (Stones et al, 1949) little attention has been paid to this element. A number of other studies, such as that of Gerdin and Torell (1969), have used K salts, but in these instances the prime objective of the studies was for another agent, such as Mn as in the study cited. The salts of K are generally very soluble and it would be surprising if K of itself should have any effect on caries. Certainly its presence in enamel would be unlikely to reduce enamel solubility. High concentrations of K in enamel would tend to increase solubility.

Sodium

As mentioned above, Na is found as a major element in human dental enamel. As such, it is not really a trace element, and neither does it appear to have any major effect on dental caries. The literature does not give indication of any research that specifically focuses on the action of Na. The element is discussed extensively with respect to its fluoride salt, NaF, whose action on caries is well known; it needs no further elaboration here.

SILICON

An abundant element geochemically, silicon (Si) comprises 25.7% of the earth's crust and is the second most abundant element after oxygen. Not found free in nature, Si occurs chiefly as the oxide (silicate). Because of its abundance, Si is one of the most useful elements to man. It is best known in dentistry as a polishing agent but has also been used as a silicate restorative material.

In metabolism Si has been listed as essential (Mertz, 1981), although the evidence is not conclusive, and debate continues as to whether the element should be included in the list of essential trace elements. Nevertheless, because it is regarded as essential by some we have included it. It has been proposed that Si is essential for bone formation and particularly for the mineralization process (Carlisle, 1972), and may be necessary for growth in rats (Schwartz and Milne, 1972).

For such an abundant element as Si, the evidence for any relationship of Si to dental caries, or to any other dental disease, is sparse. Navia (1970) commented on this lack of interest and the paucity of data, but only two other papers have been published concerning any aspect of Si and dental disease. The literature for the last 20 years shows a large number of references concerned with Si as a restorative material, and also its use as a silicon rubber impression material. However, only three papers have been concerned with Si in relation to dental disease.

Boyers et al (1963) tested silicone oil on the teeth of caries-susceptible rats by topical application with a toothbrush. By this method Si had no effect on dental caries. However, when the same author added Si oil to the cariogenic diet, at a level of 10% or 20%, a significant increase in caries was found. These findings suggest that the incorporation of high levels of Si into the enamel makes it more susceptible to dental caries. It is possible to theorize that a Si-rich apatite would be disorganized and of poor structure, which would presumably make it more susceptible to dissolution. However, this has yet to be tested, and could easily be accomplished by using synthetic apatites and X-ray crystallography.

Studies of natives living in the Kalahari desert showed they exhibited a high resistance to dental caries (Levy and Koritzer, 1976). A high level of Si, from sand, would provide a very coarse and detergent diet, which together with a diet presumably low in refined carbohydrate, would produce a low caries incidence. However, these authors noted that in these people with low caries there was a linear relationship between Si and fluoride concentrations in enamel. As the Si concentrations increased so did the fluoride. Where two trace elements appear, as in this case, to be correlated, it is possible for the relationship to be an indirect one if the two occur together geologically. Although the study quoted above showed a significant correlation, no causal relationship has been proven.

The identification of Si in enamel is difficult, as contamination is a serious problem and has not yet been adequately dealt with. All results on Si in enamel are therefore not entirely reliable. In most cases when teeth are prepared for analysis they have to be cleaned of the organic plaque and other debris found on the surface. This is usually accomplished with a prophylaxis using pumice or other Si-based abrasive. In the process of cleaning the teeth, therefore, Si is polished into the enamel and contaminates it. For example, in the study by Curzon and Crocker (1978), where care was taken not to contaminate the enamel with trace elements in general, a silicon carbide (SiC) slurry was used which was trace element-free, but inevitably contaminated the samples with Si. As a result, Si was not reported on in this study.

Si research on producing coatings on the enamel surface has attracted attention. It is presumed that these coatings can be produced with Si compounds and used to enhance the uptake and retention of fluoride in surface enamel. Using this approach, Kohlehamainen and Kerosuo (1979) tested a urethane lacquer containing silane fluoride for its effect on caries. Results showed that after one year there was no statistically significant difference between treated and control tooth surfaces. The authors concluded that the varnish application provided little or no further improvement than did a simple topical fluoride.

The data on Si in enamel show that the element is present in higher concentrations than are most other trace elements. If Si is an important factor in calcification, it would not be surprising to find high levels of Si in enamel. What role Si may play in the chemistry of enamel is as yet unknown.

REFERENCES

General

Curzon MEJ: *Trace Element Composition of Human Enamel and Dental Caries.* PhD Thesis, London, University of London, 1977.

Curzon, MEJ, Crocker DC: Relationships of trace elements in human tooth enamel to caries. *Arch Oral Biol* 1978;23:647–653.

Navia JM: Effects of minerals on dental caries, in Harris RS (ed): *Dietary Chemicals in Dental Disease.* Advances in Chemistry Series 94. Washington, DC, American Chemical Society, 1970.

Underwood EJ: *Trace Elements in Human and Animal Nutrition. New York, Academic Press, 1977.*

Arsenic

Frost DV: Arsenicals in biology-retrospect and prospect. *Fed Proc* 1967;26:194–208.

Nixon GS, Smith H: Estimation of arsenic in teeth by activation analysis. *J Dent Res* 1960;39:514–516.

Schroeder HA, Balassa JJ: Abnormal trace elements in man: Arsenic. *J Chron Dis* 1966;19:85–106.

Shaw JH: Relation of arsenic supplements to dental caries and the periodontal syndrome in experimental rodents. *J Dent Res* 1973;52:494–497.

Shintani H, Shinohara S, Ninomiya T, et al: Epidemiologic survey of dental disease in children poisoned by arsenic. *J Hiroshima Univ Dent School* 1972; 4:104–118.

Chromium

Ceen RF, Gwinnett AJ: Indelible iatrogenic staining of enamel following debonding. A case report. *J Clin Orthodont* 1980;14:713–715.

Mertz W: The essential trace elements. *Science,* 1981;213:1332–1338.

Moreira BH: Elimination of pits and fissures in permanent teeth and topical applications of 5 solutions of acetic acid and chromium anhydride for the prevention of dental caries. *Bol Fac Odontol Piracicaba* 1972;6:1–18.

Rothman KJ, Glass RL, Espinal F, et al: Dental caries in the soil content of trace metals in two Colombian villages. *J Dent Res* 1972;51:1686.

Schroeder HA, Mitchener M, Nason AP: Influence of various sugars, chromium and other trace metals on serum cholesterol and glucose in rats. *J Nutr* 1971; 101:247–258.

Schroeder HA, Nason AP, Tipton IH: Chromium deficiency as a factor in atherosclerosis. *J Chron Dis* 1970;23:123–142.

Schwartz K, Mertz W: Effects of chromium on growth. *Arch Biochem Biophys* 1959;85:292–295.

Stawinski K: Diagnostic significance of oral changes in workers exposed to chromium compounds. *Czas Stomatol* 1977;30:309–311.

Cobalt

Bird ED, Thomas WC: Effects of metal ions on apatite crystal formation. *Proc Soc Exp Biol Med* 1963;112:640–643.

Driezen S, Spies HA, Spies TD: The copper and cobalt levels of human saliva and dental caries. *J Dent Res* 1952;31:137–142.

Goldenberg H, Sobel AE: Metal ions affecting the calcification process. *Proc Soc Exp Biol Med* 1952;81:695–698.

Hendershot LC, Forsaith J: Effect of various metal salts of ethylenediamine tetra-acetic acid on dental caries in the rat. *J Dent Res* 1958;37:32–33.

Kubota J: Distribution of total and extractable forms of cobalt in morpholocally different soils of Eastern United States. *Soil Sci* 1964;99:166–174.

Schroeder HA, Nason AP, Tipton IH: Essential trace elements in man: Cobalt. *J Chron Dis* 1967;20:869–890.

Iodine

Caufield PW, Gibbons RJ: Suppression of *S. mutans* in the mouths of humans by a dental prophylaxis and topically applied iodine. *J Dent Res* 1979;58: 1317–1326.

Caufield PW: Combined effect of iodine and sodium fluoride on dental caries in rats and the variability of *S. mutans* in vitro. *Caries Res* 1981;15:484–491.

Caufield PW, Navia JM, Rodgers AM, et al: Effect of topically applied solutions of iodine, sodium fluoride, or chlorhexidine on oral bacteria and caries in rats. *J Dent Res* 1981;60:927–932.

Dale PP, Keyes PH: Inhibition of experimental dental caries in Syrian hamsters by fluorine and iodoacetic acid. *J Dent Res* 1945;24:194.

Dale PP, Powell VH: Inhibition of experimental dental caries in the rat by iodoacetic acid. *J Dent Res* 1949;22:33–36.

Hercus CE, Benson WN, Carter CL: Endemic goitre in New Zealand and its relation to the soil's iodine. *J Hyg* (Lond) 1925;24:321–402.

Kruger BJ: The effect of trace elements on experimental dental caries in the albino rat. *Univ of Queensland Papers* 1959;1:3–28.

Lundquist C: The toxicity of iodoacetic acid and its quantitative relations to inhibition of glycolysis and dental caries. *J Dent Res* 1950;30:203–213.

Maltz-Turkienicz M, Krasse B, Emilson CG: Effects of chlorhexidine and iodine on in vitro plaques of *Streptococcus mutans* and *Streptococcus sanguis*. *Scand J Dent Res* 1980;88:28–33.

McClure FJ, Arnold FA: Observations on induced dental caries in rats. *J Dent Res* 1941;20:97–105.

Powell VH, Dale PP: The mechanism by which iodoacetic acid inhibits experimental dental caries in the rat. *J Dent Res* 1943;22:257–260.

Tanzer JM, Slee AM, Kamay B, et al: In vitro evaluation of three iodine-containing compounds as anti-plaque agents. *Antimicrob Agents Chemother* 1977;12:107–113.

Iron

Bawden JW, Wennberg A, Hammarstrom L: In vivo and in vitro study of ^{59}Fe uptake in developing rat molars. *Acta Odont Scand* 1977;36:271–277.

Beighton D, McDougall WA: The influence of certain added water-borne trace elements on the percentage bacterial composition of tooth fissure plaque from conventional Spraque-Dawley rats. *Arch Oral Biol* 1981;26:419–425.

Besic FC, Bayard M, Wiemann MR, Burrell KH: Composition and structure of dental enamel: Elemental composition and crystalline structure of dental enamel as they relate to its solubility. *JADA* 1975;92:594–601.

Emilson CG, Krasse B: The effect of iron salts on experimental dental caries in the hamster. *Arch Oral Biol* 1972;17:1439–1443.

Hendershott LC, Forsaith J: Effect of various metal salts of ethylenediamine tetraacetic acid on dental caries in rats. *J Dent Res* 1958;37:32–33.

McClure FJ: Observations on induced caries in rats. VI. Summary results of various modifications of food and drinking water. *J Dent Res* 1948;27:34–40.

Torrell P: Iron and dental hard tissues. Experimental studies concerning the significance of iron in physiology and pathology of human enamel and dentine. *Odont Tidsk* 1955;63:131–172.

Wynn W, Haldi J, Law ML, Bentley KD: Studies on the cariogenicity of Emory and Harvard diets. *J Dent Res* 1958;37:33.

Nickel

Forbes JC, Smith JD: Studies on the effect of metallic salts on acid production in saliva. *J Dent Res* 1952;31:129–136.

Hendershott LC, Monsell E, Forsaith J: Metal binding, in Seven JJ, Johnson AL (eds): *Medicine*. Philadelphia, Lippincott, 1960, pp 306–311.

Rothman KJ, Glass RL, Espinal F, et al: Dental caries and soil content of trace metals in two Colombian villages. *J Dent Res* 1972;51:1686.

Schroeder HA, Mitchener M, Nason AP: Lifetime effects of nickel in rats: Survival, tumors, interactions with trace elements and tissue levels. *J Nutr* 1974; 104:239–243.

Schroeder HA, Balassa JJ, Tipton IH: Abnormal trace elements in man—Nickel. *J Chron Dis* 1961;15:51–65.

Potassium and Sodium

Gerdin PO, Torell P: Mouth rinses with potassium fluoride solutions containing manganese. *Caries Res* 1969;3:99–107.

Schamschula RG, Barmes DE, Adkins BL: Caries aetiology in Papua-New Guinea. Associations of tooth size and dental arch width. *Aust Dent J* 1972; 17:188–195.

Stones HH, Lawton FE, Bramsby ER, Hankley HD: The effect of topical applications of potassium fluoride and ingestion of tablets containing sodium fluoride on incidence of dental caries. *Br Dent J* 1949;86:263–269.

Silicon

Boyers CL, Shaw JH, Rosenthal E, et al: Effects of silicone oil on the teeth of caries-susceptible rats when applied by toothbrushing and by inclusion in the diet. *J Dent Res* 1963;42:1517–1519.

Carlisle EM: Silicon: A possible factor in bone calcification. *Science* 1970;167: 279–280.

Kolehamainen L, Kerosuo E: The clinical effect of application of a urethane lacquer containing saline fluorine. *Proc Finn Dent Soc* 1979;75:69–71.

Levy JS, Koritzer RT: Enamel silicon and fluoride relationships demonstrating a surface silicon effect that facilitates fluoride uptake. *J Dent Res* 1976;55: 733–737.

Mertz W: The essential trace elements. *Science* 1981;213:1332–1338.

Schwartz K, Milne, DB: Growth promoting effects of silicon in rats. *Nature* 1972;239:333–334.

$_{38}$ Sr
87.62
18-8-2

15 Strontium

M. E. J. Curzon

Strontium (Sr), atomic number 38, is one of the alkaline earth elements located in group IIa of the periodic table. Discovered by two Scotsmen, Cruikshank and Crawford, in the late 18th century, the element was isolated by Sir Humphrey Davy in 1808. It was found to be a major constituent of the mineral strontianite ($SrCO_3$) discovered in a Pb mine in the village of Strontian, in Scotland, from which it derives its name.

The electron distribution of the Sr atom is such as to determine the universal divalency of its compounds. Naturally occurring stable isotopes are of numbers 84, 86, 87, and 88, and give an average atomic weight of 87.63. There are several radioactive isotopes, the most notable being Sr^{90} which has attracted considerable attention because of its long half-life and association with nuclear fallout. Elements of the alkaline earths are notable bone or calcified tissue seekers, and because of this, Sr has been considered for a possible role in human enamel apatite structure. By incorporation, either developmentally or topically, Sr might be implicated with caries resistance or susceptibility.

Naturally Occurring Compounds

The naturally occurring salts of Sr are, among others:
Strontianite — $SrCO_3$ Brewsterite — $SrBaCaO$, Al_2O_3, $6SiO_2$, $5H_2O$
Celestite — $SrSO_4$ Femonite — $(Ca,Sr)_5[(P,A_5)O_4]_3F$
And $SrCl_2$, SrF_2, and SrO.

Certain large geologic deposits, usually celestite, are mined to provide elemental Sr, although this element has limited use. Principally Sr plays a role in the processing of sugar beets and manufacturing of fireworks, but also in ceramics, medicines, plastics, and iron castings (Schroeder et al, 1972).

Metabolism

The metabolic behavior of Sr, particularly in its interaction with Ca, has attracted considerable attention because of the danger of Sr^{90}, and the incorporation of this hazardous fission product in calcified tissues. The presence of natural isotopes of Sr effectively dilutes the radioactive variety and reduces the amount of Sr^{90} taken up by body tissues.

The element has an interesting relationship to Ca in animal metabolism, insofar as its absorption and subsequent deposition in bones and teeth are decided by the concentration and availability of Ca. The primary site of absorption of Sr is through the alimentary tract (Forbes and Mitchell, 1957), and it can also pass the placenta and mammary gland. Absorption is increased under fasting conditions, but decreased in the presence of high Ca levels and with advancing age. Thus, increased dietary Ca reduces Sr absorption and retention; the various aspects of this relative to bone and tooth uptake have been reviewed by Wolf et al (1973).

The biologic discrimination that occurs against Sr, as compared with Ca, is important in calcified tissue where the Sr:Ca ratios in bones and other calcified tissues in animals are lower than the Sr:Ca ratios found in their diets. A discrimination against Sr exists because of the relatively lower absorption, but higher urinary excretion, of Sr compared with Ca. Ordinary human diets generally supply 1 to 3 mg Sr/day, but it has been shown that the total intake of Sr is dependent partially upon the water quality and Sr content (Schroeder et al, 1972).

After absorption, Sr is primarily deposited in calcified tissues where it can enter the apatite crystal as a substitute for Ca (Boyde et al, 1958). While small amounts of Sr can exchange for Ca in apatites without severe dislocation, large amounts may inhibit calcification, and have been shown to induce "strontium rickets" in experimental animals (Storey, 1961). Analysis of human tissues (Table 15-1) shows that most Sr in the human body is located in the calcified tissues.

Table 15-1
Strontium Concentrations[a] in Human Tissues

Tissue	Mean ± SE	Tissue	Mean ± SE
Brain	2.3 ± 0.23	Heart	3.5 ± 0.24
Muscle	9.8 ± 0.52	Kidney	6.5 ± 0.35
Lung	9.8 ± 0.52	Liver	2.1 ± 0.19
Stomach	13 ± 0.93	Bone, rib	120 ± 5.10
Ileum	32 ± 4.10	Teeth, enamel	183 ± 15.44
		dentin	94 ± 11.47

[a]Concentrations in ppm.
Data from Schroeder et al, 1972; Curzon and Losee, 1977; Retief et al, 1978.

Although Sr has never been linked with any disease or deficiency state in man (Schroeder et al, 1972), there is evidence that it may be an essential trace element, and it was listed as a probable one by Schwartz (1972). It was described by Rygh (1949) as necessary for calcification and growth in rats. Sr has been used with reported success in the treatment of postmenopausal osteoporosis (McCaslin and Janes, 1981). It may well be that Sr has an essential role in calcification in man (Skoryna, 1981).

Strontium and Calcified Tissues

In human tissues Sr occurs in concentrations which reflect those found in drinking water supplies (Wolf et al, 1973). It is deposited in calcified tissues in man in greater quantities than are most other trace elements (Turkian and Kulp, 1956); there is some evidence that the Sr content of bones increases with age, and it varies with geographic origin (Schroeder et al, 1972). For example, higher rib ash Sr concentrations were found in Far Eastern adults (190 ppm) and children (320 ppm) than in United States adults (110 ppm) and children (96 ppm). Drinking water is considered a principal source of Sr for bones and teeth and a geographic variation in Sr content of enamel has been described by several authors (Losee et al, 1971; Wolf et al, 1973; Curzon and Losee, 1977).

Human Food and Water Sources

Schroeder et al (1972) claimed that foods make the greatest contribution to trace element intake in man. In some geographic areas, however, drinking water may contribute substantial proportions of certain elements, and the total intake of Sr is dependent partially upon the quality of the drinking water (Wolf et al, 1973). The composition of food and water, themselves related to soils, are much interrelated. In general,

Table 15-2
Strontium Concentrations in Typical Foods Used in the United States

Food Group	Item	Strontium ppm
Organ Meats	Beef liver	0.76
Meat	Beef	1.44
Bread	White	3.46
Cereals	All Bran	8.56
Milk	Whole	0.50
Vegetables	Carrot	2.65
	Kale	109.13
	Beans, green	1.09
	Tomato	0.48

Data from Schroeder et al, 1972.

foods of plant origin are richer in Sr than are animal products (except for bones); Schroeder et al (1972) have reported Sr levels for a number of foods in the United States (Table 15-2). Analysis of vegetables from different areas of the United States also shows some geographic variation with considerable differences to be noted in vegetables such as collards (kale) and other green leafy vegetables which have high Sr contents (Curzon, 1977).

Losee and Adkins (1968) found Sr to be one of the few elements which are absorbed by vegetables from cooking water. This increases the importance of the Sr content of the drinking (and cooking) water, for total ingestion of Sr. Because the Sr levels in domestic water supplies vary widely (Dufor and Becker, 1962) and there are variations in food composition, food choice and cooking practices can make large differences in Sr intake in man. Wolf et al (1973) contend that the differences in Sr content of hard tissues collected from certain areas of Israel were due to the differences in water Sr levels. Towns have been found in Wisconsin with Sr water levels as high as 39 ppm, giving a potential daily intake of about 39 mg/day from water alone. This level is considerably in excess of the estimated daily balance from food as given by Schroeder et al (1972) of 2.0 mg.

CARIES

Man

The low level of caries prevalence in Bonn (Germany) school children, when compared with that of children in Oslo (Norway), was ascribed by Lödrop (1953) to higher levels of Sr and V in the drinking water. Anderson (1969), however, could not find any reduction in caries prevalence in English children living in an area of high geologic Sr in Gloucestershire. In this English study, the influence of the composition of foodstuffs and drinking water was not adequately considered, and the intake of dietary Sr in the examination subjects was probably low. The results of an exhaustive survey of trace elements in soils, food, and water and their relationship to dental decay in New Guinea (Barmes, 1969) indicated an inverse relationship between Sr and caries. A further study was conducted in New Guinea with more emphasis on the analysis of enamel, saliva, and plaque, and factors in the geochemical environment (Schamschula et al, 1978). A less strong relationship of Sr to dental caries was found than previously reported (Barmes, 1969).

A dental survey of sailors inducted into the United States Navy (Losee and Adkins, 1969) indicated that a disproportionate number of men with no history or evidence of tooth decay originated from a small geographic area of northwest Ohio. Subsequent epidemiologic studies on school children living in this area showed them to have a significantly lower prevalence of caries than children in a control area (Curzon et al, 1970). Because the fluoride levels in the water supplies were similar in all study towns, it was suggested that the low levels of caries were due to factors other than fluoride, such as trace elements.

The identification of an area of Wisconsin where geologic deposits of celestite ($SrSO_4$) and strontianite ($SrCO_3$) occurred afforded an opportunity for further epidemiologic research. Ground water supplies in this study area were known to have Sr concentrations as high as 39 ppm (Nichols and McNall, 1957). Seven towns in the area were chosen for study, all with comparable levels of drinking water fluoride (1.0 to 1.2 ppm F), but with varying levels of Sr of from 0.02 to 34.5 ppm, as determined by atomic absorption spectrophotometry. Dental examinations were carried out on 12- to 14-year-old school children who were lifelong residents of the seven towns, and the decayed, missing, and filled permanent tooth surfaces (DMFS) recorded for each child. All examinations were carried out in the same way and by the same examiner who had completed the northwest Ohio studies (Curzon et al, 1970). Results for the mean number of DMFS scores for 12- to 14-year-old children for each town (Table 15-3) indicated that caries prevalence scores in Little Chute were the lowest. Further, the distribution of caries prevalence relative to Sr water concentrations suggested a curvilinear relationship, although this remains unproven at the present time (Curzon et al, 1978).

These epidemiologic studies on the relationship of dental caries prevalence and Sr are summarized in Table 15-4. Evidence suggests a negative association between Sr and caries—the prevalence of caries being decreased as environmental Sr increases.

Table 15-3
Mean Caries Scores (DMFS) of 12- to 14-Year-Old Lifelong Residents in Wisconsin Towns

Town	Sr in Water ppm	n	DMFS Scores[a] Mean ± SE
Eau Claire	0.02	122	6.96 ± 0.49
Beaver Dam	0.27	150	6.19 ± 0.45
Little Chute	5.37	113	3.14 ± 0.32
Kimberly	8.27	303	4.00 ± 0.24
Kewaskum	10.41	49	5.08 ± 0.75
Union Grove	15.13	37	4.93 ± 0.74
Menomonee Falls	33.91	134	6.49 ± 0.48

[a]DMFS = decayed, missing, and filled tooth surfaces.

Table 15-4
Summary of Human Studies Reporting a Relationship of Strontium to Caries Prevalence

Author	Year	Area of Study	Caries Effect
Lödrop	1953	Oslo, Norway, and Bonn, Germany	Decrease
Anderson	1969	Gloucestershire, England	Slight increase
Barmes	1969	Sepik Villages, New Guinea	Decrease
Curzon et al	1970	Northwest Ohio, United States	Decrease
Curzon et al	1978	Wisconsin, United States	Decrease
Schamschula et al	1978	Sepik Villages, New Guinea	Very slight effect (decrease)

Animals

The earliest report of any effect of Sr on dental tissues, or dental disease, was that of Rygh (1949), who studied the effects of a number of trace elements on growth and health in rats. The reduction of Sr levels in rats' diets was found by Rygh to produce poorer calcification, and also an increase in dental decay. These studies were not described in sufficient detail for clear conclusions to be drawn, but a number of studies since (Table 15-5) have generally indicated that Sr may lead to a decrease in caries in rodents.

Results have varied from significant increases in caries to significant reductions. The experimental methods used by previous researchers have differed widely, which presumably accounts for the disparity of results obtained. Thus, Shaw and Griffiths (1961) used $SrCO_3$ at doses between

Table 15-5
Summary of Studies on Strontium on Dental Caries in Rodents

	Strontium Used				
Author (year)	*Salt*	*Conc.*[a]	In	Period of Administration	Effect on Caries
Rygh, 1949	$SrSO_4$	47.4	W	Development	Decreased
Johansen and Hein, 1953	$SrCl_3$	50	W	Posteruption	No effect
Shaw and Griffiths, 1961	$SrCO_3$	0.1 or 2.0%	D	Posteruption	Moderate decrease
Grippaudo et al, 1971	$SrSO_4$	50	W	Prenatal, pre- and posteruption	Increased
Hunt and Navia, 1972, 1975	$SrCl_2$	500 BW	W	Pre- and posteruption	Increased
Gedalia et al, 1975	$SrCl_2$	25, 75	W	a. prenatal and pre-eruption b. posteruption	Decreased No effect
Meyerowitz et al, 1976	$SrCl_2$	50, 100	W	Pre- and/or posteruption	Decreased with 50
Olson et al, 1978	$SrCl_2$	50, 100 ± 10F	W	After weaning	No effect or in addition to F
Meyerowitz et al, 1979	$SrCl_2$	25, 50 100, 150 + 10F	W	Pre- or posteruption	Decreased in addition to F
Matsumoto, 1980	$SrCl_2$	0.1, 0.5g 1.0, 3.0g	D	After weaning	Decreased with > 0.5g Sr/kg

[a]Concentrations in ppm unless otherwise stated.
W = Water; D = Diet.
BW = dose/10g body weight.

Table 15-5 (continued)

Author (year)	Strontium Used				
	Salt	*Conc.*[a]	In	Period of Administration	Effect on Caries
Ashrafi et al, 1980	$SrCl_2$	5, 10, 25 50, 75	W	Pre- and/or posteruption	Decreased with 75 Sr
Curzon and Spector, 1981	Various salts	50	W	After weaning	$SrCl_2$ and SrF most effective

[a]Concentrations in ppm unless otherwise stated.
W = Water; D = Diet.
BW = dose/10g body weight.

0.1% and 2.0% of the food fed to rats and found modest but inconsistent reductions in caries. This Sr salt, $SrCO_3$, is very insoluble, and although Shaw and Griffiths used a relatively high dose, the amount of Sr absorbed must have been low. A similar question arises in considering the experiment of Grippaudo et al (1970), who used 50 ppm Sr of $SrSO_4$, another highly insoluble salt, and found an increase in caries.

Other salts of Sr, such as $SrCl_2$, dissolve far more easily; it is this salt that has been tried the most extensively. Thus, Johansen and Hein (1953) used 50 ppm Sr as $SrCl_2$ in Syrian hamsters with no reduction in caries incidence. In this study, however, the Sr was not used until the 35th day of life, well after the eruption of the molars into the mouth. Any delay between weaning and the commencement of any cariogenic challenge can reduce caries prevalence, so that significant differences become difficult to obtain when test agents are used. Later Hunt and Navia (1972, 1975), feeding $SrCl_2$ by esophageal intubation before tooth eruption, found Sr increased caries. Their dosage was 500 μg Sr/10 g body weight, which was a concentration unrelated to naturally occurring Sr in drinking waters, and comparable to about 300 ppm in human drinking water supplies. This was therefore nearly ten times higher than the highest water concentration found in Wisconsin.

Recent studies, such as those by Gedalia and co-workers (1975), have produced generally significant reductions in caries prevalence in hamsters using 25 or 75 ppm Sr pre-eruptively. A lack of posteruptive effect of the Sr was suggested as reflecting the major effect of Sr by incorporation in enamel during calcification. Gedalia's study also showed that the 25 ppm Sr was more effective than the 75 ppm concentration in preventing caries.

Preliminary studies by Meyerowitz et al (1976), subsequently substantiated in later work (Meyerowitz et al, 1979), have shown a decrease in caries prevalence in rats with $SrCl_2$, when given before or after tooth eruption. Furthermore, the maximum effect in these studies was at a dose of 50 ppm Sr, with a greater effect after tooth eruption than before. Thus, a regimen of 50 ppm Sr given posteruptively was associated with a reduction of buccolingual smooth surface lesions of from 5.5 to 1.5 (72%) compared with a reduction of from 8.3 to 5.1 (38%) for the same dose given pre-eruptively. Similar, but less marked, results were seen for other doses of Sr. This finding does not agree with that of Gedalia et al (1975) who reported significant caries reductions in hamsters, when 25 ppm Sr was given before tooth eruption (as well as prenatally). There is unlikely to be a species difference in response to the Sr between rats and hamsters, so there is no explanation of the different findings between the two studies. The only major disparity concerned the methods of administration of the Sr before weaning. In Meyerowitz's studies, esophageal intubation was used to ensure that the Sr was taken up by each rat pup, whereas in Gedalia's study the Sr was administered via the mother's drinking water. Furthermore, the use of the Sr during the prenatal period in Gedalia's study may be important, because the first molar of rodents begins to form during the intrauterine period. Recently, the findings of Ashrafi et al (1980) substantiate those of Meyerowitz, but with a maximum effect at 75 ppm Sr. The question of the effect of a trace element such as Sr during the period of tooth formation and calcification merits further study.

Various authors have obtained inconsistent results using different Sr salts. This has been partially explained recently in a report by Curzon et al (1981) who showed that varying reductions in rat caries prevalence could be obtained with different Sr salts. Lowest caries was found with SrF_2, and reductions occurred also with $SrCl_2$. No significant effect on caries was found with SrO, $SrSO_4$, $Sr(NO_3)_2$, and Sr-lactate.

MECHANISMS OF ACTION OF STRONTIUM

The association between Sr concentrations in drinking water and low caries prevalence, substantiated by animal experiments, requires explanation. The mechanism of action of the Sr effect could be associated with either a prevention of the initiation of the caries or continued changes in the surface chemistry of the enamel during a remineralizing process (Featherstone et al, 1981b). In either case, the trace element composition of the dental enamel might well be pertinent to an understanding of which trace elements affect the carious process. The trace element composition of whole enamel has, therefore, been a field of research in dentistry.

Incorporation in Whole Human Enamel

Since the early work of Drea, many studies have considered the presence of trace elements in enamel using various analytic methods. In human enamel Sr occurs in all samples analyzed, irrespective of origin, and the concentrations identified so far have ranged from 14 to 1200 ppm (Table 15-6).

Few studies have attempted to relate the concentration of Sr in enamel with the prevalence of caries. One study (Curzon and Losee, 1977) used teeth from a number of geographic areas of the United States, representing many different levels of caries prevalence. On the basis of 147 enamel samples, results were divided according to the caries prevalence of each individual tooth donor. There were significantly lower concentrations of Sr in the enamel of teeth from donors with a

Table 15-6
Strontium Concentrations in Human Whole Permanent Enamel: Summary of Studies

				Concentration Sr ppm		
Author	Year	n	Method of Analysis	*Minimum*	*Mean ± SD*	*Maximum*
Drea	1936	18	spectrographic		present	
Lowater and Murray	1937	NG	spectrographic		present	
Söremark and Samsahl	1961	15	neut. activation		93.5 ± 21.9	
Lundberg et al	1965	10	neut. activation		83 ± 32	
Cälonius and Visäpäa	1965	16	X-ray emission		10 ± 100	
Little and Steadman	1966	96	spectrographic		60 ± 100	
Hardwick and Martin	1967	2	mass spectrometry		100 ± 1,000	
Wycoff and Doberenz	1968	1[a]	X-ray spectrography		390 NG	
Retief et al	1971	7	neut. activation		111.19 ± 9.86	
Losee et al	1974[a]	93	optical emission	14	65.6 ± 5.22[b]	450
Losee et al	1974[b]	28	mass spectrometry	26	81 ± 11[b]	280
Losee et al	1974[c]	56	mass spectrometry	21	121 ± 11[b]	280
Derise and Ritchey	1974	173	atomic absorption		285.6	NG
Helsby	1974	NG	atomic absorption		NG	185
Nixon and Helsby	1976	181	atomic absorption	72	112.4	239
Curzon et al	1975	36	mass spectrometry	39[c]	92.6 ± 4.93[b]	170[c]
Curzon and Losee	1977[a]	147	mass spectrometry	21	193.04 ± 15.44	1,200
Retief et al	1978	Pooled[d]	neut. activation		140.3	NG
Vrbic and Stupar	1980	16	atomic absorption		89.7 ± 4.52	

NG = Not given.
[a]Fossil Amerindian tooth.
[b]Standard error of the mean.
[c]Minimum and maximum from unpublished data.
[d]Replicates on pooled samples.

DMFT of > 7 (high caries), than donors with a DMFT of < 4 (low caries).

Previously, Steadman et al (1958) showed that human tooth enamel from Texas had significantly higher levels of Sr than comparable samples from New England, and a number of epidemiologic surveys, reviewed by Dunning (1953), have shown these two geographic areas to be of low and high caries respectively.

Further studies relating trace elements, including Sr in human enamel, to dental caries have been carried out (Curzon and Losee, 1977b, 1978), but in this case first selecting high and low caries geographic areas and then analyzing teeth from those areas. On the basis of dental epidemiologic studies (Ludwig and Bibby, 1969), two areas were chosen for tooth collection, a high caries area with a mean DMFT of 6.92, and for comparison a lower caries area with a mean DMFT of 4.29. Results showed significantly higher Sr concentrations in enamel from the low caries area compared with the high caries area, and also higher enamel Sr in low caries individuals within each area (Table 15-7).

Table 15-7
Strontium Concentrations in Whole Enamel Samples Related to Caries Indices of Tooth Donors, Irrespective of Geographic Origin

	Concentration of Sr $\mu g/g$			
Group	*n* =	*Mean* ± *SE*	*t*	*p*
High caries	31	111.29 ± 15.45		
Los caries	96	204.52 ± 19.86	3.70	< 0.001

The simultaneous increase in the Sr and fluorine concentrations in enamel developed in a low fluoride water environment, such as New England and South Carolina, becomes interesting since in earlier work a similar pattern was found in enamel developed in an optimally fluoridated (1.0 to 1.2 ppm F) environment. The 12- to 14-year-old children from Ft. Recovery and Delphos, Ohio, had significantly lower caries prevalence, and enamel samples from persons living in these communities had greater Sr and fluoride concentrations than were found in enamel from comparable people in Portsmouth, Ohio (1.0 ppm F).

An analytical study carried out in South Africa by Retief and coworkers (1978) showed significantly higher concentrations of Sr in whole enamel samples derived from black subjects with low caries than from white subjects with high caries. Not only does this study add more evidence of an association of Sr in tooth enamel with low caries prevalence, but it also makes an interesting comparison with the earlier study of Curzon and Losee (1977a). Concentrations of Sr in enamel from high-versus-low caries donors were 103.0 and 177.7 ppm Sr respectively in the South African study. Corresponding values in the United States study were 104.6 and 183.0 ppm Sr. The similarity of these findings is interesting, even though different racial and cultural populations were used, as well as different sampling and analytic methods.

In the Dalmatian area of Yugoslavia, a recent report (Vrbic and Stupar, 1980) has again shown a relationship of Sr in soil, water, and human enamel to low caries prevalence. In the town of Novigrad, 3.9% of school children, aged 8 to 15 years, were caries-free, and analysis of enamel gave mean concentrations of 83 ppm Sr. By comparison, 29.6% of school children in Zamunik were caries-free with mean concentrations of 113 ppm Sr. The differences in Sr concentrations were different to a high degree of significance. If these Yugoslavian results are compared with those for the United States (Curzon and Losee, 1977a) and South African (Retief et al, 1978) studies, it is seen that the mean Sr concentrations are rather low. This is no doubt because of the lower Sr concentrations in the drinking water of Zamunik and Novigrad, which were 0.44 and 0.23 ppm Sr respectively. These drinking water Sr concentrations were markedly lower than those found in northwest Ohio, 5.2 ppm Sr (Curzon et al, 1970). Nevertheless, it is of interest that significant inverse relationships of Sr to caries have been found in a population with lower water Sr concentrations than previously were reported.

Analysis of whole enamel may only give results indicating the intake of Sr during the limited period of tooth development. The Sr concentrations in the outermost layers of enamel may be more important, since caries is initiated at the enamel surface.

Strontium in Surface Enamel

The concentrations of Sr found in permanent teeth surface enamel by a number of authors are summarized in Table 15-8. The range varies widely, but may only reflect the disparity of populations studied. Thus, the samples of Spector and Curzon (1978) were largely from the high Sr areas of northwest Ohio and Wisconsin, while those of Brudevold et al (1963), at the other end of the scale, were from metropolitan areas of Massachusetts.

Cutress (1972) suggested that there may be an increase in both Sr and fluoride in the outer enamel layers of some teeth obtained from low caries individuals. Although Steadman and co-workers (1958) found an even distribution of Sr across the depth of enamel, this has not been a consistent finding. The work of Steadman et al (1958) was not designed to consider caries prevalence in relation to Sr, and no information was given as to the caries history of the tooth donors used.

Table 15-8
Strontium Concentrations in Human Surface Enamel: Results of Previous Reports

			Concentration Sr ppm		
Author	Year	n	*Minimum*	*Mean ± SD*	*Maximum*
Cutress	1972	8	97	222 ± 145	495
Brudevold et al	1963	35	26	67 ± 20	132
Little and Barrett	1976	84	NG	449 ± 238	NG
Spector and Curzon	1979	439	5	366 ± 435	2077
Schamschula et al	1978	301	NG	100 ± NG	NG

NG = Not given.

Little and Barrett (1976) used an acid etch analytic technique on enamel samples collected for the whole enamel studies previously discussed (Curzon and Crocker, 1978). For this study, the caries history of each individual tooth donor was known as well as the caries prevalence in the geographic area of tooth origin. Analytic results showed a gradient of Sr concentrations from the enamel surface inward, indicating that Sr may be acquired by the surface enamel after tooth eruption, which may possibly affect the ability of a tooth to resist carious attack. Little and Barrett (1976) concluded that teeth derived from areas of high caries prevalence had less Sr overall (as well as fluoride) in the surface enamel than those from low caries areas. When enamel samples of individuals from high and low caries groups were compared, there was at least twice as much Sr (and fluoride) in the surface enamel of the low caries group. These studies identified significant differences of Sr in enamel by comparing groups of individuals. However, Sr concentrations in enamel and caries indices within groups vary widely. Spector and Curzon (1979) could find no significant relationship beween surface enamel Sr and individual caries prevalence of the tooth donors.

If Sr by incorporation in enamel, and most likely surface enamel, has an effect on dental caries, then the physicochemical action of the Sr ion must be important. Any possible effect upon enamel dissolution may play a role in whether or not the surface enamel becomes carious.

Effects on Enamel Dissolution

Sr has been identified in virtually every sample of whole enamel analyzed, by whatever method, and in most samples of surface enamel. By its close similarity to Ca in chemical and physical properties, Sr can theoretically replace Ca in hydroxyapatite (ionic radii Sr = 0.112 and Ca = 0.099 nm). This is true, and up to four of the Ca ions in human enamel apatite may be replaced by Sr. Changes in the physical and chemical properties of enamel apatites may be brought about by the replacement of Ca by Sr, and the effect of their substitutions upon caries needs to be considered.

Since the early enamel lesion is the result of dissolution of enamel by acids produced by bacterial fermentation of carbohydrates, the resistance to such dissolution is important. Dedhiya et al (1974) studied how the addition of Sr and fluoride affected the dissolution of synthetic hydroxyapatite. The incorporation of either Sr or fluoride retarded apatite dissolution, although it was much greater with fluoride. When the Sr and fluoride were used in combination, the effect was greater even than for fluoride alone, and greater than would have been expected from the addition of the Sr and fluoride separately. This suggests a synergistic relationship for the two trace elements.

Recently, Herbison and Handelman (1975) looked at the effects of Sr, fluoride, B, Mo, and Li on bacterial growth of *S. mutans,* acid production, and hydroxyapatite dissolution. This work is of interest because the concentrations and combinations of trace elements used were based on those identified in the communal water supplies of the low caries towns in northwest Ohio (Curzon et al, 1970). This was, therefore, an important attempt to relate in vitro experiments to epidemiologic findings. The results showed no effect of any of the trace elements, alone or in whatever combination, on bacterial growth or acid production. On hydroxyapatite dissolution, however, Sr and fluoride in combination had a significant effect in reducing dissolution, which was not due to changes in the buffering capacity of the medium, or to bacterial growth or acid production. It appeared that in the case of the best combination, 1.5 ppm F and 17 ppm Sr, the effect was produced by reducing the solubility of synthetic hydroxyapatite.

It is of note that in the studies of both Dedhiva et al and of Herbison and Handelman, synthetic apatites were used. Obviously, it is easier, and it reduces the number of variables, to use synthetic material, but one must question how applicable these results are to human enamel, an imperfect apatite with many impurities as well as an organic component. On the other hand, Gedalia and co-workers who used powdered human enamel to test the effects of Sr and fluoride on enamel dissolution found that Sr alone had no effect, while fluoride alone reduced dissolution considerably. A combination of Sr and fluoride gave no better dissolution reduction than fluoride alone. Their method involved pretreating the powdered enamel with Sr, fluoride, or both, and then challenging the enamel with 0.1 mol HCl. (Gedalia et al, 1975).

It would seem, therefore, that although an effect of Sr on dissolution can be demonstrated with synthetic hydroxyapatite, this does not take place with human enamel. Recently Ladrigan et al (1979) showed that there is an effect of Sr and a synergistic effect of Sr and fluoride on human enamel dissolution, when these ions are added to a challenging acetate buffer solution. Use of powdered whole enamel with high Sr and fluoride content shows no effect on enamel dissolution rate. These findings suggest that a cariostatic effect of Sr and/or fluoride takes place at the enamel surface by a mechanism acting through the solutions present at the enamel-plaque interface, thus supporting the findings from rat caries studies of a posteruptive effect of Sr.

Effects on Mineralization

There is also evidence from a number of animal studies which indicates that Sr in excessive levels, when administered during enamel and dentin formation, produces defective calcification. Weinmann (1942) demonstrated that ameloblasts were affected during tooth development when $SrCl_2$ was injected into growing rats. Hunt and Navia (1972) reported macroscopic and microscopic evidence of rachitic lesions in the ribs, and proliferative hypomineralized foci in the dentin of the developing molars when a dosage of 500 mg Sr/10 g body weight were fed to rats in their diet. Other authors have also reported this effect by high dosages of Sr when injected into rats (Irving, 1944; Yaeger, 1966; Eisenman and Yaeger, 1969). Later work by Castillo-Mercado and Bibby (1973) showed also that Sr produced defects in rat enamel identified as banding of the enamel in the rat incisor, which corresponded to the time of injection. It appears that Sr interferes with the production of normal mineralized organic matrix.

In man there has been one report of an association of defects in enamel (mottling) with Sr in drinking water (Curzon and Spector, 1977). At the time of the dental examinations for caries in Wisconsin, records were kept of the prevalence of mottling, opacities, and pigmentation in permanent anterior teeth. The percentage of lifelong resident children affected by mottling significantly increased as the Sr concentrations in drinking water increased from 0.02 to 33.9 ppm Sr. This finding has not been substantiated or refuted by any other research to date.

There is evidence that structural modification of the apatite crystal may occur by Sr incorporation. The studies of LeGeros et al (1977) indicated that incorporation of Sr in synthetic hydroxyapatite increased the size of each crystallite. There was a systematic increase in lattice parameters of apatite formed from solutions with increasing amounts of Sr (Table 15-9). When Sr and fluoride were present together, the resulting apatite had lattice parameters lower than fluoride-free, Sr-containing apatites, but higher than the Sr-free fluoride apatites. This suggested a simultaneous substitution of Sr for calcium and fluoride for

Table 15-9
Effects of Incorporation of Strontium on Lattice Parameters of Synthetic Apatite

Sample	Sr/Ca in Solution	Lattice Parameters a-axis	($\pm$ 0.003nm) c-axis
1	0/100	0.9438	0.6880
2	1/99	0.9455	0.6898
3	2/98	0.9469	0.6895
4	5/95	0.9475	0.6911
5	10/90	0.9480	0.6924
6	20/80	0.9495	0.6953
7	30/70	0.9543	0.6994
8	50/50	0.9639	0.7034

After LeGeros et al, 1977.

the hydroxyl ion. The simultaneous presence of both carbonate and Sr in the solutions used to produce the synthetic apatites also showed interesting results. It has been thought for some time that the presence of carbonate in human apatites rendered them more susceptible to dissolution. Studies by LeGeros et al (1976) have shown that the presence of Sr limits the incorporation of carbonate. So the increased incorporation of Sr may bring about a reduction in dissolution, either by increasing the size of the apatite crystal (and hence its surface area) or by reducing the incorporation of carbonate.

The criticism of dissolution studies that results for synthetic apatites may or may not be applicable to human enamel also holds true for X-ray crystallographic studies. However, using pieces of enamel from the same teeth used in the studies of Curzon and Losee (1977a), LeGeros (1977) found increases in the c-axis of the apatites with increasing Sr. This series also included the tooth containing 1200 ppm Sr mentioned above. The increases in crystallite size were similar to those found for synthetic hydroxyapatite.

Recent work by Featherstone and co-workers (1981b) has studied the role of metal ions, including Sr, on the processes of demineralization and remineralization of natural and synthetic apatites. Enamel apatites are far from pure structures but contain Ca-deficient areas. Carbonate inclusions may be involved in these defects and may also influence the subsequent dissolution of the enamel. Metal ions, or metal-fluoride complexes, are apparently incorporated in Ca-deficient areas (Featherstone et al, 1981a). Experiments on remineralization or artificially induced lesions have shown that there are solutions with components which will diffuse rapidly in combination into carious enamel (Featherstone et al, 1981b). A major component of such solutions is Ca, but in conjunction with Sr as well as Zn, it appears to reduce enamel solubility as well as to improve apatite structure (Featherstone et al, 1979).

Changes in Tooth Morphology

It has been demonstrated in man that one action of fluoride may be to change not only tooth size, but also fissure patterns and cuspal slope. In any of these cases initiation of the caries could be reduced by making tooth configurations less susceptible to food impaction and stagnation. Previous authors (Kruger, 1959, 1966; Castillo-Mercado and Bibby, 1973) have reported on the effects of trace elements on molar morphology in the rat. In the study by Castillo-Mercado and Bibby an increase in fissure width and thickness of dentin was associated with intraperitoneal injections of 50 ppm Sr. In a later study Curzon et al (1981) fed rats $SrCl_2$ (0, 50, and 150 ppm Sr) via esophageal intubation from day 3 to day 19 of life. After sacrifice on day 65, mandibular molars were sagittally sectioned and measurements made of crown diameter, fissure widths, and thicknesses of enamel and dentin. Results showed a significant increase in dentin thickness, both horizontally and vertically, at the base of the first molar sulci. This finding confirmed the earlier result of Castillo-Mercado and Bibby. An increase in dentin (and possibly enamel thickness) could be important in relation to the rate of progress of caries.

Interactions with Other Trace Elements

From the foregoing discussion of the relationship of Sr to dental caries, the reader will have noted the concern with fluoride. It would appear that there is an interaction of Sr with fluoride which may affect dental caries. Some of the rat studies, such as those of Meyerowitz et al (1976, 1979) showed a greater reduction in caries by a combination of Sr with fluoride than by fluoride alone. Similarly in the dissolution studies of Dehiya et al (1974), synergism was evident in reducing apatite solubility by Sr with fluoride. Furthermore, the work of LeGeros et al (1977) has suggested simultaneous substitution of Sr for Ca and fluoride for hydroxyl ions in synthetic hydroxyapatite. These relationships merit further study with an emphasis on the physicochemistry of Sr and fluoride in dental enamel.

Another obvious interaction is that of Sr and Ca. Although this interaction has been extensively studied with regard to bone calcification and the prevalence of Sr^{90}, the relationship has not been well looked at in teeth. Other interactions of Sr with other trace elements that should be considered are the suggested fluoride-Al-Sr found in the enamel studies of Curzon and Crocker (1978), and the relationships of all of the alkaline earths—Mg, Ca, Sr, and Ba—to each other in tooth enamel.

SUMMARY

The evidence for a role of Sr in the dental caries process appears now to be strong enough to merit serious consideration of Sr as a cariostatic agent. Found regularly in human whole and surface enamel, Sr is in concentrations related to those occurring in drinking water supplies, and may or may not be related to caries. Epidemiologic studies in man and animal studies with rats have indicated an association of Sr in drinking water to low caries prevalence at concentrations of about 5 and 50 ppm Sr respectively. Evidence for a posteruptive effect of Sr derives from the findings in rats where greater caries reductions have been obtained when Sr was given after weaning. Incorporation of Sr in the apatite crystal as a substitution of Ca may be one possible mechanism of action whereby the physicochemical characteristics of apatite could be changed by Sr. This may be related to changes in resistance to dissolution, either by Sr alone or by synergism with fluoride.

REFERENCES

Anderson RJ: The relationship between dental conditions and the trace element molybdenum. *Caries Res* 1969;3:75–87.

Ashrafi MH, Spector PC, Curzon MEJ: Post-eruptive effect of low doses of strontium on dental caries in the rat. *Caries Res* 1980;14:341–346.

Barmes DE: Caries aetiology in Sepik Villages—Trace element, micronutrient and macronutrient content of the soil and food. *Caries Res* 1969;3:44–59.

Boyde J, Neumann WF, Hodge HC: On the mechanism of skeletal fixation of strontium. University of Rochester, Atomic Energy Project UR 512:1–16, 1958.

Brudevold F, Steadman LT, Spinelli MA, et al: Distribution of strontium in teeth from different geographic areas. *Arch Oral Biol* 1963;8:135–144.

Calonius PEB, Visäpaä A: Inorganic constituents of human teeth and bone by X-ray emission spectrography. *Arch Oral Biol* 1965;10:9–13.

Castillo-Mercado R, Bibby BG: Trace element effects on enamel pigmentation, incisor growth and molar morphology in rats. *Arch Oral Biol* 1973;18: 629–635.

Curzon MEJ, Adkins BL, Bibby BG, Losee FL: Combined effect of trace elements and fluorine on caries. *J Dent Res* 1970;49:526–529.

Curzon MEJ, Losee FL, McAlister AD: Trace elements in the enamel of teeth from New Zealand and the USA. *NZ Dent J* 1975;71:80–83.

Curzon MEJ, Losee FL: Strontium content of enamel and dental caries *Caries Res* 1977a;11:321–326.

Curzon MEJ, Losee FL: Trace element composition of whole human enamel and dental caries. Part I. Eastern USA. *JADA* 1977b;94:1146–1150.

Curzon MEJ: *Trace Element Content of Human Enamel and Dental Caries.* PhD. Thesis. University of London, London. 1977.

Curzon MEJ, Spector PC: Enamel mottling in a high strontium area of the USA. *Community Dent Oral Epidemiol* 1977;5:243–247.

Curzon MEJ, Losee FL: Trace element composition of whole human enamel and dental caries. Part II. Western USA. *JADA* 1978;96:819–822.

Curzon MEJ, Spector PC, Iker HP: An association between strontium in drinking water supplies and low caries prevalence. *Arch Oral Biol* 1978;23: 317–321.

Curzon MEJ, Crocker DC: Relationship of trace elements in human tooth enamel to dental caries. *Arch Oral Biol* 1978;23:647–653.

Curzon MEJ, Spector PC: Effect of using different strontium salts on dental caries in the rat. *Caries Res* 1982;15:296–301.

Curzon MEJ, Ashrafi MH, Spector PC: Effects of strontium administration on rat molar morphology. *Arch Oral Biol* 1981; In press.

Cutress TW: The inorganic composition and solubility of dental enamel from several specified population groups. *Arch Oral Biol* 1972;17:93–109.

Dedhiya MG, Young F, Higuchi WI: Mechanism of hydroxyapatite dissolution. The synergistic effects of solution fluoride, strontium and phosphate. *J Phys Chem* 1974;78:1273–1279.

Derise NL, Ritchey SJ: Mineral composition of normal human enamel and dentin and the relation of composition to dental caries. II. Microminerals *J Dent Res* 1974;53:853–858.

Drea WF: Spectrum analyses of dental tissues for "trace" elements. *J Dent Res* 1936;15:403–406.

Dunning JM: The influence of latitude and distance from the sea coast on dental caries. *J Dent Res* 1953;32:811–829.

Dufor CN, Becker E: *Public Water Supplies of the 100 Largest Cities of the US.* United States Geological Survey Water Supply, paper #1812, 1962.

Eisenman DR, Yaeger JA: Alterations in the formation of rat dentin and enamel induced by various ions. *Arch Oral Biol* 1969;14:1045–1064.

Featherstone JDB, McGrath MP, and Smith MW: The effect of trace elements in some New Zealand water supplies on synthetic apatite structure. *NZ Dent J* 1979;75:206–211.

Featherstone JDB, Nelson DG, McClean JD: An electron microscope study of the modifications to defect regions in dental enamel and synthetic apatites. *Caries Res* 1981a;15:278–288.

Featherstone JDB, Rodgers BE, Smith MW: Physico-chemical requirements for rapid remineralization of early carious lesions. *Caries Res* 1981b;15:221–235.

Forbes RM, Mitchell HH: Accumulation of dietary boron and strontium in young and adult albino rats. *Arch Indust Health* 1957;16:489–492.

Gedalia I, Anaise J, Laufer E: Effect of prenatal, preeruptive and posteruptive strontium administration on dental caries in hamster molars. *J Dent Res* 1975;54:1240.

Grippaudo G, Cattabriga M, Valfre F: L'influenza degli elementi traccio molibdeno, vanadio e stronzio sulla carie sperimentale. *Ann Stomatol* 1970;8: 223–236.

Hardwick JL, Martin CJ: A pilot study using mass spectrometry for the estimation of trace element content of dental tissues. *Helv Odont Acta* 1967;11:62–70.

Helsby CA: Determination of strontium in human tooth enamel by atomic absorption spectrometry. *Anal Chim Acta* 1974;69:259–265.

Herbison RJ, Handelman SL: Effect of trace elements on dissolution of hydroxyapatite by cariogenic streptococci. *J Dent Res* 1975;54:1107–1114.

Hunt CE, Navia JM: Effects of Sr, Mo, Li, and B in developing teeth and other tissues of neonatal rats, in *Trace Substances in Environmental Health,* vol 6. Columbia, MO, University of Missouri, 1972.

Hunt CE, Navia JM: Pre-eruptive effects of Mo, B, Sr and F on dental caries in the rat. *Arch Oral Biol* 1975;20:1–5.

Irving JT: "Fluorine-like" action of various substances on the teeth. *Nature* 1944; 154:149–150.

Johansen E, Hein JW: Effect of strontium chloride on experimental caries in the Syrian hamster. *J Dent Res* 1953;32:703.

Kruger BJ: The effect of trace elements on experimental dental caries in the albino rat. *University of Queensland Papers* 1959;1:3–28.

Kruger BJ: Interaction of fluoride and molybdenum on dental morphology in the rat. *J Dent Res* 1966;45:714–725.

LeGeros RZ, Miravite MA, Bonel G: CO_3 in apatites: Influence of other ions on carbonate incorporation. IADR Abst #955, *J Dent Res* 1976;55:300.

LeGeros RZ, Miravite MA, Quirologico GB, et al: The effect of some trace elements on the lattice parameters of human and synthetic apatites. *Calcif Tissue Int* 1977;22:362–366.

Little MF, Steadman LT: Chemical and physical properties of altered and sound enamel. IV Trace elements. *Arch Oral Biol* 1966;11:273–278.

Little MF, Barrett K: Strontium and fluoride content of surface and inner enamel versus caries prevalence in the Atlantic coast of the USA. *Caries Res* 1976; 10:297–307.

Lödrop H: The low rate of dental decay in Bonn am Rhein and the conclusions that can be drawn from it. *Den Norske Tannl Tid* 1953;63:35–50.

Losee FL, Adkins BL: Anti-cariogenic effect of minerals in food and water. *Nature* 1968;219:630–631.

Losee FL, Adkins BL: A study of the mineral environment of caries-resistant Navy recruits. *Caries Res* 1969;3:23–31.

Losee FL, Little-McClellan MF, Orbell GM: Strontium content of teeth related to geologic environment. IADR Abst #638. *J Dent Res* 1971;50:123.

Losee FL, Curzon MEJ, Little MF: Trace element concentrations in human enamel. *Arch Oral Biol* 1974a;19:467–471.

Losee FL, Cutress TW, Brown R: Natural elements of the periodic table in human dental enamel. *Caries Res* 1974b;8:123–134.

Losee FL, Cutress TW, Brown R: Trace elements in human dental enamel, in *Trace Substances in Environmental Health VII.* Columbia, MO, University of Missouri, 1974c.

Lowater F, Murray MM: Chemical composition of teeth: Spectrographic analyses. *Biochem J* 1937;31:837–843.

Ludwig TG, Bibby BG: Geographic variations in the prevalence of dental caries in the USA. *Caries Res* 1969;3:32–43.

Lundberg M, Söremark R, Thilander H: The concentrations of some elements in the enamel of unerupted (impacted) human teeth. *Odont Revy* 1965;16:8–11.

Matsumoto S: The effects of strontium upon experimental dental caries in rats. *Shika Gakuho* 1980;80:53–75.

McCaslin FE, Janes JM: Effects of stable strontium in treatment of osteoporosis, in Skoryna SK (ed): *Handbook of Stable Strontium.* New York, Plenum Press, 1981.

Meyerowitz C, Little MF, Curzon MEJ: Sr in rat enamel and caries. A preliminary report. IADR Abst #259, *J Dent Res* 1976;55:B126.

Meyerowitz C, Spector PC, Curzon MEJ: Pre- or post-eruptive effects of strontium alone or in combination with fluoride on dental caries in the rat. *Caries Res* 1979;13:203–210.

Nichols MS, McNall DR: Strontium content of Wisconsin municipal waters. *J Am Water Works Assoc* 1957;49:1493–1501.

Nixon GS, Helsby CA: The relationship between strontium in water supplies and human tooth enamel. *Arch Oral Biol* 1976;21:691–695.

Olson BL, McDonald JL, Stookey GK: The effect of strontium and fluoride upon in vitro plaque and rat caries. *J Dent Res* 1978;57:903.

Retief DH, Cleaton-Jones, PE: The quantitative analysis of Sr, Au, Br, Mn and Na in normal human enamel and dentin by neutron activation and high resolution gamma spectrometry. *J Dent Res* 1971;26:63–69.

Retief DH, Turkstra J, Cleaton-Jones PE, et al: Mineral composition of enamel from population groups with high and low caries. IADR Abst #302, *J Dent Res* 1978;57:150.

Rygh O: Recherches sur les oligo-elements. I. De l'importance du strontium, du beryum et du zinc. *Bull Soc Chim Biol* 1949;31:1052–1057.

Schamschula RG, Adkins BL, Barmes DE, et al: *WHO Study of Dental Caries Aetiology in Papua-New Guinea.* WHO Publication #40, Geneva, 1978.

Schroeder HA, Tipton IH, Nason AP: Trace metals in man: Strontium and barium. *J Chron Dis* 1972;25:491–517.

Schwartz K: New essential trace elements (Sn, V, F, Si): Progress report and outlook, in Hoekstra WG (ed): *Trace Element Metabolism in Animals.* 2. Baltimore, University Park Press, 1974.

Shaw JH, Griffiths D: Developmental and postdevelopmental influences on incidence of experimental dental caries resulting from dietary supplementation by various elements. *Arch Oral Biol* 1961;5:301–322.

Skoryna SC: *Handbook of Stable Strontium.* New York, Plenum Press, 1981.

Söremark R, Samsahl K: Gamma ray spectrometric analysis of elements in normal human enamel. *Arch Oral Biol* 1961;6:275–279.

Spector PC, Curzon MEJ: Relationship of strontium in drinking water and surface enamel. *J Dent Res* 1978;57:55–58.

Spector PC, Curzon MEJ: Surface enamel fluoride and strontium in relation to caries prevalence in man. *Caries Res* 1979;13:227–230.

Steadman LT, Brudevold F, Smith FA: Distribution of strontium in teeth from different geographic areas. *JADA* 1958;57:340–344.

Storey E: Strontium rickets, bone calcium and strontium changes. *Aust Ann Med* 1961;10:213–222.

Turkian KK, Kulp JL: Strontium content of human bones. *Science* 1956;124: 405–407.

Vrbic V, Stupar J: Dental caries and the concentration of aluminum and strontium in enamel. *Caries Res* 1980;14:141–147.

Weinmann JP: Effect of strontium on the incisor in the rat. I. Ingestion of small doses of strontium chloride as a means of measuring in the rat of incremental dentine apposition. *J Dent Res* 1942;21:497–499.

Wolf N, Gedalia I, Yariv S, et al: The strontium content of bones and teeth of human foetuses. *Arch Oral Biol* 1973;18:233–238.

Wyckoff RWG, Doberenz AR: The strontium content of fossil teeth and bones. *Geochim Cosmochim Acta* 1968;32:109–115.

Yaeger JA: Recovery of rat incisor dentin from abnormal mineralization produced by strontium and fluoride. *Anat Rec* 1966;154:661–673.

$_{56}$ **Ba**
137.34
18-8-2

16 Barium

M. E. J. Curzon

In 1808 Sir Humphrey Davy added to the periodic table a number of new elements which included barium (Ba). This element he named after the Greek word 'barys' meaning heavy, hence "barium." As Ba occurs in relatively high concentrations in the earth's crust, as well as being present in sea water, it is found in the tissues of most living things, but at much smaller concentrations than Mg, Ca, or Sr.

The most common geologic stratum containing Ba is barite of $BaSO_4$, which is also the main source of the element for industrial use. Barite is used as a lubricant in drilling oil wells, while other Ba compounds are used in glass, ceramics, and television picture tubes, and as a paint pigment among many other uses. The extensive use of the element in modern industry has meant that man is exposed to increasing concentrations of Ba. In 1968 over two million tons were used in the world in comparison with Sr, of which only 12,500 tons were used in the same period.

There has been much research on Ba in dentistry, and the number of publications concerning the element have been limited, despite a report more than ten years ago (Healy and Ludwig, 1968) which suggested that dietary differences in Ba intake are reflected in contents of bone, enamel, and dentin. The element merits further investigation, particularly since it is ever-present in the geochemical environment, and readily incorporated into calcified tissues.

Metabolism of Ba and its Presence in Calcified Tissues

The whole body contains about 22 μg pf Ba (Schroeder et al, 1972) of which some 91% is located in the bones. Other areas where Ba is found are fat, skin, and lungs. According to Schroeder there are marked geographic differences in tissue Ba concentrations in various parts of the world; he suggested that highest exposures were found in the Near and Far East and Africa, probably from dust.

Municipal drinking waters in the United States contain an average of 43 μg/l of Ba (range 1.7 to 380) which, if the average adult consumes 2 liters of water per day, provides about 0.086 mg or 9% of the average daily intake. The bulk of the Ba intake is, therefore, from foods except where waters are particularly hard, where the Ba concentration would be much higher. The major sources of Ba in the diet have not been reported. Schroeder et al (1972), in the major paper on Sr and Ba, gave extensive lists of the Sr composition of foods, but nothing on Ba.

In living organisms Ba occurs in everything that has been tested from marine animals to man. Some animals depend upon the element, such as the rhizopod Xenophyofera where the exoskeleton is made up of $BaSO_4$. For this animal, therefore, the element is essential. In general, however, evidence shows that Ba is not essential for life, and nothing is known of any biochemical function in calcified and soft tissues. No catalytic or electrochemical roles for Ba have been identified, although it is possible the element might have a role in calcification. Its place in the periodic table makes this unlikely.

Clearance of Ba in the urine in test subjects was 9% of that of Ca, and about 5% of that of Sr, indicating preferential excretion of Sr (Schroeder et al, 1972). Absorption of Ba compounds from the gut appears poor both from food and from water.

Taken orally, the alkaline earths have a low order of toxicity to living organisms in the order, Be $>$, Ba $>$, Sr $>$, Mg $=$ Ca. For Ba the toxicity seems to depend upon the compound taken, eg, the carbonate is not as toxic as the chloride. Insoluble $BaSO_4$ is well tolerated and, indeed, forms the compound for Ba meals or enemas used in radiology. Instances of acute toxicity of Ba in man appear to be unknown. However, baritosis, a disease of miners, is caused by Ba inhalation, but apart from this condition there is no evidence that Ba, in deficient or excess amounts, causes any chronic disease. A more detailed discussion on Ba in human metabolism is given by Schroeder et al (1972).

BARIUM IN HUMAN ENAMEL

Because of its position in the periodic table, together with Ca and Sr, it is not surprising to find Ba in most samples of calcified tissues including enamel. Over the years, a number of analyses have been carried out on enamel, the results of which are summarized in Table 16-1. From this review, it is apparent that the concentration of Ba can vary considerably within the range of 0 to 500 ppm.

The mean concentration of Ba varies in the reports listed. Geographic factors probably explain the variation of Ba incorporation in enamel. Thus, the analyses of Losee et al (1974 a,b,c) were all carried out on United States tooth enamel, while those of Retief et al (1970) were on South African teeth, and those of Cutress (1972) on a mixed group of South Pacific enamel samples. As noted above, Schroeder et al (1972) reported on the higher exposures of Ba in the Near and Far East and Africa, and this could be reflected in human tooth enamel.

Nevertheless, differences in analytic techniques used by the different researchers no doubt contributed to the variation observed. Ba analyses of enamel have been by spark source mass spectrometry, which show that Ba is readily detectable and uniform, but present in low concentrations. Analysis with neutron activation or emission spectrography produces high mean concentrations.

Attempts to relate Ba in human enamel to dental caries have been few and amount only to the report of higher Ba in enamel from a high caries area (Curzon and Losee, 1977), a report concerned with many trace elements; there remains a need for a specific study of the role of Ba in human tooth enamel.

BARIUM IN DENTAL DISEASE

The presence of Ba in teeth was noted early (Drea, 1936) and has been found in most samples of enamel analyzed (Curzon and Crocker, 1978). The element was suggested by Rygh (1949) as impairing calcification in rats, and associated with an increase in dental caries. Shaw and Griffiths (1961) considered Ba to only have a minor influence on caries in rats, and no other work on experimental caries and Ba has been reported.

Only three studies have considered links between Ba and dental caries. One of those studies concerned the Ba precipitability criterion for

Table 16-1
Reported Concentrations of Ba in Human Whole Enamel

Author	Year	n	Method	Mean Conc. ppm ± SE	Range
Drea	1936	18	Spectrographic	present	NG
Lowater and Murray	1937	NG	Spectrographic	present	NG
Hardwick and Martin	1967	3	Spark source mass	NG	10–100
Retief et al	1970	7	Neutron activation	125.1 ± 8.97	NG
Cutress et al	1972	11	Emission spectrography	48.6 ± 5.59[a]	11–500
Losee et al	1974a	28	Spark source mass	4.2 ± 0.60	0.8–13
Losee et al	1974b	93	Optical emission	15.3 ± 0.75	4.2–44
Losee et al	1974c	56	Spark source mass	5.6 ± 0.51	0.8–17
Curzon et al	1975	36	Spark source mass	13.9 ± 2.25	NG
Curzon and Crocker	1978	451	Spark source mass	18.8 ± 0.83	0.0–510

[a]Calculated from data given in paper.
NG = not given.

the presence of polyphosphate formation by caries-conducive *Streptococcus mutans*. Schamschula et al (1978) looked for Ba as one of the trace elements in the Papua-New Guinea study. Although identified in a number of samples analyzed, the only trend of a correlation of Ba with dental caries was in surface enamel, where a positive association between Ba in enamel and dental caries was reported. Similarly, as part of the large study of whole human enamel trace elements and caries, Curzon and Losee (1977) indicated significantly higher Ba concentrations in enamel from a high caries area in New England compared with a low caries area of South Carolina.

It could be argued that being well above Ca in the periodic table in group IIa, Ba would not normally be able to fit into the apatite structure. Ba is considerably bigger than Ca and therefore would cause distortion of the lattice parameters to a much greater extent than Sr, with an unknown effect on the physical and chemical properties of the apatite.

SUMMARY

Research on Ba and dental disease has been very limited, and is insufficient for any conclusions of any possible role for the element. From a knowledge of the physicochemistry of Ba relative to other elements of the alkaline earth group of the periodic table, it may be that the element has a detrimental effect on enamel and is thus cariogenic.

REFERENCES

Curzon MEJ, Losee FL, MacAlister AD: Trace elements in the enamel of teeth from New Zealand and in the USA. *NZ Dent J* 1975;71:80–83.

Curzon MEJ, Losee FL: Trace element composition of whole human enamel and dental caries. Part I. Eastern USA. *JADA* 1977;94:1146–1150.

Curzon MEJ, Crocker DC: Relationships of trace elements in human tooth enamel to dental caries. *Arch Oral Biol* 1978;23:647–653.

Cutress TW: The inorganic composition and solubility of dental enamel from several specified population groups. *Arch Oral Biol* 1972;17:93–109.

Drea WF: Spectrum analysis of dental tissues for "trace" elements. *J Dent Res* 1936;15:403–406.

Hardwick JL, Martin CJ: A pilot study using mass spectrometry for the estimation of the trace element content of dental tissues. *Helv Odont Acta* 1967;11: 62–70.

Healy WB, Ludwig TG: Barium content of teeth, bone and kidney of twin sheep raised on pastures of differing barium content. *Arch Oral Biol* 1968;13: 559–563.

Losee FL, Curzon MEJ, Little MF: Trace element concentrations in human enamel. *Arch Oral Biol* 1974a;19:467–471.

Losee FL, Cutress TW, Brown R: Natural elements of the periodic table in human dental enamel. *Caries Res* 1974b;8:123–134.

Losee FL, Cutress TW, Brown R: Trace elements in human dental enamel, in *Trace Substances in Environmental Health.* VII. Columbia, MO, University of Missouri, 1974c.

Lowater F, Murray MM: Chemical composition of teeth: Spectrographic analysis. *Biochem J* 1937;31:837–843.

Retief DH, Cleaton-Jones PE: The quantitative analysis of Cr, Ba, Sb, Ag, Zn, Co, and Fe in normal human enamel and dentin by neutron activation and high resolution gamma spectrometry. *J Dent Assoc SA* 1970;25:370–375.

Rygh O: Recherches sur les oligo-elements. I. De'importance du strontium, du beryum et du zinc. *Bull Soc Chim Biol* 1949;31:1052–1057.

Schamschula RG, Adkins BL, Barmes DE, et al: *WHO study of Dental Caries Aetiology in Papua-New Guinea.* WHO Publication #40. Geneva, 1978.

Schroeder HA, Tipton IH, Nason AP: Trace metals in man: Strontium and barium. *J Chron Dis* 1972;25:491–517.

Shaw JH, Griffiths D: Developmental and postdevelopmental influences on incidence of experimental dental caries resulting from dietary supplementation by various elements. *Arch Oral Biol* 1961;5:301–322.

3 Li

6.94

2-1

17 Lithium

A. D. Eisenberg

The lightest of all metals, lithium (Li) has a density only half that of water. It is highly reactive and never occurs naturally as a free element. One of the alkali metals, Li occurs in group one of the periodic table together with Na, K, Rb, Cs, and Fr. Small amounts of Li are to be found in igneous rocks and also in gypsum. Its association with the latter means it usually is found in many mineral springs and artesian wells in areas where gypsum is found.

In commercial use, Li is an alloying agent, but it is also used widely in special glasses and ceramics. The stearate of Li is used as a lubricant. Of greater interest here, however, is the use of the carbonate in the treatment of manic-depressive syndromes in man (Maletzky and Blachley, 1971). Large doses of $LiCO_3$ are given on a daily basis to these patients, and an increase in dental caries has been noted. However, the increase in caries may not be due solely to increased Li intake. Recently, several other epidemiologic studies have described an association of low caries to Li in drinking water.

Metabolism

No clear proof of an es sential role for Li has yet been demonstrated (Mertz, 1981), and little is known of its metabolic behavior in man or animals (Underwood, 1977). All human tissues contain some levels of Li (Table 17-1), and it appears to be readily absorbed from food and water but only small quantities are retained. Dietary intakes are variable and would appear to be dependent on the degree of Li in water. Mineral springs, or sources of hard water contain high levels of Li. The average daily intake of Li has been estimated at 2 mg per day but this figure is not widely accepted and a much lower intake of 20 μg has been reported (Underwood, 1977).

Table 17-1
Lithium Concentrations[a] in Human Tissues

Tissue	Mean ± SE	Tissue	Mean ± SE
Brain	0.004 ± 0.001	Blood	0.006 ± 0.002
Muscle	0.005 ± 0.002	Lymph nodes	0.2 ± 0.07
Lung	0.06 ± 0.01	Liver	0.007 ± 0.003
		Enamel	0.92 ± 0.10

[a]Concentrations in ppm (μg/g).
After Underwood, 1977; Curzon and Crocker, 1978.

The mechanism of the pharmacologic action of Li in manic depressives is still not well understood. The large doses used for manic-depressive psychosis, following an acute attack, may reach 1800 mg $LiCO_3$ (50 mEq Li) and can produce toxic symptoms which include polyuria, ataxia, and hypothyroidism. By contrast, deficiency states in both man and animals are not known. It would appear that if the element is essential, the daily requirement is low and is easily obtained through a normal diet.

Recently, interest in Li has increased because of a possible association with atherosclerosis in man (Voors et al, 1975). Atherosclerotic heart disease appears to have a negative correlation with hardness of drinking water. Epidemiologic studies coupled with trace element analysis of water supplies showed that only Li levels were significantly inversely correlated with the atherosclerotic heart disease in Caucasian men in 99 cities of the United States. High concentrations of Li occur in several parts of the United States, notably Texas, but these are not areas where the incidence of heart disease is necessarily the lowest. Because of the recent finding of hard water and Li as related to low dental caries (Schamschula et al, 1981), the interaction of trace element hardness of water and disease merits further attention.

DENTAL CARIES

Man

Recently, several epidemiologic surveys have implicated Li as being associated with low caries prevalence. Interest in Li arose about twenty years ago mainly as a result of research on trace elements in general; the findings are reflected in a number of subsequent animal studies.

An early indication of a possible association of Li to caries in man was in the Naval recruit studies of Losee and Adkins (1969) which identified an area of low caries prevalence in Ohio (USA) as related to high levels of Li, B, Mo, and Sr. Following these studies the emphasis was placed upon Sr (see Chapter 15) but Li was also present in the Ohio drinking waters at a very high level of about 50 μg/l.

As described in Chapter 2, studies in Papua–New Guinea (Barmes, 1969) suggested a caries-protective effect of Li in man on the basis of associations of Li in soils and food. Further studies in the same area of Papua–New Guinea gave more evidence of significant inverse relationships, or consistent trends, between low caries and Li concentrations in surface enamel, saliva, and plaque (Schamschula et al, 1978).

Studies on possible relationships of hardness of water to low dental caries can be studied in terms of ground (well) and surface water (rivers and lakes). A recent study (Schamschula et al, 1981) found that where low caries was associated with artesian water in New South Wales (Australia) the most outstanding difference in water constituents was that of Li concentration (Table 17-2). There were 660 to 1000 times higher concentrations of Li in the artesian well water compared with the river (surface) water. In this study the high Li concentration was 0.132 ppm (0.019 mmol). This compares with a concentration of about 0.05 ppm (0.0072 mmol) in the Ohio waters. By contrast the Li concentrations in the Papua–New Guinea study were quite low.

Table 17-2
Concentrations in River and Artesian Water Supplies in Relation to Dental Caries in Australian Children

	Caries (DMFT)	Trace Element Conc.			(Mean ± SE ppm)	
	Mean ± SD	*Li*	*Ca*	*Mg*	*Sr*	*F*
River water	3.8 ± 2.5	0.0002	19.4	11.07	2.12	0.20
Artesian water	2.3 ± 1.9	0.132	1.1	0.42	0.04	0.59

After Schamschula et al, 1981.

These three studies of northwest Ohio (Losee and Adkins, 1969), Papua–New Guinea (Schamschula et al, 1978a, 1978b), and in New South Wales (Schamschula et al, 1981) comprise the evidence in man for considering Li as a possible cariostatic trace element. On the other hand, there have been reports pointing out that manic depressives treated with $LiCO_3$ show increased caries but it is known that salivary function is reduced by high levels of Li, and the increase in caries may be due to decreased flow of saliva. When multiple trace elements have been studied, such as in Colombia (South America), (Glass et al, 1973) Li was not associated in any way with low caries. It must be pointed out however that analysis of Li is difficult and is often neglected.

Animals

There have been several studies on the effects of Li on caries in animals. The equivocal results are summarized in Table 17-3. Except for the most recent experiments the concentration of Li used has been quite high, or even toxic, which may be a reason why results have not been encouraging.

Early work by Malthus et al (1964) used LiCl but at a concentration of 25 mg/l (ppm) in drinking water, which can be compared with a concentration of 0.13 ppm in the artesian water from New South Wales. It should also be noted that usual Li levels in water supplies in the United States are in the range of 0 to 132 μg/l (ppb). A level of 25 ppm would therefore be at least 200 times that found in natural drinking waters. As shown in Table 17-3 most authors have also used very high levels.

In a recent series of animal experiments (Curzon, 1982), where concentrations of Li were selected on the basis of those found in naturally occurring drinking water, results were again equivocal. Although a trend for a decrease in caries was noted with Li, the variance of the data for most of the caries scores in rats was high, which prevented a statistical significance being reached. This trend was noted in all three experiments.

No clear conclusion can therefore be made on the basis of animal experiments reported so far. There are trends in the data that indicate a reduction in caries by the use of Li, but perhaps the animal model is inappropriate to test this effect of Li.

MECHANISM OF ACTION

Lithium in Enamel

As with most other trace elements, a role of Li in enamel has been considered. In the Papua–New Guinea study one of the correlations

Table 17-3
Summary of Reports on Lithium and Dental Caries in Rats

Author (year)	Lithium Used			Effect on Caries	Remarks
	Salt	*Conc.*	*In*		
Wisotsky and Hein, 1958	$LiSO_4$	7 mg/l	W	No effect	
Shaw and Griffiths, 1961	Li_2CO_3	0.1%	D	Reduced[b]	Low weight gain
Hunt and Navia, 1972	LiCl	100 mg[a] 125 mg[a]	W	Reduced Reduced	Toxic effects at high doses
Olson et al, 1979a	LiCl	50 mg/l	W	No effect	
Curzon, 1981	LiCl	5, 20, 40, 60 125, 250 μg/l	W	No significant effect	Wide variance of the data

[a] = mg/10 g body weight.
[b] = barely significant.
W = water; D = diet.

showing significance was a low caries prevalence with relatively high concentrations of Li in surface enamel. Li is found in enamel in the concentration range of 0.01 to 13.2 ppm, and is not found in all samples examined (Curzon and Crocker, 1978). A distinction needs to be made between whole and surface enamel. Caries is initiated at the enamel surface, and here increased concentrations of a trace element may affect the initiation or severity of caries. Most studies have concerned whole enamel while the only study correlating Li in enamel significantly with low caries dealt with surface enamel (Schamschula et al, 1978).

Lithium and Oral Microbiology

It has been shown that group I monovalent cations are involved in microbial cation, sugar, and amino acid transport in non-oral bacteria. Relatively little is known about the direct metabolic effects of Li on oral micro-organisms.

Plaque Following the identification of low caries areas in northwest Ohio that contained drinking water high in Li (among many other elements) Gallagher and Cutress (1977) investigated the antimicrobial effects of 26 elements on the oral bacteria *S. mutans, S. sanguis, S. salivarius, Actinomyces viscosus,* and *A. naeslundi.* At a high concentration of 1000 μg/ml, (ppb or 144 mmol), Li did not inhibit cell growth but did render *S. mutans* more sensitive to acid. Cell growth ceased at a higher pH (5.07) than in control cultures (4.60). The other organisms tested were not significantly affected by Li at a concentration that was unspecified.

Eisenberg et al (in preparation, 1982) found little effect of Li on cell growth or acid production by *S. mutans,* AHT, BHT, FA-1, GS-5, LM-7, SL-1 or 6715 or for *A. viscosus* 15987 or M100. Moreover, Li, in concentrations up to 1 ppm, had little or no effect on fluoride-mediated decreases in cell growth or acid production. Similarly no significant changes in acid production were detected when suspensions of pooled plaque were incubated with glucose, compared with non-Li controls.

Handelman and Losee (1971) and Herbison and Handelman (1975) investigated the combined effects of elements found in the water supplies of communities in northwest Ohio where caries prevalence is low. Fourteen strains of human oral streptococci were used to determine the effect of specific Ohio water, which contained the following minerals significantly more concentrated than in waters of other communities: fluoride (1.5 ppm), B (0.0054 ppm), Li (0.35 ppm), Mo (0.12 ppm), and Sr (17.0 ppm). Microbiologic media were prepared with the highly mineralized "Ohio Water" or distilled water with 1.5 ppm of added fluoride.

In one set of studies, pour plates were prepared from appropriate dilutions of bacteria. Those media were prepared with the incorporation of powdered commercial hydroxyapatite (HA) so that the pour plate appeared cloudy. Around colonies that produced sufficient acid to dissolve the HA, circular clear zones were observed and their diameters could be measured. It was found that the usual hydroxyapatite dissolution zones were absent or greatly reduced for all 14 strains of oral streptococci tested when "Ohio Water" was used rather than distilled water. The reduction in enamel dissolution was 46.5 or 60.7% for *S. mutans* LM-7 and *S. mutans* GS-5. These reductions were almost twice as great as for distilled water plus 1.5 ppm fluoride alone, although in vitro there was more extracellular polysaccharide produced when *S. mutans* was grown in media made with the highly mineralized Ohio water. It was concluded that media made with highly mineralized water from the low caries area of northwest Ohio inhibited the dissolution of synthetic HA and that effect could be explained only partially by fluoride. In a further study, Herbison and Handelman (1975) attempted to separate the effect of Li from the elements tested above. They found that although Li may have had an effect with the other cations it had little if any inhibitory effect alone or in combination with fluoride. It was tested, however, at low (0.0035 to 0.35 ppm) concentration.

Olson et al (1979b) investigated the effects of Li in combination with fluoride on bacterial growth and plaque formation in vitro. Combinations of fluoride (0 or 25 ppm; 0 or 1.3 mmol) and Li (0, 50, 100, or 200 ppm; 0, 7.2, 14.4 or 28.8 mmol) were tested as antiplaque agents on *S. mutans* 6715 in Jordan's medium containing 5% sucrose. There were no significant reductions in the dry weight of plaque formed on glass microscope slides between the control or any of the test groups at the 5% probability level. However, there was a numerical reduction of 27% in all tests that contained 25 ppm fluoride. Fluoride significantly decreased acid production. Even at 200 ppm, Li did not alter cell growth characteristics, nor did it significantly alter the effects of fluoride on *S. mutans* 6715. Fluoride was observed to increase growth rate significantly, but final cell mass was reduced. Although Olson et al (1979) found no effect of Li in plaque accumulation in vitro, 100 and 200 ppm Li significantly reduced total cell accumulation in a standard growth assay. At 50 ppm Li was without effect.

In an epidemiologic study of 72 schoolchildren in Australia (Schamschula et al, 1978) Li, Sr, Zn, and F content in individual plaque samples were inversely related to plaque accumulation. Of these elements, only Sr and fluoride were inversely related to DMFT. Concentrations of Li in plaque have been determined in only two studies (Schamschula et al, 1976, 1977) where values were 0.13 (SD = 0.13, n = 72) or 0.31 (SD = 0.46, n = 288) ppm, dry weight respectively. This correlates to means of 0.026 to 0.062 ppm wet weight assuming the dry-weight to wet-weight proportion of plaque is 1 to 5.

Curzon (1981) found, in one of three rat experiments, a significant effect of Li^+ on plaque extent. The lowest plaque severity was associated with 0.02 ppb (μg/l) Li^+. However, this finding could not be reproduced in further experiments.

The effects of F, Li, and Sr on extracellular polysaccharide production (EPS) in *S. mutans* and *A. viscosus* was studied by Treasure (1981). None of the above elements altered the activity of cell free glucosyl- or fructosyl-transferase, although cells grown in the presence of F had more EPS per gram cell protein. Both Li and Sr decreased the fluoride-mediated enhancement of total EPS production. Concentrations of Li up to 1 ppm were used, whereas the F and Sr were tested at levels up to 100 ppm.

Intracellular polysaccharide (IPS) accumulation was unaffected by Li in three strains of *S. mutans* (BHT, FA-1 and GS-5) and in strain RC-45 of *A. viscosus.* Nor did Li modify an observed fluoride-mediated decrease in IPS producton (Eisenberg et al, in preparation, 1982).

Cation transport Both procaryotic and eucaryotic cells generate electrical potentials and gradients of ions across ion-permeable membranes. These gradients are used to perform the metabolic work necessary for life itself. For reviews on the importance of ion currents and potentials in ion regulation, sugar, and amino acid transport the reader is referred to Harold (1972, 1977), Boyer et al (1977), and Mitchell (1981). Protons, Na or K ion gradients, are commonly used by micro-organisms to transport both sugars and amino acids into the cell and waste products of metabolism out. Substances that interfere with these processes generally perturb normal metabolism. Ionophores, or compounds that render the cell membrane permeable to specific ions, are one such group of inhibitors of these transport systems. The inhibitors allow ion movements without doing the "work" of metabolic transport.

Actively growing microbial cells regulate the cation content of their interior. They generally accumulate K^+ and extrude Na^+ and often pump out H^+. These reactions are membrane mediated, and their importance in animal cell physiology and the treatment of manic-depressive illness has been reviewed (Ehrlich and Diamond, 1980). On the other hand, the possibility of Li, a monovalent group I element, acting to disturb the delicate $K^+/Na^+/Ha^+$ balance in the micro-organisms needs consideration. If cation balance is adversely affected, then Li could act to decrease the efficiency of cell metabolism.

Rodriguez-Navarro (1981) has studied Li efflux from yeast cells which are known to exchange cellular K^+ for H^+, accumulating K^+ against steep gradients. The efflux of Li^+ was found to depend on proton motive force (Mitchell, 1979) and the ATP content that regulated the efflux was sensitive to a decrease in cellular pH. They therefore proposed an H^+/Li^+ antiport as the mechanism for efflux. An alkaline external pH drove the influx uphill whereas an acidic external pH drove the efflux uphill. Thus, the transmembrane pH difference was used as the "energy" for transmembrane Li^+ translocation; the cells' proton motive force is used to move Li^+ in or out of the cell. Experiments with ATPase inhibitors and proton ionophores demonstrated that ATP is an important regulator of the Li^+ transport system. The Li concentrations used in the above experiments were generally 30 mmol (208 ppm) or less.

In *E. coli* (Sorensen and Rosen, 1980) and *S. lactis* (Kashket, 1979) Na^+ was shown to stimulate K^+ uptake via the TrkA and Kdp transport system, whereas Li^+ inhibited K^+ uptake via TrkA but stimulated K^+ uptake via the Kdp system. Although several models are plausible, it appears that in these bacteria as well as in yeast, a proton motive force is required, and ATP may be involved in the regulation of intracellular cation content. It is possible that Li would be detrimental to the process of Na^+/K^+ regulation and cell metabolic energy might be wasted.

Carbohydrate transport The ability of bacteria to survive, grow, and reproduce is dependent on their ability to utilize nutrients from the surrounding medium. Generally, the transport of available carbohydrates from the extracellular space to the interior is required (Dills et al, 1980). Of importance here is that transport of those sugars which are linked to proton transport movements appears to be modified by Li ions.

The lactose permease system in *E. coli* appears to be coupled to transport energy by a proton gradient (Mitchell, 1963). The lactose peroxidase system is involved in the uptake of the β-galactosides lactose and melibiose, among others. Lactose transport has been shown to be coupled to proton movements in *E. coli* (West, 1970) and in *S. lactis* (Kasket and Wilson, 1973). Furthermore, the major end product of fermentation by oral streptococci, lactic acid, appears to be transported out of the cell by a process that is dependent on the difference in proton concentration across the cell membrane in *S. faecalis* (Harold and Levin, 1974). Melibiose transport in both *E. coli* and *Salmonella typhimurium* (Tanaka et al, 1980; Niiya et al, 1980) is coupled to proton cation transport. Lactose, the sugar in milk, and melibiose utilization could be affected by Li.

Although it is likely that Li ions will affect those systems dependent on $H^+/Na^+/K^+$ movements there has been little investigation of this phenomenon. A Na-melibiose transport system in *Sal. typhimurium* was suggested by Niiya et al (1980) because they found that (1) influx of Na^+ was induced by melibiose influx, (2) the efflux of protons induced by melibiose influx was observed only in the presence of Na^+ or Li^+, which demonstrated the absence of H^+-melibiose cotransport, and (3) an artificially imposed Na^+ gradient, or membrane potential, could drive melibiose uptake. Furthermore, they found that the Na^+ gradient that could be produced by these cells was coupled to proton movements. It was found that Li was inhibitory for cell growth when melibiose was the sole carbon source, even though Li stimulated this transport system. It was suggested that high intracellular Li^+ may be harmful. As little as 10 mmol (69.4 ppm) of Li^+ was very stimulatory for melibiose uptake, but the same level of Li^+ inhibited growth of cells on melibiose. However, the same level of Na^+ was without effect. When Li^+-melibiose transport system functions, Li^+ enters the cell together with the melibiose and then accumulates to levels toxic to the cell.

In *E. coli* (Tanaka et al, 1980, Tsuchiya et al, 1977) it appears that Na^+ stimulates melibiose transport and Li^+ inhibits it. In addition K^+ may be essential for the inhibitory effect of Li^+. Should Li^+ and melibiose be transported, then toxic levels of Li^+ may be reached intracellularly, or growth may be reduced simply by the inhibition of uptake of this carbohydrate.

In the presence of melibiose or lactose it is possible that Li may be toxic for some bacteria. However, higher concentrations are required than are normally present in drinking water supplies. The lack of demonstrated inhibition of oral bacteria, by Li, to date, may be due to the generalized use of glucose or sucrose as the carbohydrate source.

Amino acid transport Monovalent ions, including protons, are involved in the uptake of a number of amino acids, such as α-aminoisobutyrate in *S. lactis* (Thompson, 1976). Thus the involvement of cations in amino acid transport is widespread, but it has not been studied extensively in oral bacteria. Nonetheless it is likely that the uptake of amino acids by some species of oral bacteria is regulated by monovalent cations. The regulation by Li^+ has been studied for *E. coli* on a number of occasions (Kawasaki and Kayama, 1973; Kayama and Kawasaki, 1976, Kayama-Gonda and Kawasaki, 1979). The stimulation in the uptake of proline by whole cells was specific for Li^+, and the effect was synergistically enhanced by the presence of a carbon source. The Li gradient may be based on a proton-Li antiport as demonstrated by West and Mitchell (1974). Evidence has been presented by Kayama-Gonda and Kawasaki (1979) for a proline-Li synport and a H^+:Li^+ antiport. Cell growth in the presence of proline and Li was not tested. It is possible that Li^+ uptake with proline during growth could result in toxic intracellular Li concentrations, as was the case for melibiose uptake in *S. typhimurium* (Niiya et al, 1980).

SUMMARY

Although epidemiologic evidence implicated Li as a naturally occurring cariostatic agent, studies on the effects of Li on rat caries have yielded equivocal results. In combination with other elements, Li reduced hydroxyapatite dissolution by oral bacteria, but Li did not appreciably affect plaque formation on glass slides, nor growth and acid production. Nor did Li reduce the inhibitory effects of fluoride on oral bacteria. A reduction of the fluoride-mediated increase in extracellular polysaccharide production caused by Li in *S. mutans* and *A. viscosus,* was affected by Li.

Little is known about the effects of Li on oral bacteria, but Li could penetrate the cell membrane and accumulate intracellulary. Further work on this possible action of Li is required to explain the findings of significant inverse relations of Li in plaque to dental caries. At the low levels of Li to be found in water supplies it is unlikely that the element could be solely responsible for the reductions in caries noted in epidemiologic surveys. Studies on the interaction of Li with other elements and on the combinations with sugars that may be cotransported into the bacterial cell are indicated.

REFERENCES

Barmes DE: Caries etiology in Sepik villages—Trace element and micronutrient content of soil and food. *Caries Res* 1969;3:44–59.

Boyer PD, Chance B, Ernster L, et al: Oxidative phosphorylation and photphosphorylation. *Ann Rev Biochem* 1977;46:955–1026.

Curzon MEJ, Crocker DC: Relationships of trace elements in human tooth enamel to dental caries. *Arch Oral Biol* 1978;23:647–653.

Curzon MEJ: Lithium and dental caries in the rat. *Arch Oral Biol* 1982;27: 573–576.

Dills SS, Apperson A, Schmidt MR, et al: Carbohydrate transport in bacteria. *Microbiol Rev* 1980;44:385–418.

Ehrlich BE, Diamond JM: Lithium membranes and manic-depressive illness *J Memb Biol* 1980;52:187–200.

Facklam RR: Physiological differentiation of viridans streptococci. *J Clin Microbiol* 1977;5:184–201.

Gallagher IHC, Cutress TW: The effect of trace elements on the growth and fermentation by oral streptococci and actinomyces. *Arch Oral Biol* 1977;22: 555–562.

Glass RL, Rothman KJ, Espinal F, et al: The prevalence of human dental caries and water-borne trace elements. *Arch Oral Biol* 1973;18:1099–1104.

Hamada S, Slade HD: Biology, immunology and cariogenicity of *Streptococcus mutans. Microbiol Rev* 1980;44:331–384.

Handelman SL, Losee FL: Inhibition of enamel solubility in a highly mineralized water. *J Dent Res* 1971;50:1605–1609.

Harold FM: Conservation and transformation of energy by bacterial membranes. *Bacteriol Rev* 1972;36:172–230.

Harold FM: Ion currents and physiological functions in microorganisms. *Ann Rev Microbiol* 1977;31:181–203.

Harold FM, Levin E: Lactic acid-translocation: Terminal stepinglycolysis by *Streptococcus faecalis. J Bacteriol* 1974;117:1141–1148.

Herbison RJ, Handelman SL: Effect of trace elements on dissolution of hydroxyapatite by cariogenic streptococci. *J Dent Res* 1975;54:1107–1114.

Holloway Y, Schaareman M, Dankert F: Identification of viridans streptococci on the minitek miniaturized differentiation system. *J Clin Pathol* 1979;32: 1168–1173.

Hunt CE, Navia JM: Pre-eruptive effects of Sr, Mo, Li and B in developing teeth and other tissues of neonatal rats, in *Trace Substances in Environmental Health,* vol 6. Columbia, MO, University of Missouri, 1972.

Kashket ER: Active transport of thallous ions by *Streptococcus lactis. J Biol Chem* 1979;254:8129–8131.

Kashket ER, Wilson TH: Proton-coupled accumulation of galactoside in *Streptococcus lactis.* 7962. *Proc Natl Acad Sci* USA 1973;70:2866–2869.

Kawasaki T, Kayama Y: Effect of lithium on proline transport by whole cells of *Escherichia coli. Biochem Biophys Res Commun* 1973;55:52–59.

Kayama Y, Kawasaki T: Stimulatory effect of lithium ion on proline transport by whole cells of *Escherichia coli. J Bacteriol* 1976;128:157–164.

Kayama-Gonda Y, Kawasaki T: Role of lithium ions in proline transport in *Escherichia coli. J Bacteriol* 1979;139:560–564.

Losee FL, Adkins BL: Anti-cariogenic effect of minerals in food and water. *Nature* 1968;219:630–631.

Maletsky B, Blackly PH: *The Use of Lithium in Psychiatry.* Cleveland, OH, CRC Press, 1971.

Malthus RS, Ludwig TG, Healy WB: Trace elements and dental caries in rats. *NZ Dent J* 1964;60:291–297.

Mertz W: The essential trace elements. *Science* 1981;213:1332–1338.

Mitchell P: Molecule group and electron translocation through natural membranes. *Biochem Soc Symp* 1963;22:142–168.

Mitchell P: Keilin's respiratory chain concept and its chemiosmotic consequences. *Science* 1979;206:1148–1159.

Niiya S, Moriyama Y, Futal M, et al: Cation coupling to melibiose transport in *Salmonella. J Bacteriol* 1980;144:192–199.

Olson BL, MacDonald JL, Stookey GK: Influence of lithium upon dental caries in the rat. *J Dent Res* 1979a;58:1123–1126.

Olson BL, MacDonald JL, Stookey GK: The effect of lithium and fluoride upon in vitro plaque and bacterial growth. *J Dent Res* 1979b;58:1428.

Ottina K, Lopilato J, Wilson TH: Membrane transport of p-nitrophenyl-oc-galactoside by melibiose carrier of *Escherichia coli. J Memb Biol* 1980;56: 169–175.

Rodriguez-Navvaro A, Sancho ED, Perez-Lloveres C: Energy source for lithium efflux in yeast. *Biochem Biophys Acta* 1981;640:352–358.

Schamschula RG, Agus H, Bunzel M, et al: The concentration of selected major and trace minerals in human dental plaque. *Arch Oral Biol* 1977;22:321–325.

Schamschula RG, Adkins BL, Barmes DE, et al: WHO *Study of Dental Caries Aetiology in Papua-New Guinea.* WHO Publication #40. Geneva, 1978a.

Schamschula RG, Bunzel M, Agus HM, et al: Plaque minerals and caries experience: Associations and interrelationships. *J Dent Res* 1978b;57:427–432.

Schamschula RG, Cooper MH, Agus HM, et al: Oral health in Australian children using surface and artesian water supplies. *Commun Dent Oral Epidemiol* 1981;9:27–31.

Setterstrom JA, Gross A, Stanko RS: Comparison of minitek and conventional methods for the biochemical characterization of oral streptococci. *J Clin Microbiol* 1978;10:409–414.

Shaw JH, Griffiths D: Developmental and postdevelopmental influences on incidence of experimental dental caries resulting from dietary supplementation by various elements. *Arch Oral Biol* 1961;5:301–322.

Sorenson EN, Rosen BP: Effects of sodium and lithium ions on the potassium ion transport systems of *Escherichia coli. Biochemistry* 1980;19:1458–1462.

Tanaka K, Niiya S, Tsuchiya T: Melibiose transport in *Escherichia coli. J Bacteriol* 1980;141:1031–1036.

Thompson J: Characteristics and energy requirements of an α-aminioisobutyric acid transport system in *Streptococcus lactis. J Bacteriol* 1976;127:719–730.

Tosteson DC: Lithium and mania. *Sci Am* 1981;244:164–166.

Tsuchiya T, Raven J, Wilson TH: Co-transport of Na^+ and methyl-β-D-thiogalactopyranoside mediated by the melibiose transport system of *Escherichia coli. Biochem Biophys Res Commun* 1977;76:26–31.

Tsuchiya T, Copilato J, Wilson TH: Effect of lithium on melibiose transport in *Escherichia coli. J Membr Biol* 1978;42:45–59.

Treasure P: Effects of fluoride, lithium and strontium on extracellular polysaccharide production by *Streptococcus mutans* and *Actinomyces viscosus. J Dent Res* 1981;60:1601–1610.

Underwood EJ: *Trace Elements in Human and Animal Nutrition,* ed 4. New York, Academic Press, 1977.

Voors AW, Shuman MS, Gallagher PN: Atherosclerosis and hypertension in relation to some trace elements in tissues. *World Rev Nutr Diet* 1975;20: 299–326.

West JC: Lactose transport coupled to proton movements in *Escherichia coli. Biochem Biophys Res Commun* 1970;41:655–661.

West JC, Mitchell P: Proton/sodium antiport in *E. coli. Biochem J* 1974;144: 87–90.

Wisotsky J, Hein JW: Effects of drinking solutions containing metallic ions above and below hydrogen in the electromotive series on dental caries in the Syrian hamster. *JADA* 1958;57:796–800.

81 Ti

204.37

32-18-3

18 Titanium

Buddhi M. Shrestha

Titanium (Ti), discovered in 1791 by William Gregor, is the ninth most abundant element in the earth's crust, in which it is widely distributed. It is present in most soils in concentrations of 0.5% to 10% (Gilbert, 1957), and in sea water, from 1 to 9 ppb (μg/l) (Schroeder, 1965).

When purified, Ti is a lustrous white metal having low density, good strength and excellent corrosion resistance properties. It is resistant to dilute sulfuric, hydrochloric, and most organic acids. Alloys of Ti are used in building aircraft and missiles where lightweight, strength, and ability to withstand extremes of temperature are important. Because of its excellent tissue tolerance (Meachim and Williams, 1973; Lemons et al, 1976), Ti and its alloys are also used as implants in various orthopedic, plastic reconstructive, and cardiovascular surgeries as well as various aspects of oral surgery, (Zallen and Fitzgerald, 1976; Williams, 1977; Schettler et al, 1979; Young et al, 1979).

A number of organotitanium compounds such as tetraisopropyl titanate and tetrabutyl titanate have been used as catalysts for the polymerization of various resins, such as silicons, epoxy resin, and polyester, because they permit rapid curing at lower temperatures and provide improved resin properties. Organic titanates can also be used to deposit (by either pyrolysis or hydrolysis) a thin layer of titanium dioxide on a variety of surfaces to improve adhesion, impart scratch resistance to glass, improve electrical properties, or act as dispersing aids.

Titanium dioxide is extensively used as a pigment in various house paints since it is permanent and provides a good cover. Titanium tetrafluoride (TiF_4) has been used recently as a topical cariostatic agent in the prevention of dental caries in animals as well as in humans.

Despite increasing industrial and medical usage, there is little information available as to the biologic effects of Ti on animals or humans. This lack of interest may, in part, be due to Ti, as such, long being considered physiologically inert (Clark et al, 1975; Williams, 1977) and nontoxic. However, the present indication, based on studies such as those of Bernheim and Bernheim, (1939), Kato and Gözsy (1955), Köpf and Köpf-Maier (1979), and Shrestha (1980), is that several Ti compounds could be considered physiologically active.

Distribution and Essentiality

The presence of trace amounts of Ti in plant, animal, and human tissues was reported as early as a century ago. Baskerville and his associates (1899) found 0.1088% titanic oxide in wild strawberries, while beef and human flesh contained 0.0130% and 0.025%, respectively. These findings have been supported in recent years by a number of studies (Schroeder, et al, 1963; Tipton and Cook, 1963; Schroeder, 1965) which have also shown varying amounts of Ti in plant and animal tissues. In plants, Ti is present mainly in leaves and is thought to be directly associated with the chlorophyl concentration (Browning, 1961). In animals, Ti appears to be present mainly in calcified tissues. Baskerville et al (1899) reported the presence of titanic oxide in both beef and human bones. Maillard and Ettori (1936) found Ti at 0.05 and 0.02 ppm in costal cartilage and bone marrow, respectively. A number of studies (Schroeder et al, 1963; Tipton and Cook, 1963) have shown the presence of higher amounts of Ti in human lungs and skin tissues than in the kidney, liver, spleen, or heart, believed to be mainly due to an accumulation through airborne sources.

A trace amount of Ti has also been found in human dental tissues. Hardwick and Martin (1967) reported that human enamel, dentin, and plaque all contained Ti in concentrations ranging from approximately 10 to 100 ppm and that dentin contained a higher amount of Ti than either enamel or plaque. Losee et al (1974) found 0.1 to 4.8 ppm Ti in human tooth enamel samples obtained from persons raised in different geographic areas of the United States. Despite these findings, there is no direct evidence to support the possibility that Ti performs any vital function in the body or that it is a dietary essential for any living organisms.

Metabolism

Absorption of Ti through the gastrointestinal (GI) tract is generally regarded as poor. Schroeder et al (1963) reported the presence of Ti in most human foods and in all vegetation analyzed; they estimated that about 300 μg per day were ingested in an average diet, of which 3% was absorbed. According to Miller et al (1976) only 4% of the ingested Ti was absorbed following intraoral administration of the radioisotope ^{44}Ti as $TiCl_4$ in experimental lambs. Because of poor absorption, TiO_2 has been frequently used as markers in various physiologic and metabolic studies (Kotb and Luckey, 1972; Huggins and Froehlich, 1977; Ferin and Leach, 1976). However, a recent study by Shrestha (1980) has indicated that the absorption of Ti may vary, depending on the type of compound used. Accordingly, the highest Ti absorption, as indicated by serum Ti concentration three to seven hours following gastric intubation, was observed with TiF_4 followed by sodium hexafluorotitanate, potassium hexafluorotitanate, and TiF_3, whereas titanium lactate and $TiCl_3$ both failed to show any absorption in experimental rats. An interesting finding of this study was that the administration of either titanium lactate or $TiCl_3$, when preceded by a dose of NaF, resulted in varying amounts of Ti absorption which indicated the possibility that fluoride may have an enhancing effect on Ti absorption through the GI tract. This is consistent with the previous finding by Ruliffson et al (1963) in which a significant increase in blood ^{59}Fe uptake was observed in rats when gastric administration of radioactive iron (^{59}Fe) was immediately followed by a dose of NaF solution.

Most human and animal tissues contain a trace amount of Ti. It is not known, however, whether this Ti serves any specific function or functions in these tissues. Miller et al (1976) observed the highest amounts of ^{44}Ti (about 25% of the original dose) in the skeletal and cartilage tissues, 48 hours following intravenous administration of the radioisotope titanium-44 in the lamb. They also found considerable amounts of ^{44}Ti in the skeletal muscle and various internal organs. Shrestha (1980) observed significant amounts of Ti uptake by rat enamel following gastric administration of TiF_4 in neonatal rats during tooth development.

Excretion

Schroeder et al (1963) reported that most of the Ti absorbed from daily diets, which was found to be about 9.0 μg Ti/day, was excreted in the urine. In a balance study, Tipton et al (1966) noted that human test subjects excreted almost twice as much Ti as they ingested and suggested that some Ti might have come from unknown deposits within the body. According to Perry and Perry (1959), normal human urine contains an average of 10.2 μg/Ti per liter.

All unabsorbed ingested Ti is expected to be excreted in the feces. However, it is possible that some of the fecally excreted Ti comes from biliary excretion. The clearance of unabsorbed Ti particles in the lung, inhaled through airborne sources, involves mainly the alveolar macrophages, the mucociliary transport, and lymphatic drainage systems (Ferin and Leach, 1976).

Other Biologic Activity

Certain Ti compounds are known to enhance growth and nitrogen fixation in plants (Schroeder et al, 1963). Sodium pertitanate has been found to inhibit both the conversion of cysteine to cysteic acid and the oxidation of ethyl mercaptan and thioglycolic acid by rat liver (Bernheim and Bernheim, 1939). Kikkawa et al (1955) found that Ti formed a complex with melanin to produce yellow hair in animals. Cavanaugh et al (1955) reported that triethanolamine titanate and titanium lactate form complexes, in vitro, with several catechols, including 3, 4, dihydroxyphenylalanine (DOPA), to produce an intense orange-yellow color. Apparently, the triethanolamine titanate and the titanium lactate inhibit the action of tyrosinase on DOPA by competing with the enzyme for the substrate.

According to Köpf and Köpf-Maier (1979), titanocene dichloride, a metallocene Ti complex, given intraperitoneally (IP), is an effective antitumor agent in experimental mice. Intraperitoneal implantations of Ehrlich ascites tumor cells in mice followed after 24 hours by IP injection of titanocene dichloride in doses of 25 to 65 mg/kg body wt produced no sign of tumor development in 83% of the treated mice up to 30 days after implantation. They further observed that titanocene dichloride was not only a more effective antitumor agent but also was less toxic than cis-diaminedichloro-platinum, a well-known antitumor agent. Kato and Gözsy (1955) observed increased phagocytic activity of endothelial cells of skin capillaries, when skin medications containing titanium peroxide, titanium tannate, or titanium salicylate were applied topically in experimental mice. Titanium oxide, however, was ineffective in inducing such phagocytic activity.

Toxicity

The metal Ti as such is generally considered nontoxic (Schroeder et al, 1963). Attempts to produce toxicity in experimental animals have usually failed (Williams, 1977). According to Browning (1961), Ti is regarded as "not highly toxic" either to animals or to human beings and is well tolerated by animals when given by mouth as well as by injection.

It is important to bear in mind, however, that in most of these studies, Ti compounds of a low solubility, such as titanium dioxide, were used. Therefore, Ti compounds which are more soluble and which allow increased Ti absorption can have varying degrees of physiologic or toxicologic effects in animals as well as in human subjects.

Single-dose oral toxicity studies on a number of organic titanates using laboratory rats have all indicated a low order of toxicity ranging from slightly toxic tetrabutyl titanate (ALD 7500 mg/kg body wt) to almost nontoxic tetrastearic tetraisopropyl titanate (ALD 17,000 mg/kg body wt).

On the other hand, the acute oral LD_{50} for rats dosed with 10% TiF_4 has been estimated to be 55 mg/kg body wt (Shrestha, 1980), which means that TiF_4 could be considered toxic. TiF_4-related toxicity is believed to be primarily due to its F content rather than the Ti.

CARIES

Man

Reed and Bibby (1976) conducted a three-year study on the topical effects of TiF_4 and APF in two groups of 11- to 13-year-old school children. The treatment group received an annual, one-minute topical application with 1% TiF_4 to the teeth on one side of the mouth. A similar treatment with APF (1.23% F) was given on the other side of the mouth, with the exception that the treatment was carried out for four minutes instead of one minute. A second group of untreated children served as control. The results at the end of the three-year study showed that the caries increment in the TiF_4-treated teeth was 33% less ($p = < 0.04$) than in the APF-treated teeth and 50% less ($p = < 0.004$) than in the untreated control teeth. The authors concluded that a large measure of protection against dental caries can be provided by making a single, annual one-minute topical application of a 1% aqueous solution of TiF_4 to the teeth of children.

Recently, Langer and Galon (1979) reported the effect of mouth rinsing with TiF_4 solutions on caries incidence in children. One hundred and fifty 10- to 11-year-old school children, previously unexposed to fluoridated water, were randomly divided into two groups. One group was treated with the TiF_4 solution and the other with distilled water (control). The treatment consisted of rinsing the mouth with the test solution or water for 30 seconds, after which the children were instructed not to eat or drink for a period of one hour. Caries examinations were carried out every six months. The results, at the end of a two and one-half year experimental period, indicated a significant reduction in caries incidence for the TiF_4-treated group, as compared with the control group.

To date, there is no report on the systemic effect of TiF_4 on human dental caries. In experimental animals, systemic TiF_4 was found to be effective in producing caries-resistant teeth when administered during tooth development (Shrestha, 1980). Whether TiF_4 will have a similar effect on human teeth remains to be determined.

Animals

The indication that Ti may have some cariostatic effect dates back to the 1940s when Manly and Bibby (1949) reported a reduction in acid enamel dissolution following topical treatment with titanium nitrate. However, studies relating to animal caries appeared only in the 1970s when Shrestha (1970), Orbell (1973), and Regolati et al (1974) all reported caries reductions in rats following topical treatments with TiF_4. Shrestha (1970) observed a reduction of 52% in the number of smooth surface carious lesions in rats, following daily one-minute topical treatment with 1.5% TiF_4 for seven days, as compared with those treated with distilled water. Orbell (1973) reported 53% and 29% caries reduction in rats receiving daily one minute topical treatment with 1% TiF_4 for four days, as compared with those treated with distilled water and APF (1.23% F), respectively. Regolati et al (1974) reported that both TiF_4 and NaF were effective in reducing rat caries. There was, however, no statistical difference in reductions of caries between the TiF_4- and NaF-treated groups.

There is some indication that Ti may enhance the cariostatic effect of F when administered systemically during tooth development. In a recent study by Shrestha (1980), a significant reduction in caries was observed in neonatal rats following gastric administration of TiF_4 during tooth development. He noted that TiF_4, when administered intragastrically (IG) in daily dosages of 7.5 mg F/kg body weight from the 2nd through the 17th day of age, produced a significantly ($p = < 0.05$) lower amount of caries than NaF at the same F dose level. This enhanced cariostatic effect of TiF_4 is believed to be due to increased Ti and F uptake by the rat enamel during tooth development.

In any case, based on the various animal and clinical studies cited above, it may be concluded that TiF_4 is an effective cariostatic agent which reduces dental caries in animal as well as human subjects when applied topically. Although the extent to which the Ti is directly involved in the overall caries reductions produced by TiF_4 is not known, an indirect inference can be made in this regard from the systemic effect study by Shrestha (1980), in which the IG administration of a TiF_4 solution, given in the same way and at the same F dose level as the NaF solution, was nearly twice as effective in resisting dental caries in neonatal rats as the NaF. A similar inference can also be made from the clinical study by Reed and Bibby (1976) in which TiF_4, containing less F, administered in shorter application periods, was significantly more effective in preventing dental caries in school children than APF.

MECHANISMS OF ACTION

The mechanism by which Ti may produce a cariostatic effect is not fully understood. As most of the Ti-related caries studies in the past were carried out using either TiF_4 or a Ti compound in conjunction with another compound, the exact role of Ti in each of these instances is difficult to discern. The weight of evidence based on these studies seems to indicate that Ti and F ions act synergistically to impart caries resistance to teeth following either topical or systemic administration.

Enamel Dissolution Effect

Several studies indicated that TiF_4 is one of the most effective topical agents in reducing enamel dissolution. Shrestha et al (1972) tested the effects of 1% TiF_4 along with a number of other F agents, including ZrF_4, HfF_4, Sn_3BrF_5, $SnF_2 \cdot 3ZrF_4$, $SnZrF_6$, APF, SnF_2, and NaF, on acid dissolution of intact bovine enamel. Treated enamel blocks were subjected to a series of seven 15-minute post-treatment decalcification runs using 0.1 mol/l acetate buffer pH 4.0.

The results not only showed the highest solubility reduction for TiF_4 as compared with the other fluorides tested, but also that the antisolubility effect of TiF_4 was sustained at the 90% level throughout seven consecutive post-treatment decalcification runs (Figure 18-1). The antisolubility effects of APF, SnF_2 and NaF, on the other hand, completely disappeared after the fourth post-treatment decalcification run.

Wei et al (1976) studied the effects of four-minute topical application of TiF_4 or APF on enamel dissolution by exposing the treated human enamel to 0.5M perchloric acid ($HClO_4$) for 15 seconds. The amount of enamel dissolved from the TiF_4-treated tooth sections, as indicated by their mean depths, was significantly less ($p < 0.05$) than from either the APF or the untreated control enamel. Regolati et al (1974) observed a 41% reduction in enamel dissolution in rat enamel following a five-day topical treatment with TiF_4 (2000 ppm F) as compared with a 12% reduction for NaF, both groups having received the same amount of F.

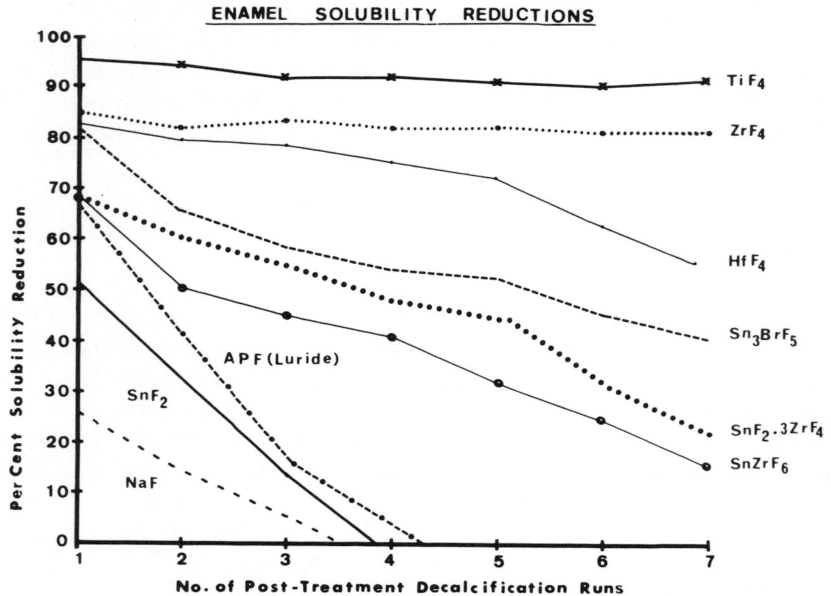

Figure 18-1 Enamel solubility effects of various fluoride compounds tested by repeated decalcification.

So far, TiF_4 has been shown to be more effective in reducing enamel dissolution than conventional topical F compounds. It should be pointed out, however, that the enamel dissolution-reducing effects vary from one Ti compound to another. For instance, trivalent Ti compounds such as titanium trifluoride (TiF_3) have been found to be less effective in reducing enamel dissolution than tetravalent TiF_4, when applied topically at the same F concentration levels (Shrestha, unpublished data, 1981). Tetravalent Ti compounds such as lithium hexafluorotitanate (Li_2TiF_6) have also been found to be less effective than TiF_4 (Shrestha, 1970). Furthermore, the enamel dissolution-reducing effects of Ti decrease considerably in the absence of F. Among the various non-F Ti compounds tested thus far, only titanium nitrate ($TiNO_3$) has been shown to be effective in reducing enamel dissolution when applied topically (Manley and Bibby, 1949). Others, such as titanium trichloride ($TiCl_3$) or titanium tetrachloride ($TiCl_4$), have often been found to be ineffective in reducing enamel dissolution without concomitant F treatment. Thus, based on currently available information, the indication is that both Ti and F react with the enamel synergistically to produce acid-resistant enamel and the role of Ti appears to be secondary to the effect of F.

Formation of an Acid-resistant Coating in the Surface Enamel

When applied topically TiF_4 is known to produce a highly acid-resistant coating (ARC) in the surface enamel. Shey (1970) noted that when pre-etched bovine enamel was treated with TiF_4, a thin membrane-like structure formed that could be floated off by immersing the enamel block in 5% hydrochloric acid. When examined under the light microscope, the undersurface of this membrane revealed a replica-like impression of the enamel rod pattern with the preservation of the inter-rod matrices (Figure 18-2).

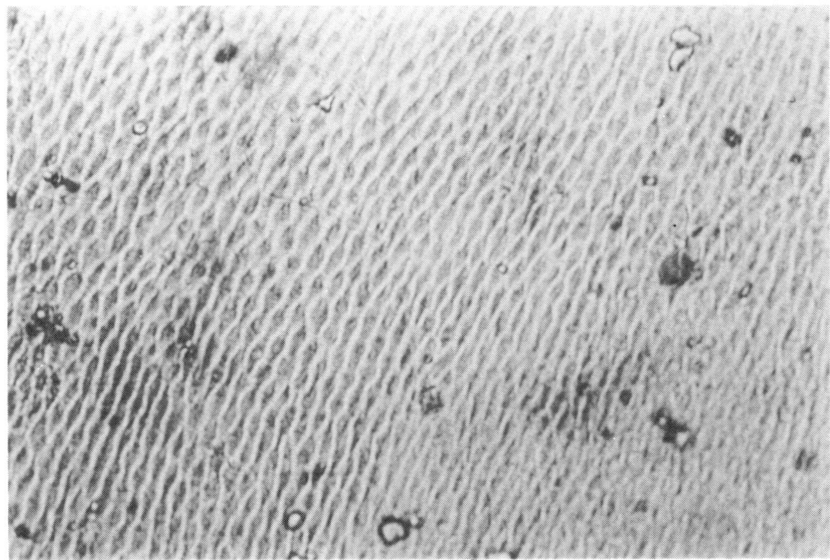

Figure 18-2 Acid-resistant coating from TiF_4-treated bovine enamel surface. The under surface revealed a replica-like impression of the enamel rod pattern with the preservation of the inter-rod matrices. (Photomicrograph, magnification X450, courtesy Dr Z. Shey.)

Mundorff et al (1972) reported the formation of a "glaze" on the enamel surface of teeth treated topically with 1% TiF_4 followed by an application of a number of organic agents, such as acetic or formic acid buffer. This surface coating is believed to be due to an organometallic complex formation, involving the organic materials of the surface enamel. Bibby and Little (1975) suggested that formation of such a coating with TiF_4 may involve polymerization of organic materials on the enamel surface.

Wei et al (1976) indicated that organometallic complex formation may involve the formation of a titanium dioxide coat that may also act as a barrier to F loss from the enamel. Shrestha (1980) reported the formation of an ARC following TiF_4 topical treatment, in both the presence and absence of the organic matrix of the enamel. Although these studies have confirmed the formation of an ARC on TiF_4-treated enamel, detailed information as to the chemical and physical nature of the coating is lacking.

Enamel Fluoride Uptake and Retention Effects

Tetravalent titanium ions (Ti^{4+}), when present in enamel, have been found to enhance the uptake of topically applied F. McCann (1969) observed increased uptake and retention of F from APF solutions when intact human enamel blocks were pretreated with a 0.05 mol/l Ti^{4+} solution as compared with those treated with APF only. It is known that TiF_4 produces higher enamel F uptake when applied topically to teeth, as compared with NaF, SnF_2 or APF, containing the same or higher amounts of F (Mundorff et al, 1972; Wei et al, 1976; Clarkson and Wefel, 1979). In addition, TiF_4 appears to enhance F retention in the enamel, following topical treatment, possibly by minimizing the F loss as a result of surface coating formation (Wei et al, 1976; Clarkson et al, 1981). Mundorff et al (1972) observed a 40% retention of enamel F, when TiF_4-treated bovine enamel blocks were subjected to a 48-hour washing in 0.05 mol/l acetate buffer, pH 5.5, containing 1 mM calcium and 5 mM phosphate. Wefel et al (1977) observed a significantly ($p = < 0.05$) higher F retention in enamel treated with TiF_4 as compared with a control, following 24-hour post-treatment washing in synthetic saliva.

It should be pointed out, however, that none of the above studies directly compared the F retention effects of TiF_4 with other fluorides, such as NaF, SnF_2, or APF. It will therefore not be possible to make a proper assessment of the effects of TiF_4 until such data are obtained.

That Ti may also enhance the F uptake by enamel, when administered systemically during tooth development, has been indicated by a report by Shrestha (1980), in which he observed increased enamel F concentration in neonatal rats following IG administration of TiF_4, as compared with rats treated with NaF at the same F dose level. This increased F uptake in the TiF_4-treated rat enamel appeared to be associated with increased Ti uptake since no Ti was found in the NaF-treated rat enamel.

A recent study by Hals et al (1981) showed a fine layer of Ti-containing material covered specimens of root surfaces, from human teeth, after treatment with TiF_4. In a comparison study TiF_4 and APF solutions, with the same F molarity and pH, showed different effects on root surfaces. Results showed an increase in F following treatment with TiF_4 that was much greater than with APF. The topical application of TiF_4 on root surfaces shows some promise as a preventive agent for the growing problem of root caries in the older patient.

Anti-microbial Effect

There are preliminary indications that Ti may have some antibacterial properties. Pape et al (1975) reported that restoration of carious lesions in human subjects, with amalgam containing 0.05% TiF_4, resulted in marked inhibition of *S. mutans* in plaque samples obtained from marginal areas of the restorations. Recently, Beierle and Grenoble (1981) observed reduced adherence of plaque when teeth were treated with a liquid dentifrice containing 0.001 to 0.05% TiF_4. It should be noted, however, that no attempt was made in these studies to determine if the effect was due to Ti, or F, or both.

It is of interest that in the collection of the enzyme dextranase, titanium dioxide has been found to be effective in immobilizing the dextranase when used as column packing material (Kennedy and Kay, 1977). Since the ARC which forms as a result of topical TiF_4 treatment is believed to contain some amount of titanium dioxide Wei et al, 1976, it appears possible that this ARC may enhance the accumulation of dextranase on the tooth surface which in turn may inhibit the formation of dextran in the plaque.

SUMMARY

Although Ti has long been known to be present in various plant, animal, and human tissues, it has been only in recent years that information has become available regarding its various biologic functions. Contrary to previous beliefs, Ti has been found to be absorbed through the GI tract, particularly in the presence of F, under certain conditions.

During the past decade the role of Ti in the prevention of dental caries has received increased attention. The compound TiF_4 has been studied for its cariostatic effect, in topical as well as systemic applications. Results of these investigations have shown that TiF_4 could be used advantageously as an effective cariostatic agent when administered either topically or systemically.

TiF_4 is unique compared with conventionally used F compounds, in that an ARC is formed, following topical application, that protects the enamel against acid dissolution and enhances the uptake and retention of F. Although Ti and F are believed to be acting synergistically, the extent to which Ti is directly involved in the observed cariostatic effect of TiF_4 is not known and requires further study. In addition, there is some preliminary indication that TiF_4 may also provide an antibacterial effect against cariogenic microorganisms.

REFERENCES

Baskerville C: On the universal distribution of titanium. *J Am Chem Soc* 1899; 21:1099.

Beierle JN, Grenoble DE: Method for limiting adherence of plaque and dental composition therefore. U.S. Patent #4291017, September 22, 1981.

Bernheim F, Bernheim MLC: The effect of titanium on the oxidation of sulfhydryl groups by various tissues. *J Biol Chem* 1939;127:695.

Bibby BG, Little MF: Fluorine and titanium and the organic material of dental enamel. *J Dent Res* 1975;54 (Special Issue B):B137.

Browning E: Titanium, in *Toxicity of Industrial Metals,* London, Butterworths, 1961, p 296.

Cavanaugh DJ, Harris J, Hearon JZ: Enzyme inhibition by complexing of substrates: Inhibiting of tyrosinase by titanium compounds. *J Am Chem Soc* 1955;77:1531.

Clark RJH, Bradley DC, Thornton P: *The Chemistry of Titanium, Zirconium and Hafnium.* Pergamon Texts in Inorganic Chemistry, vol 19. New York, Pergamon Press, 1975.

Clarkson BH, Wefel JS: Titanium and fluoride concentration in titanium tetrafluoride and APF treated enamel. *J Dent Res* 1979;58:600.

Clarkson BH, Wefel J, Edie J, et al: SEM and microprobe analyses of enamel-metal-fluoride interactions. *J Dent Res* 1981;60:1912.

Ferin J, Leach LJ: The effect of amosite and chrysotile asbestos on the clearance of TiO_2 particles from the lung. *Environ Res* 1976;12:250.

Hals E, Tveit AB, Total B, et al: Effect of NaF, TiF_4 and APF solutions on root surfaces in vitro and with special reference to uptake of fluoride. *Caries Res* 1981;15:468–472.

Gilbert FA: *Mineral Nutrition and the Balance of Life.* Norman, OK, University of Oklahoma Press, 1957.

Hardwick JL, Martin CJ: A pilot study using mass spectrometry for the estimation of the trace element content of dental tissues. *Helv Odont Acta* 1967;11:62.

Huggins CB, Froehlich JP: High concentration of injected titanium dioxide in abdominal lymph nodes. *J Exp Med* 1967;124:1099.

Kato L, Gözsy B: Stimulation of the cell-linked deference forces of the skin: Mechanism of action of certain topical agents. *Can Med Assoc J* 1955;73:31.

Kennedy JF, Kay IM: The use of titanium (IV) oxide coated with diazotized 1, 3-diaminobenzene for the immobilization of carbohydrate-directed enzymes. *Carbohydr Res* 1977;59:553.

Kikkawa H, Ogita Z, Fujito S: Nature of pigments derived from tyrosine and tryptophan in animals. *Science* 1955;121:43.

Köpf H, Köpf-Maier P: Titanocene dichloride—The first metal-locene with cancerostatic activity. *Angew Chem Int Ed Engl* 1979;18:477.

Kotb AR, Luckey TD: Markers in nutrition. *Nutr Abst Rev* 1972;42:28.

Langer M, Galon H: The effect of mouth rinsing with titanium tetrafluoride solutions on caries incidence in children. *J Dent Res* 1979;58: (Special Issue C) Abst. #1 (Israel), 1277.

Lemons JE, Niemann KM, Weiss AB: Biocompatibility studies on surgical grade titanium-, cobalt-, and iron-base alloys. *J Biomed Mater Res* 1976;10:549.

Losee FL, Cutress TW, Brown R: Trace elements in human dental enamel, in Hemphill D (ed): *Trace Substances in Environmental Health,* vol 7. Columbia, MO, 1974, University of Missouri, p 19.

Maillard LC, Ettori J: Titanium as a new constituent of the human body. *Bull Acad Med Paris* 1936;115:631.

Manley RS, Bibby BG: Substances capable of decreasing the acid solubility of tooth enamel. *J Dent Res* 1949;28:160.

McCann HG: The effect of fluoride complex formation on fluoride uptake and retention in human enamel. *Arch Oral Biol* 1969;14:521.

Meachim G, Williams DF: Changes in nonosseous tissue adjacent to titanium implants. *J Biomed Mater Res* 1973;7:555.

Miller JK, Madsen FC, Hansard SL: Absorption, excretion, and tissue deposition of titanium in sheep. *J Dairy Sc* 1976;59:2008.

Mundorff SA, Little MF, Bibby BG: Enamel dissolution: II. Action of titanium tetrafluoride. *J Dent Res* 1972;51:1567.

Orbell GM: *The Effect of Topical Applications of Titanium Tetrafluoride and Other Agents on Dental Caries in Rats.* M.S. Thesis, University of Rochester, NY, 1973.

Pape HR Jr, Charbeneau GT, Loesche WJ: Effects on *Streptococcus mutans* of titanium tetrafluoride in amalgam restorations. *J Dent Res* 1975;54(Special issue A):129, Abstr. #336.

Perry HM Jr, Perry EF: Normal concentrations of some trace metals in human urine: Changes produced by ethylenediaminetetracetate. *J Clin Invest* 1959; 38:1452.

Reed AJ, Bibby BG: Preliminary report on effect of topical application of titanium tetrafluoride on dental caries. *J Dent Res* 1976;55:357.

Regolati B, Schait A, Schmid R, Muhlemann HR: The effect of titanium aluminum and fluoride on rat caries. *Helv Odont Acta* 1974;18:92.

Ruliffson WS, Burns LV, Hughs JS: The effect of the fluoride ion on Fe^{59} iron levels in blood rats. *Trans Kans Acad Sci* 1963;66:52.

Schettler D, Baumgart F, Bensmann G, Haasters J: Method of alveolar bracing in mandibular fractures using a new form of fixation made from memory alloy. *J Maxillofac Surg* 1979;7:51.

Schroeder HA: The biological trace elements. *J Chron Dis* 1965;18:217.

Schroeder HA, Balassa JJ, Tipton IH: Abnormal trace metals in man:Titanium. *J Chron Dis* 1963;16:55.

Shey Z: *Bonding of Adhesive Resins.* M.S. Thesis, University of Rochester, Rochester, NY, 1970.

Shrestha BM: *In vitro Chemical Effect of Different Fluoride and Non-Fluoride Compounds on Enamel Solubility.* M.S. Thesis, University of Rochester, Rochester, NY, 1970.

Shrestha BM, Mundorff SA, Bibby BG: Enamel dissolution: I. Effects of various agents and titanium tetrafluoride. *J Dent Res* 1972;51:1561.

Shrestha BM: Titanium tetrafluoride in the prevention of dental caries: Effects of systemically administered titanium tetrafluoride on rat caries. Ph.D. Thesis, University of Rochester, Rochester, NY, 1980.

Tipton IH, Cook MJ: Trace elements in human tissue, Part II. *Health Phys* 1963; 9:103.

Tipton IH, Stewart PL, Martin PG: Trace elements in diets and excreta. *Health Phys* 1966;12:1683.

Wefel JS, Valesio J, Wei SHY: TiF_4 as a topical fluoride agent. *J Dent Res* 1977; 56:(Special Issue B) Abst. #321.

Wei SHY, Soboroff DM, Wefel JS: Effects of titanium tetrafluoride on human enamel. *J Dent Res* 1976;55:426.

Williams DF: Titanium as a metal for implantation. Part 2: Biological properties and clinical applications. *J Med Eng Technol* 1977;1:266.

Young FA, Kresch CH, Spector M: Porous titanium tooth root: Clinical evaluation. *J Prosthet Dent* 1979;41:561.

Zallen RD, Fitzgerald BD: Treatment of mandibular fractures with use of malleable titanium mesh: Report of 56 cases. *J Oral Surg* 1976;34:748.

B, Al,

Y, Rb, Zr, Sn,

Cl, Br, Pl, Au, Ag, Be,

and others

19 Other Nonessential Trace Elements

M. E. J. Curzon

This chapter completes the information on all trace elements not discussed in the other chapters of Part III. Chapters on Pb and Cd follow. There are a very large number of elements in the periodic table still to be covered, but the task is simplified because there is no information on most of these elements. Such data as there are concern the trace elements: boron, aluminum, yttrium, rubidium, zirconium; these elements will be discussed as far as is possible. All remaining elements will be considered together in the final part of this chapter.

BORON

Boron (B) is essential for good growth of plants, particularly those used in agriculture. There is no evidence of its essentiality in animals or man although a number of attempts have been made to demonstrate that B is necesary for metabolism (Underwood, 1977). Widely distributed in all tissues and organs, the concentrations in human enamel vary between

0.5 to 69 ppm (Losee et al, 1974), and there is considerable geographic variation in B levels (Losee and Little, 1973).

There is now widespread use of B as a food additive, usually in the form of sodium borate or boric acid. Most of this B is completely absorbed and excreted via the urine. Large intakes have also been noted from the use of boric acid for treatment of burns (Underwood, 1977). Toxic effects have been reported but generally B has a low order of toxicity when taken orally. A concentration of 5 ppm B, as sodium metaborate, when given to mice via drinking water, did not affect lifespan, longevity, or health (Schroeder and Mitchener, 1975).

The limited research on B has generally shown the element to have a negative association with dental caries. In the low caries area of northwest Ohio drinking water supplies contained high levels of B (Curzon et al, 1970) which were reflected in tooth enamel levels (Curzon, 1977). However, in the various epidemiologic studies that have been carried out in this area of Ohio, statistical analyses have not shown any association of B in the drinking water to caries (Curzon et al, 1978).

Other human studies have indicated that there was a relationship between soil concentrations of minerals, including B, to caries prevalence (Barmes et al, 1970), and that this may be related to tooth size (Schamschula et al, 1972). In Schamschula's report there was a significant direct association between B content of village garden soils and the mesiodistal and buccolingual crown diameters of the second molars. There was also a consistent, but not significant, similar trend for the first molars. There appear to be no other reports on the effect of B on dental disease in man.

Animal studies, summarized in Table 19-1, have shown varying effects of B on dental caries and tooth morphology in animals. The earlier studies of Kruger (1959), Pappalardo (1968), and Malthus et al (1964) showed B as reducing caries, while the later studies of Hunt and Navia (1973, 1975), Liu (1975), and the early study by Shaw and Griffiths (1961) found either no effect of B or an increase in caries.

Concentrations of B used in the animal experiments have generally been quite high, 1 to 283 ppm, or even as percentages of the diet, and much higher than those encountered in drinking water supplies such as those in northwest Ohio of about 0.03 ppm B. The study by Losee et al (1976) which used water solids derived from the drinking waters of northwest Ohio reduced caries in rats.

Kruger (1959) also showed that a daily injection of 25 μg B, between the seventh and tenth day of life, gave significant changes in tooth morphology. It was interesting that in 25% of the rats the mesial fissure was absent, and in the other rats the fissure was partially eliminated or wider and shallower. Such a change in fissure dimensions would discourage food retention and cause less susceptibility to tooth decay. This finding is

Table 19-1
Summary of Experiments of Reports on the Effect of Boron on Dental Caries in Rats

	Boron Used			
Author (year)	*Salt*	*Conc.*	Vehicle	Effect on Caries
Kruger, 1959	Boric acid	5 μg/day, 25 μg	Injection	Reduced caries
Shaw and Griffiths, 1961	Sodium borate	0.5%, 1.0%	Diet	No effect
Malthus et al, 1964	NG	25 ppm	Water	25% reduction in caries
Pappalardo, 1968	Sodium borate + magnesium fluorsilicate	1%	Water	Reduced caries
Hunt and Navia, 1973	Sodium borate	25,125	Water	Increased caries pre-eruptively
Hunt and Navia, 1975	Sodium borate	150 μg per 10g BW	Water	Increased caries pre-eruptively
Liu, 1975	Sodium borate	a) 1–283 ppm b) + 10, 25F ppm	Water	No effect on caries Antagonized F effect
Losee et al, 1976	–	Water solids[a]	Diet	Reduced caries

[a]Containing B, Sr, Li, and F from northwest Ohio drinking water.
NG = not given.

of great interest in the light of the later report of significant associations of tooth size and caries in Papua-New Guinea (Schamschula et al, 1978) and merits further study.

Little attention has been devoted to other possible effects of B, such as incorporation of B in enamel, dissolution, or remineralization. Similarly, it is not known whether B has any effects on oral bacterial metabolism. The report of Herbison and Handelman (1975) found an effect for trace element solutions, based upon the trace element content of northwest Ohio drinking water, on dissolution of hydroxyapatite by cariogenic bacteria. These solutions contained B as well as F, Sr, and Mo. These authors found no effect of the trace elements on bacterial metabolism, but concluded that the major effect was for Sr and F and on hydroxyapatite dissolution.

In summary, therefore, B is an interesting trace element but its action on dental caries has barely been investigated. Tentatively, it has a cariostatic effect in animals and possibly also in man, which may be related to changes in tooth morphology.

ALUMINUM

A relatively abundant element in the earth's crust, aluminum (Al) is widely used by man in its elemental form as various products from airframes to cooking pots. Ingested poorly, Al does not appear to perform any essential function in plants, animals, or micro-organisms (Underwood, 1977).

The element has an interesting interaction with F that is important in human metabolism, but also has significant dental implications. Absorption of Al from food is related to fluoride intake, and it has been shown that administration of 1 mg F to rats gives an increase in Al elimination. Simultaneously giving 50 mg Al/kg and 1 mg F resulted in decreased levels of Al in rat tissues, as shown by Ondreicka et al (1971). These same authors suggested that a readily soluble complex (AlF_6) is formed by an interaction between Al and F. This complex is far more soluble than the usual fluoride salt, CaF, and so the Al is removed from the body as the fluoride intake increases. The reverse process, whereby Al might lessen the intake of fluoride, is of more interest to us here.

Animal experiments have shown that the Al counteracts dental fluorosis as a result of decreased deposition of F in the calcified tissues (Wadhawani, 1954). When adult mice were given solutions of NaF at 150 μg F/day by gastric intubation, fluorosis of the incisors, characterized histologically, was seen. Simultaneous administration of ammonium hexafluoraluminate reduced the incidence of fluorosis by 100% (Ruzicka et al, 1974). Interestingly the F content of the incisor ash was only reduced to 93% of the control by use of the Al-complex. These results suggest that the Al-F complexes could usefully be used as fluoride supplements with less likelihood of occurrence of fluorosis. Alternatively they might be tested where high fluoride occurs (drinking water fluoride levels over 2 ppm) to offset the excessive fluorosis found in those areas.

The use of Al cookware affects the availability of F′ from the water. As the amount of fluoride in water modifies the concentrations in foods cooked in that water, so the influence of Al cooking pots may bring about transfer, or complexing, of the F′ onto the walls of the vessels used. Shannon (1977) demonstrated a considerable drop in F′ concentration from 1.0 ppm down to less than 0.01 ppm in ten hours of boiling. However, the initial drop in F′ content was greatest in the first 30 minutes, which would be an average cooking time for vegetables. Full and Parkins (1975) had found that in just 15 minutes the loss of F′, presumably complexed as AlF_3, was about 80%.

Besides this effect of interacting with F, there is the question of what effect Al has on dental caries, either by itself or by its interaction with fluoride. Reports on the relationship of Al to dental caries in man are rare. In the Papua-New Guinea study soil samples were analyzed for Al but no biologic material such as enamel or plaque was analyzed (Schamschula et al, 1978). However, in the Colombian dental survey Al was shown to be at significantly higher concentrations in the soil and water of the high caries village (Rothman et al, 1972). There are no other epidemiologic reports in the literature on the effect of Al on dental caries.

Collecting and analyzing soil and water samples from high and low caries areas in Yugoslavia, Vrbic and Stupar (1980) found Al concentrations in water of 330 and 30 μg/l (ppb), where the percentage of children with carious teeth was 9.4 and 27.0% respectively. However, there was no difference in the concentrations of Al between tooth enamel samples collected from the same communities.

Nearly all samples of dental enamel contain Al, and Little and Steadman (1966) reported a range of 95 to 220 ppm. Most other studies since then have recorded lower values than these. Attempts to relate Al concentrations in enamel to dental caries have not been successful. The lack of such a correlation in the Yugoslavian study has already been noted. In an earlier study by Derise and Ritchey (1974), Al was not one of trace elements showing a significant caries relationship. In the regression analysis of trace elements in enamel and caries by Curzon and Crocker (1978), a significant negative relationship of Al to caries prevalence in the tooth donors was found. However, the partial *t* effect, that part of the caries index that could be explained by the Al, was only -0.30×10^{-1}. The Al therefore accounted for only a very small increment of the caries. In a further stage of this multiple regression analysis an interaction of Al and F was found to be significant, but again the partial t effect was $+ 0.18 \times 10^{-3}$. The positive sign indicated a cariogenic association with caries. This finding would support the suggestion that Al may offset the beneficial effect of F.

A number of animal studies, summarized in Table 19-2, have shown varying effects of Al on caries in rats. In most studies Al, when tested alone, had no effect on caries, or even in some cases increased the degree of acid erosion on teeth (Regolati et al, 1975). When used in combination with fluoride, there was generally a greater reduction in caries than by use of the fluoride alone. Because of the indication of a detrimental effect by Al with F in human enamel further experiments are required.

Uptake of Al by rat molars has been studied in a few of the animal experiments listed above. As would be expected, there is an uptake of Al by rat enamel that seems to affect the enamel dissolution rate (Regolati et al, 1969).

Several researchers have studied the interactions of Al and F and their effects on enamel dissolution. Thus, McCann (1969) showed that maximum retention of F in surface enamel was achieved after first treating the surface enamel with a solution of an Al salt. Enamel treated for one minute with 0.05M $Al(NO_3)_3$ and then for three minutes with acidulated phosphate fluoride (APF), retained 1800 ppm F compared with 800 ppm F for treatments with APF alone. Similarly Vrbic and Brudevold (1970) found that the use of a polishing paste containing 0.2 mmol/l $Al(NO_3)_3$, followed by exposure to fluoride solutions, gave an increased deposition of F in enamel. Finally Gerhardt and Windler (1972) demonstrated that a pretreatment of enamel tooth crowns with a Zr-silicate prophylaxis mixed with 0.02 M solution of $Al(NO_3)_3$ gave an increased F uptake from topical solutions of both APF and SnF_2. Further interest in this promising effect of Al is lacking, and there have not been any clinical trials.

Various salts of Al have recently been used as abrasives in toothpastes. Clinical trials of dentifrices, such as that of Andlaw and Tucker (1975) using AlO with 0.8% MFP, have shown good reductions in caries. The effect of the presence of an Al salt on caries by the Al of itself is not known.

Research on any bacterial effect of Al has been limited. Olsson et al (1978) found that adherence of oral streptococci was inhibited with a pretreatment of enamel with $AlCl_3$. However, tests on the bacterial composition of plaque in rats, treated with $AlCl_3$ via their drinking water, showed no effect of the Al on the percentage bacterial composition of fissure plaque (Beighton and McDougall, 1981).

In summary, therefore, little is known about the effects of Al on enamel and on dental caries. There are indications that there is potentially useful interaction of Al with F to affect fluoride uptake and perhaps lead

Table 19-2
Summary of Reports on the Effect of Aluminum on Dental Caries in the Rat

Author (year)	Aluminum Used		Vehicle	Effect on Caries
	Salt	*Conc.*		
Wynn and Haldi, 1954	$AlCl_3$ 20 ppm	0.16, 2.0	Diet	No effect
Kruger, 1958	A. Al-Ac	0.25, 0.08 mg	Injection	No effect
	B. Al-Ac plus NaF	0.25, 0.08 mg 0.108, 0.054 mg		Significant reduction
Van Reen et al, 1967	$AlKSO_4$	10, 50, 100 ppm	Water	No effect
Regolati et al, 1969	A. $AlCl_3$	0.05M	One topical	Reduced caries
	B. $AlCl_3$ plus NaF	0.05M 0.1%	Topical for five days	Additive reduction in caries
Regolati et al, 1974	A. $AlCl_3$	0.0264M	Daily topical for 5 days	No effect
	B. $AlCl_3$ plus NaF	0.0264M 2000 ppm	Daily topical for 5 days	Caries slightly lower than F alone
Regolati et al, 1975	A. $AlCl_3$	1%	Daily topical for 21 days	Increased erosion
	B. $AlCl_3$	1%	Twice daily topical 5 days	No effect
Riethe, 1979	A. Al-lact	NG	Used as a toothpaste daily	Significant reduction in all test groups
	B. Al-lact plus $AlF_x(OH)_y$	NG 0.068%		
	C. Al-lact plus $AlF_x(OH)_y$	NG 0.136%		

NG = not given.
Al-Ac—aluminum acetate.
Al-lact—aluminum lactate.

to changes in the chemistry of the enamel. Adherence of bacteria to surface enamel may also be reduced by pretreatment with Al salts.

YTTRIUM

Yttrium (Y) occurs in nearly all of the rare earth minerals; the metal is produced commercially by reduction of the fluoride by Ca metal. The earth is widely used in making Y_2O_3: Europium phosphors to impart the red color on television screens.

A report by Thomassen and Leicester (1964) showed that ionic Y, when injected peritoneally, could be incorporated into the enamel of rats when given during the period of tooth development. Where and how it is incorporated, or what effect it has on the chemistry of the enamel, is unknown. Castillo-Mercado and Bibby (1973) produced significant alterations in rat molar morphology by injection of Y during the period of tooth development in rats. In an associated study, Castillo-Mercado and Ludwig (1973) gave rats 0.1 mg Y/day, again by injection, or 50 ppm Y in drinking water. All animals were fed a cariogenic diet, and results showed significantly lower ($p = < 0.001$) caries scores in the Y-injected animals. Although the rats receiving Y via the drinking water also had lower caries prevalence, the differences were not statistically significant.

In human enamel Y is present at low concentrations. Losee et al (1974) gave a mean value of 0.007 ppm with a minimum of 0.01 and a maximum of 0.17. However, although 28 teeth were used, Y was found in only 2 samples.

RUBIDIUM AND CESIUM

Occurring in the same group of the periodic table as Li, K, and Na, rubidium (Rb) and cesium (Ce) should be further investigated because of their close physicochemical relationship to K. Thus, for a variety of physiologic processes, Rb can affect the action of other members of the group, such as the neutralization of the toxic effect of Li on fish larvae (Underwood, 1977). Intake of Rb in man is about 2 to 4 mg/day with the major sources being foods of plant origin and meats.

Rubidium

Resembling K in its distribution and excretion in the animal body, Rb does not have any known physiologic function. The element is not normally taken up by calcified tissues any more than by other parts of the body. The concentration of Rb in human dental enamel, 0.39 ppm, is much lower than that of K, 401 ppm (Losee et al, 1974).

Cesium

In enamel Ce was found at a very low concentration of 0.04 ppm by Losee et al (1974). In this study, however, it was of interest that Ce was found in only 27 out of the 28 samples used. Cesium may occur in enamel frequently, but at such low concentrations that would probably not affect apatite chemistry.

A survey of the dental literature shows that apart from the reporting the presence of these two elements in dental tissues little work has been done on their possible effect on dental disease. With the present interest in Li, and its role in dental caries, it would seem that a new look should be taken at the alkali metals, Na, K, Li, Rb, and Ce. Many of the comments made on Li in Chapter 17 can also be applied to these other group I elements and would merit further attention.

ZIRCONIUM

Zirconium (Zr) is relatively abundant as a trace element and is well known in dentistry as a polishing agent. The metabolic action of Zr, if any, has not been directly studied (Underwood, 1977). It is absorbed from ordinary diets and, therefore, presumably absorbed from any swallowed polishing pastes. Schroeder and Balassa (1966) listed Zr as a "biologically neglected" element. Its concentrations in human tissues are comparable or larger than Cu.

The levels found in enamel are quite low. In the study of Curzon and Crocker (1978) a mean concentration of 0.07 ppm was reported, but with no significant relationship to dental caries. When the data given in the above paper were broken down by geographic origin a significant difference was noted between enamel samples from high and low caries donors. Concentrations were 0.27 and 0.16 ppm Zr respectively. These concentrations were significantly different at the $p = < 0.02$ level (Curzon, 1977). This finding has not been supported by any other work.

Polyvalent ions such as Zr, as well as Al, Ti, and Sr have been shown to be effective in increasing topical F deposition in enamel (McCann, 1969). By this approach the fluoride is deposited in the enamel as a metal complex. Other studies, such as that of Vrbic and Brudevold (1970), have also shown zircate polishing pastes to give a slight gain in fluoride uptake. There is no evidence of any action of Zr, of itself, on dental caries.

TIN

Although tin (Sn) has been shown to be essential for growth in rats, this has not been demonstrated in other animals, although deficiency states in other animals have been reported (Underwood, 1977). Large amounts of Sn can be ingested from foods when they are contaminated by prolonged contact with tinplate, but humans are very tolerant of such high concentrations because of poor Sn absorption.

In dentistry Sn has been extensively studied because of its widespread use as SnF_2 as a topical fluoride agent; it was listed by Navia (1970) as having a doubtful effect on caries. However, it is not clear whether Sn has any effect on caries by itself, as a separate entity from the effect of F. Thus Muhler and Day (1950) showed a cariostatic action of 10 ppm SnF_2, but no effect of $SnCl_2$ on dental caries in rats. However, in a subsequent experiment, Muhler (1957) showed that $SnCl_2$ did not reduce caries but that SnF_2 was better than NaF in inhibiting caries. This effect was greater than would be expected from the fluoride alone.

Binding of Sn to enamel can be considerable; it was suggested by Meckel (1962) as due to uptake in "enriched" areas of enamel. Other studies showed that Sn could be slowly released with time (Hoerman et al, 1966). Such a slow release of Sn^{2+} could have an inhibitory effect on oral bacteria. Lilienthal (1956) noted that SnF_2 inhibited acid production in vitro more than NaF, proportional to the Sn content rather than to the fluoride content.

Further studies have shown SnF_2, as a mouth rinse, inhibits acid formation in plaque (Svatun and Attramadal, 1978), and also lessens the formation of plaque (Svatun et al, 1977). The major effect was thought to be due to Sn. Bacteria such as *S. mutans* show a rapid uptake of Sn that has been shown to be greater than other metal cations (Attramadal and Svatum, 1980). The presence of Sn^{2+} may have an inhibitory effect on growth rate and metabolism of bacteria (Aikin and Dean, 1976) which could account for the indicated effect of Sn on caries.

A recent study on the effect of Sn has shown that a 0.4% SnF_2 experimental toothpaste significantly reduced both plaque and gingivitis (Bay and Rolla, 1980). However, in longer term trials the plaque inhibition effect of SnF_2 has been shown by Leverett et al (1981, and personal communication, 1981) to be only of short duration (less than four months). It has also recently been demonstrated, in a preliminary report, that prerinsing with metal ions, Zn and Sn, reduced the antiplaque effect of chlorhexidine (Wales and Rolla, 1980).

The dental literature of Sn is much more extensive than is indicated here, mainly because of the many studies that have been carried out on SnF_2. However, data on the effect of Sn itself are still limited, although there has been considerable interest recently. If this trend continues, much valuable information will be available on the relationships of Sn to dental disease.

CHLORINE AND BROMINE

Apart from F, the other halogens have not been seriously studied for their effects on dental disease. The element I has already been discussed; thus the remaining elements to consider are chlorine (Cl), bromine (Br), and astatine (At). All of these elements have marked antibacterial action and some of them, as salts of other elements, are used as bacteriocides or disinfectants in medicine. For At there is no information on any effect on dental disease; because of its position in the periodic table and high molecular weight, it is most unlikely that it has any effect. The element was not listed by Losee et al (1974) in their study on trace elements occurring in enamel.

Chlorine

Navia (1970) did not list Cl (as chloride) in his review of trace elements and dental caries. The element occurs in relatively large concentrations in enamel (see Chapter 3) but has never been shown to have any relationship to dental caries, although it must have a bacteriocidal effect on oral bacteria if present at high enough levels. The element has not been shown to have any important role in animal metabolism nor has any effect on growth or reproductive performance been demonstrated (Underwood, 1977).

There is an indication of a limited effect of Cl on the uptake of fluoride. Ericsson (1962) showed that the presence of Cl enhanced the uptake of fluoride by intact enamel surfaces although it had the opposite effect on powdered enamel. Ericsson's conclusion was that the results supported the use of kitchen salt, NaCl, as a vehicle for caries-preventive administration of fluoride. Subsequent studies have continued to show a reduction in caries prevalence by the use of fluoridated salt, but this is by action of F and not of Cl.

Bromine

There is no substantial evidence from human studies of any effect of Br on dental disease. In enamel Br is frequently found at about 1 ppm (Losee et al, 1974), but in the various epidemiologic studies concerned with multiple trace elements, Br has not been mentioned. In water supplies Br occurs only at low concentrations with little variation.

In animal studies, Sognnaes (1949) fed bromine to rats during the period of tooth development and found an increase in caries. However, there was a tendency to reduce caries with Br when given after tooth eruption. This might be explained by a posteruptive, topical, antibacterial effect of Br. Such an action does not seem to have been studied further, but in the light of findings on I, it would be interesting to see if Br does have such an effect.

PLATINUM, GOLD, SILVER, BERYLLIUM, AND OTHERS

After considering all the elements discussed in previous chapters, as well as Cd and Pb (Chapters 20, 21), there are still a large number of trace elements left, such as the rare earths and lanthanide series, but there are very few reports in the literature associating them in any way to dental diseases. As Losee et al (1974) stated, there are 39 elements regularly incorporated into enamel during development, and a further 34 elements are not present or only in exceptionally low concentrations. In Table 6 of Losee et al there are 31 elements not detected above a limit of approximately 0.1 ppm, all with an atomic number below 81. From this and other enamel studies it appears that Pb, atomic number 82, is the heaviest element to occur regularly in enamel. On this basis, we shall not consider further any element with an atomic weight greater than this.

Platinum

Sometimes used in small quantities in restorative materials, platinum (Pt) when present in enamel is probably there only as a contaminant. In a series of rat studies Wisotsky and Hein (1956, 1958) found that $PtCl_4$, when given to rats at 1 mEq, caused excessive levels of caries. Similarly Hein et al (1958) found Pt to be markedly cariogenic.

Gold

The dental literature contains many articles concerning gold (Au) as a restorative material. Any therapeutic effect of Au itself on dental caries or periodontal disease is largely unknown. The noble metal is often found in human enamel but at low levels, indicating that its presence is as a result of contamination (Losee et al, 1974). Incorporation of Au into apatite probably does not effect the properties of the enamel. Wisotsky and Hein (1958) tested $AuCl_3$ in hamsters and found that it inhibited caries. Metal coating of Au onto the teeth of Rhesus monkeys was tried by Cutwright et al (1977). Using $AuCl_3$ as well as AgF and $AgNO_3$, the solution was reduced in situ, using hydrazine after previous etching of the enamel surfaces with H_3PO_4 (O'Keefe and Christie, 1975). After six months all pits and fissures remained coated and free of caries. Further studies indicated possible antibacterial and antiplaque properties of the process. Although this was an interesting finding, the appearance of the metal-coated teeth was hardly esthetic and the use of such a procedure in man seems doubtful.

Silver

There are some interesting interactions of Ag with other metals such as Cu and Se in animal metabolism. Deficiency of Cu is accentuated by Ag and can lead to depressed growth rate in poultry (Underwood, 1977). There is no evidence of any such interactions affecting dental caries.

In addition to Au, silver (Ag) is a major restorative material in dentistry through its role in Ag-amalgam. Also, as with Au, the dental literature contains an extensive list of references on the use of this metal for fillings. However, Ag has been used as a possible anticaries agent for a long time as Ag-nitrate. For many years Ag-nitrate solutions were used on carious dentin, particularly in deciduous teeth, to arrest or prevent dental decay. Studies to demonstrate this effect were carried out by Klein and Knutson (1942). Kuroiwa (1979) has suggested the use of ammonium AgF_3 for the prevention of caries.

Recently studies on therapeutic effects of Ag in dentistry have concentrated on antibacterial and cytotoxic actions. Sterilization of carious dentin has been proposed by Rose (1976) by using Ag-nitrate; Lan (1977) used ammoniacal Ag-nitrate to sterilize root canals. Another approach to the use of Ag for sterilization has been to use the generation of Ag^{2+} by low intensity direct current (Thibodeau et al, 1978). This has the effect of inhibiting or killing of oral bacteria, and has been tried with some success for sterilizing root canals. Leirskar (1974) showed that Ag and Cu amalgams, by the release of free ions, were cytotoxic in cell cultures.

Beryllium

An alkaline earth, beryllium (Be) is toxic when inhaled, (causing berylliosis), or when injected or implanted into animal tissues. The very small quantities that are ingested do not seem to be a problem. The element has been shown experimentally to produce changes in bones and teeth when used in high concentrations. Maynard et al (1950) produced rachitic lesions in rats when these animals were fed very high doses of 5%

$BeSO_4$ or $BeCO_3$ in the diet. A similar finding was reported by Wentz (1955) who observed rickets in rats when they were fed 6% $BeCO_3$, and enamel hypoplasia was noted in the incisors. This action of Be on calcification also includes an effect on calculus. Sherman and Sobel (1965) found that a concentration of 0.01 ppm Be inhibited $CaPO_4$ crystal growth in vitro. Formation of calculus was affected by a solution of 1 ppm $BeCl_2$, by prevention of calcification of the organic precursor that forms the nidus for initiation of calculus (Sobel et al, 1966). Although this work suggested a beneficial use for $BeCl_2$ the concentration of Be was, however, felt to be too toxic for human use. Leicester et al (1954) obtained a highly significant reduction in caries using 0.02% $BeCl_2$ either in the diet or via the drinking water. However, because of the known toxicity of Be, it is unlikely that this element will have any future as a caries-preventive agent.

Palladium

Another metal used in dental restorative materials, palladium (Pd) has been identified in enamel (Losee et al, 1974), but is again probably there as a contaminant. Wisotsky and Hein (1956) found Pd had no effect on caries in hamsters. This element has attracted no further research interest.

Other Elements

Of the remaining elements none appears to be of any importance in dental caries or periodontal disease. Some of the remaining list, such as mercury (Hg), are widely used and therefore are found in enamel and other oral tissues. In the case of Hg this is undoubtedly because of contamination from amalgam that has been used as a filling material. The elements niobium (Nb) and germanium (Ge) have been found regularly in enamel but no research has been carried out on any possible dental implications. It would seem unlikely that these remaining elements have any effect on dental disease.

REFERENCES

General

Curzon MEJ: Trace element composition of human enamel and dental caries. PhD Thesis. University of London, London, 1977.

Curzon MEJ, Crocker DC: Relationships of trace elements in human tooth enamel to dental caries. *Arch Oral Biol* 1978;23:647–653.

Losee FL, Cutress TW, Brown R: Trace elements of the periodic table in human dental enamel. *Caries Res* 1974;8:123–134.

Navia JM: Effect of minerals on dental caries, in Harris RS (ed): *Dietary Chemicals: Dental Caries.* Advances in Chemistry Series 94. Washington, American Chemical Society, 1970.

Schamschula RG, Adkins BL, Barmes DE, et al: WHO *Study on Dental Caries Aetiology in Papua*-New Guinea. WHO Publication #40, Geneva, 1978.

Underwood EJ: *Trace Elements in Human and Animal Nutrition.* ed 4. New York, Academic Press, 1977.

Boron

Barmes DE, Adkins BL, Schamschula RG: Caries aetiology in Papua-New Guinea. Associations in soil, food and water. *WHO Bulletin* 1970;43:6, 769–784.

Curzon MEJ, Adkins BL, Bibby BG, et al: Combined effect of trace elements and fluorine on caries. *J Dent Res* 1970;49:526–529.

Curzon MEJ, Spector PC, Iker HP: An association between strontium in drinking water supplies and low caries prevalence. *Arch Oral Biol* 1978;23: 317–321.

Herbison RJ, Handelman SL: Effect of trace elements on dissolution of hydroxyapatite by cariogenic streptococci. *J Dent Res* 1975;54:1107–1114.

Hunt CE, Navia JM: Pre-eruptive effects of Mo, B, Sr and F on dental caries in the rat. *Arch Oral Biol* 1975;20:497–501.

Hunt CE, Navia JM: Effects of Sr, Mo, Li and B on developing teeth and other tissues of neonatal rats, in *Trace Substances in Environmental Health* VI: 159–167, Columbia, MO, University of Missouri, 1973.

Kruger BJ: The effect of trace elements on experimental dental caries in the albino rat. *University of Queensland Papers* 1959;1:1–28.

Liu FT: Post-developmental effects of boron, fluoride and their combination on dental caries activity in the rat. *J Dent Res* 1975;54:97–103.

Losee FL, Little MF: Boron content of human teeth. IADR Abstract #273, *J Dent Res* 1973;52:127.

Losee FL, Adkins BL, Curzon MEJ: Effect of water solids on dental caries in the rat. *Caries Res* 1976;10:382–386.

Malthus RS, Ludwig TG, Healy WB: Effect of trace elements on dental caries in rats. *NZ Dent J* 1964;60:291–297.

Pappalardo G: Effects of a combination of borate and magnesium fluorosilicate on experimental caries in the rat. *Riv Ital Stronatal* 1968;23:509–518.

Schroeder HA, Mitchener M: Effect of boron on life span of mice. *J Nutr* 1975; 105:421–424.

Shaw JH, Griffiths D: Developmental and post-developmental influences on the incidence of experimental dental caries resulting from dietary supplements by various elements. *Arch Oral Biol* 1961;5:301–322.

Aluminum

Andlaw RJ, Tucker GJ: A dentifrice containing 0.8% MFP in an aluminum oxide trihydrate base. *Br Dent J* 1975;138:426–432.

Beighton D, McDougall WA: The influence of certain added water-borne trace elements on percentage bacterial composition of tooth fissure plaque from Spraque-Dawley rats. *Arch Oral Biol* 1981;26:419–425.

Derise NL, Ritchey SJ: Mineral composition of normal human enamel and dentin and the relation of composition to dental caries. II. Microminerals. *J Dent Res* 1974;53:853–858.

Full CA, Parkins FM: Effect of cooking vessel composition on fluoride. *J Dent Res* 1975;54:192–193.

Gerhardt DE, Windler AS: Fluoride uptake in natural tooth surfaces pretreated with aluminum nitrate. *J Dent Res* 1972;51:870.

Kruger BJ: The effect of trace elements on experimental dental caries in the rat. *Aust Dent J* 1958;58:298–302.

Little MF, Steadman LT: Chemical and physical properties of altered and sound enamel. IV Trace element composition. *Arch Oral Biol* 1966;11:273–278.

McCann HG: The effect of fluoride complex formation on fluoride uptake and retention in human enamel. *Arch Oral Biol* 1969;14:521–531.

Olsson J, Odham G: Effect of inorganic ions and surface active organic compounds on the adherence of oral streptococci. *Scand J Dent Res* 1978;86: 108–117.

Ondreicka P, Kortus J, Ginter E: in Skoryna SC, Waldon-Edwards D (eds): *Intestinal Absorption of Metal Ions, Trace Elements and Radionuclides.* Oxford, Pergamon Press, 1971.

Regolati B, Schait A, Schmid R, et al: Effects of aluminium and fluoride on caries, fluorine content and dissolution of rat molars. *Helv Odont Acta* 1969;13:59–64.

Regolati B, Schait A, Schmid R, et al: The effect of titanium, aluminium and fluoride on rat caries. *Helv Odont Acta* 1974;18:92–96.

Regolati B, Schait A, Schmid R, et al: Effect of enamel solubility-reducing agents on erosion in the rat. *Helv Odont Acta* 1975;19:31–36.

Riethe VP: Uber die Karieshemmede Werking eines Aluminiumfluoridkomplexes und des Aluminiumlactats in Tierexperiment. *Dtsch Zahnärtzl* 1979;34: 16–18.

Rothman KJ, Glass RL, Espinal F, et al: Dental caries and soil content of trace elements in Colombian villages. *J Dent Res* 1972;51:1686.

Ruzicka JA, Mrklas L, Rokytova K: Incorporation of fluoride in bones and teeth and dental fluorosis in mice after administration of complex fluoride. *Arch Oral Biol* 1974;19:947–950.

Shannon IL: Effect of aluminum and teflon cooking vessels on fluoride content of boiling water. *J Mo Dent Assoc* 1977;57:26–28.

Van Reen R, Ostrom CA, Berzinkas VJ: Trace elements and dental caries.: Molybdenum, aluminum and titanium. *Helv Odont Acta* 1967;11:53–58.

Vrbic V, Brudevold F: Fluoride uptake from treatment with different fluoride prophylaxis pastes and from the use of pastes containing soluble aluminum salts followed by topical application. *Caries Res* 1970;4:158–167.

Vrbic V, Stupas J: Dental caries and the concentration of aluminium and strontium in enamel. *Caries Res* 1980;14:147–157.

Wadhawani TK: Effects of aluminium on dental fluorosis. *J Indian Instit Sci* 1954;36:64–69.

Wynn W, Haldi J: Effects of aluminium on caries in the rat. *J Nutr* 1954;54: 285–290.

Yttrium

Castillo-Mercado R, Bibby BG: Trace element effects on enamel pigmentation, incisor growth and molar morphology in rats. *Arch Oral Biol* 1973;18: 629–635.

Castillo-Mercado R, Ludwig TG: Effect of yttrium on dental caries in rats. *Arch Oral Biol* 1973;18:637–640.

Thomasson PR, Leicester HM: Uptake of radioactive beryllium, vanadium, cerium and yttrium in the tissues and teeth of rats. *J Dent Res* 1964;43: 346–352.

Zirconium

McCann HG: The effect of fluoride complex formation on enamel fluoride uptake and retention. *Arch Oral Biol* 1969;14:521–531.

Schroeder HA, Balassa JJ: Abnormal trace metals in man: Zirconium. *J Chron Dis* 1966;19:573–586.

Vrbic V, Brudevold F: Fluoride uptake from treatment with different fluoride prophylaxis pastes from the use of pastes containing a soluble aluminum salt followed by topical application. *Caries Res* 1970;4:158–167.

Tin

Aiken RM, Dean ACR: Action of stannous and stannic chlorides on bacteria. *Experimentia* 1976;32:1040–1041.

Attramadal A, Svatun B: Uptake and retention of tin by *S. mutans. Acta Odont Scand* 1980;38:349–354.

Bay I, Rolla G: Plaque inhibition and improved condition by use of stannous fluoride toothpaste. *Scand J Dent Res* 1980;88:313–315.

Hoerman KC, Klima JE, Birho LS, et al: Tin and fluoride uptake in human enamel in situ: Electron probe and chemical microfilm. *JADA* 1966;73: 1301–1305.

Leverett DH, McHugh WD, Jensen OE: The effect of daily mouth rinsing with stannous fluoride on dental plaque and gingivitis. Four months results. *J Dent Res* 1981;60:781–784.

Lilienthal B: Inhibition of acid formation from carbohydrates by stannous fluoride and stannous chlorofluoride. *Aust Dent J* 1956;1:165–173.

Silver

Klein H, Knutson JW: Studies on dental caries. XIII. Effect of ammoniacal silver nitrate on caries in first permanent molars. *JADA* 1942;29:1420–1426.

Kuroiwa M: Use of silver ammonium fluoride for the prevention of caries development. *Dent Outlook* 1979;53:85–100.

Lan WH: Efficacy of ammoniacal silver nitrate in root canal therapy. *Bull Tokyo Med Dent Univ* 1977;24:169–176.

Leirskar J: On the mechanism of cytotoxicity of silver and copper amalgams in cell culture system. *Scand J Dent Res* 1974;82:74–81.

Rose TP: Sterilization of carious dentin with silver nitrate. *Chron Omaha Dist Dent Soc* 1976;39:151–159.

Thibodeau EA, Handelman SL, Marquis RM; Inhibition and killing of oral bacteria by silver ions generated with low intensity direct current. *J Dent Res* 1978;57:922–926.

Beryllium

Leicester HM, Thomassen PR, Nicholas LW: Effect of beryllium upon hamster caries. *J Dent Res* 1954;33:670.

Maynard EA, Downs WL, Scott JK: Effects of beryllium on calcification. *Fed Proc* 1950;9:338–341.

Sherman BS, Sobel AE: Inhibition of tertiary calcium phosphate crystal growth with low concentrations of beryllium. *J Dent Res* 1965;44:454–461.

Sobel AE, Eilberg RG, Gould D: Inhibition of calculus formation by beryllium chloride. *Dent Surv* 1966;42:64–65.

Wentz J: Alterations in dental tissues following beryllium carbonate feeding. *J Dent Res* 1955;34:735.

$_{82}$ **Pb**

207.2

32-18-4

20 Lead

M. V. Stack

Lead (Pb) is a relatively heavy metal, atomic number 82, placed in group IVa of the periodic table. However, because of the availability of 6s and 6p electrons its properties resemble those of group IIa metals rather than those of the other group IVa elements.

The metal is widely used in storage batteries and radiation shielding, and its inorganic compounds, such as oxides, sulfate, chromate, and titanate, are constituents of pigments, paints, and varnishes; minor uses in plastics and insecticides are found for the arsenate and borate. It is the very widespread use of an organic Pb compound, in the form of a tetra-alkyl derivative, as a gasoline (petrol) additive that has resulted in Pb as a major environmental pollutant. Also, its use for many years in house painting has left a legacy of potential Pb poisoning, particularly causing risk to children when they live in old houses in deprived urban areas.

Metabolism

Stable complexes of Pb are formed with free carboxylate-, phosphate-, and thiol-bearing ligands of biopolymers and membranes (Venugopal and Luckey, 1978). Great stability of Pb nucleoside has been noted, especially of the cytidine complex. The solubility of Pb phosphate is very low, with a solubility product of 10^{-30} mol/l and microcrystalline precipitates occur in cytoplasm when the concentration of Pb exceeds the rather high level which can be attained in physiologic media (0.1 mmol/l). Under certain conditions Pb can be stimulatory, eg, in DNA synthesis.

Typical diets contain about 0.1 ppm Pb, and daily intakes of 0.2 to 0.4 mg are relatively high compared with intakes from water, air, and smoking (Moore, 1979; Nishimura, 1978). However, much of the intake from urban air can be ascribed to fuel tetra-alkyl derivatives, and the unburned proportion can be taken up in brain tissue. The object of detoxification is to convert the Pb intake into a nondiffusible form. A protein of the thionein type may be involved, as is certainly the case for Cd; Pb is known to have a high affinity for such protein. However, characteristic nondiffusible states for Pb are as large intranuclear bodies formed in proximal tubular cells of kidneys and also in osteoblast nuclei.

Lead and Neurophysiologic Impairment

A number of studies of groups of children in hospitals, population surveys, and industrial locations where there are Pb smelters and battery factories have sought to relate the Pb hazard to intelligence. The more disadvantaged of pairs of groups tested showed the same mean IQ scores from seven studies in which tooth Pb was determined as from eight studies featuring blood Pb differences (Lawther, 1980, Table 14). It has been objected that social and other variables may not be accounted for adequately in such group comparisons, although some 40 potential correlates were studied by Needleman et al (1979). Significant findings in a multivariate follow-up reassessment of 63 urban children by Ernhart and Schell (1981) arose because of methodologic difficulties. The authors concluded that if there are behavioral and intellectual sequelae of low levels of Pb burden independent of other aspects of parental and social influences on development, these effects are minimal.

In the best known study of impairment in relation to subclinical Pb burdens (Needleman et al, 1979) the IQ scores differed by four points in groups of children representing the extreme deciles of a distribution ordered according to tooth Pb values (dentin). Evaluation according to ratings by teachers showed that the frequency of nonadaptive classroom behavior increased in a dose-related fashion with dentin Pb levels. Further study of smaller groups of children, also representing the extreme deciles in the same population (Burchfield et al, 1980), using electroencephalography and psychologic tests, generated data that, together with that of Needleman et al (1979), confirmed that the function of the developing brain could be impaired by everyday (urban) exposure to Pb. They concluded that the effects were not disclosed unless highly sensitive measurements were made, and that the significance of the results would be likely to be enhanced by combining measures of behavior with those obtained by electrophysiologic methods.

It appears that IQ deficits can be related to differences in blood Pb levels exceeding 10 μg/dl. Thus, a much greater deficit might be expected when blood Pb levels fall within the rather high range of 40 to 80 μg/dl. This does not seem to be the case, however, for the same IQ deficits of 4 to 5 points are observed also in comparisons of children showing these higher blood Pb levels with those in the normal range (Bryce-Smith and Stephens, 1980). In an investigation of children with elevated blood Pb levels (de la Burdé and Choate, 1975) the circumpulpal dentin Pb was found to be twice that in the controls, and the mean IQ difference was 5 points.

In another study (Hrdina and Winneke, 1978) the mean IQ score was about 6 points higher in the control low Pb subgroup (tooth-Pb 2.4 ppm) than in the higher Pb subgroup (9.2 ppm). However, background factors may not have been sufficiently comparable in these subgroups, which represented less than one-tenth of the total number of children studied.

Lead in Calcified Tissues

Depositions of some trace elements are much greater in skeletal tissues than in soft tissues. Whereas the Pb content of the skeleton is more than 0.1 g, the soft tissues contain only one-tenth as much. An even more marked ratio characterized Sr (100:1), since Sr and Pb, as well as F, are all bone-seeking elements. It is well known that the Pb content of bone increases with age, and there is a typical 1% retention of the Pb ingested daily. The rate of accumulation can reflect not only dietary habits but also smoking activity. Concentrations of Pb in tissues of children, including bone, have recently been reported (Barry, 1981), and these confirm a number of those values already reported, eg, those of Schroeder and Tipton (1968) and of Casey and Robinson (1978). For three types of bone (rib, tibia, and calvaria) mean values were given by Barry as: 0.7, 2.9, and 15 ppm, for age groups 1 to 11 months, 1 to 16 years, and adults, respectively. This author has reviewed the literature on bone Pb and concluded that Pb levels are three times greater in bones of older children than in infants, and 10 to 40 times greater in adults than in infants. The bones of stillbirth cases were lower in Pb level than those of newborn infants in the series studied by Barry, but the reverse was reported in a brief communication by Bryce-Smith et al, (1977).

Exposure over a period of years is indicated by Pb levels in bones and teeth. With bone, especially in the young, problems arise because of deposition and resorption, and specimens are only rarely available. With teeth there can be uncertainty arising because of the degree of wear, which could give rise to a loss of relatively Pb-rich regions of the outer enamel surface, and of root resorption of dentin. Nevertheless, the analysis of primary tooth crowns has generally been used, exfoliated teeth being comparatively easily obtained for population surveys.

It appears that the youngest children are most at risk from Pb intoxication. Water supplies sometimes contain Pb at levels close to the permitted maximum; "first-drawn" water is particularly liable to provide excessive Pb (Moore et al, 1980). Then, if infants are reared on proprietary (low-Pb) dried milk preparations made up with such water they may receive relatively higher doses of Pb than at any other stage in their lives. Infants have an efficient utilization of Pb, as of other minerals from their diet, and so these factors should be added to their greater susceptibility to airborne and dustborne Pb.

The importance of Pb as an environmental pollutant has led to the accumulation of evidence from many sources on body burdens, particularly during the last decade. Teeth have been recognized as providing a useful long-term record of uptake (Needleman and Shapiro, 1974) (discussed in Chapter 3). However, care has to be taken in the collection of samples when studying teeth (discussed in the section on the analysis of trace elements in teeth).

DENTAL CARIES

Man

Environmental excess of Pb has been associated, during the last quarter of a century, with increased incidence of caries. Earlier in this period there were observations resulting from medical and dental care of industrial workers. Similar caries scores were reported in workers in Pb plants and other industrial settings (Aston, 1952). It has earlier been noted that higher caries incidence was associated with a probable Pb intake above normal (Kehoe, 1941; Cantarow and Trumper, 1944).

In a recent study (Anda, 1976) an increased Pb level in teeth was found to be associated with increased industrial exposure and higher caries incidence. But a negative trend has also been observed (Proud, 1976; Moses et al, 1976), according to preliminary reports. Deleterious effects on the periodontium and oral mucosa in Pb miners were also noted in a study by Kobylańska et al (1966).

From a thorough study of micronutrients and macronutrients in soil, food, and water in neighboring village populations in the hinterland of Papua–New Guinea (Barmes, 1969) it was shown that reduced caries scores could be related to an "alkaline earth factor" in soils. The only metal, among numerous minor elements analyzed, that appeared to overcome this factor was Pb.

Soil levels high in Pb (over 100 ppm) and extremely high in Zn (more than 25,000 ppm) prevailed in a community in which eruption of teeth was retarded and caries incidence was greater by 40%, compared with a reference community in the same state of the United States (Curzon and Bibby, 1970). Water Pb levels were 3 ppm in the reference community but up to 10 ppm in the study area with heavy metal pollution.

An above average caries level was seen in children living in a mineralized district in the southern portion of a border region between two southwestern counties in the United Kingdom (Anderson et al, 1976). Soil Pb was thought to be more available for uptake in this area than in another survey area on the opposite side of the adjoining county (Anderson et al, 1979); this opinion is consistent with the higher soil pH noted in the more recent survey. Similar caries scores (DMF 3.5 to 4.8 at age 12) were noted in the five villages visited, one of which has been the subject of a medical investigation arising from the identification of Cd pollution during routine geochemical survey (see also Anderson and Davies, 1980).

Urban children separated into two groups by Brudevold et al (1977), by values of enamel Pb biopsies, were of the same mean age but they differed in three other respects—higher F, one more erupted tooth, and a DFT (decayed and filled teeth) score less than one unit different.

A more extensive survey of enamel Pb values showed no firm relationship to caries scores (Curzon, 1977). Table 20-1 shows a much greater difference associated with geographical grouping, but a trend toward higher enamel Pb with lower caries scores.

There has not been as many reports of a relationship between Pb and caries in man as might have been expected in view of its long history as a hazard to health. The element remains a health hazard, and limited dental research continues, but attention to the role of Pb in promoting or retarding caries remains limited. This may be because of the difficulty of achieving accurate Pb analyses or perhaps because of Pb's likely role as a cariogenic trace element with a lack of potential role in prevention.

Animals

An early note of dental decay in Pb-intoxicated cats suggested that Pb might increase tooth decay in experimental animals (Aub et al, 1925).

Table 20-1
Lead Concentrations in Whole Enamel Samples Related to Geographic Distribution of Caries Prevalence

Region	DMFT Score	n	Pb conc. in enamel (Mean μg/g ± SE)
New England	Above 7	39	2.41 ± 0.46
	Below 4	25	2.54 ± 0.74
South Carolina	Above 7	15	0.97 ± 0.42
	Below 4	47	1.51 ± 0.31
California	Above 7	12	5.78 ± 0.97
	Below 4	11	6.63 ± 1.64
Oregon	Above 7	13	4.79 ± 0.68
	Below 4	12	4.89 ± 1.01

After Curzon, 1977.

Since then, the effects of Pb on dental tissues of experimental animals have been studied by a number of investigators, but the focus has not always been upon caries. Wisotzky and Hein (1958) observed a caries-promoting effect of Pb in male, but not female, hamsters. A caries reduction in hamsters was demonstrated, according to a preliminary report (Lazansky, 1947), by the brushing of teeth thrice weekly, using a mascara brush charged with Pb-fluoride solution. The proportion of caries-affected areas on molars was reduced by three-quarters but an even greater effect was obtained by using a Na-fluoride solution. Some effect followed brushing with water alone.

Just as with the epidemiology of Pb and caries in man, so there are few reports on experimental animal groups, no doubt because the lack of interest in studying the Pb/caries relationship in man has not led to incentive to investigate animal response. At present the indications are that there is a positive association between Pb and caries, but the mechanisms are little understood or studied. The great interest in Pb from the dental aspect has been the use of teeth to monitor the Pb environmental effect of uptake, and this interest has extended into work with animals.

Lead in animal teeth The distribution of Pb in organs and tissues of animals resembles those in man. However, radiotracer Pb can be employed in animal investigations, making possible a determination of percentages of the total dose at various sites (Momčilović and Kostial, 1974; Strehlow, 1974; Maruta, 1978; Kaplan et al, 1980). Dietary changes other than increases in minerals and fibers generally have been found to promote Pb absorption in rodents given diets with Pb levels (1 ppm, Shah et al, 1980; 1.5 to 2.0 ppm, Wesenberg et al, 1979) similar to those in human diets. Similar Pb levels are also seen in the teeth of rats (Maruta, 1978; Niwa, 1980). There was a 15-fold increase in Pb levels in incisors and molars of rats after supplementation of drinking water with 25 ppm $PbCl_2$, together with 5 ppm $CdCl_2$ (Wesenberg et al, 1979). Increases by factors of 200 (incisors) or even 300 (molars) have been induced in rats at doses of 1 mmol/kg body weight (Niwa, 1980).

MECHANISMS OF ACTION

Effects of Lead on Teeth

Tooth type Levels of Pb in teeth decrease from the anterior to the posterior teeth (incisors to molars) according to most surveys (Mackie et al, 1977; Proud, 1976; Lockeretz, 1975; Pinchin et al, 1978; Kelsall and Hunter, 1978) excluding that of Fosse and Berg-Justesen (1978). The differences may depend to some extent on the proportions of circumpulpal dentin in the various types of teeth, but the Pb concentrations in this region of the dentin do not seem to depend upon tooth type (de la Burdé and Shapiro, 1975).

Table 20-2 shows values found for Pb in incisors, together with values for other types of teeth expressed as percentages of the data for the incisors. These observations indicate that tooth type should, if possible, be standardized in tooth surveys. Since canines tend to become available from primary dentitions with less resorption than other teeth and roots contain more Pb than crowns, it would appear that there are several advantages to the analysis of canines in tooth Pb surveys (Stephens and Waldron, 1976).

When more than one tooth is available from a donor, there is a choice of averaging the values to represent the tooth Pb status, or of making an assessment based upon general findings for the survey population on ratios of Pb values for the various types of teeth. When contralateral pairs of teeth are available, correlations are often high (eg, $r = 0.87$, Ewers et al, 1979), but correlations of Pb concentrations in central and lateral incisors are often not impressive. When all four teeth (eg, canines) from a child have been analyzed it has often been found that there are two tending to have lower values and two having higher total Pb concentrations (Stephens, personal communication 1981). Possibly this is related to an observation (Proud, 1976) that teeth from the upper jaw show Pb values greater than those in the lower jaw.

Age Most of the tooth Pb studies in which age has been considered show a positive trend, sometimes recognized only up to adulthood; Zakson (1968) reported no significant changes in eight trace metals in teeth representing the last three decades of life. Table 20-3 lists some of the findings derived from these studies, and it can be seen that the typical increase with age is of the order of 1 ppm/yr. In Table 20-4 is

Table 20-2
Lead Concentrations in Deciduous Teeth Related to Tooth Type

Author (year)	Location	Pb Conc. in Teeth			
		Incisors (μg/g or ppm)	Conc. as % of Pb of Incisors		
			Canines	*Molars*	*n*
Stephens and Waldron, 1976	Birmingham, United States	18.3	64	49	1392
Lockeretz, 1975	St. Louis, United States[a]	18.8	94	55	324
	St. Louis, United States[b]	10.6	89	60	
Oshio, 1973	Tokyo/Toyama, Japan	8.2	92	76	795

[a]High Pb area.
[b]Low Pb area.

Table 20-3
Summary of Reports on Changes in Lead Concentrations in Teeth with Age

Author (year)	Location (age)[a]	Tooth Lead Levels			
		n	*m*	*K*	*r*
Deciduous Teeth					
Stewart, 1974	N. Ireland (3–9)				
	Rural	69	0.61	0.0	0.89
	Suburban	86	0.74	1.5	0.92
	Urban	141	0.81	2.7	0.98
Jones et al, 1981	United Kingdom (4–7)				
	School clinic	323	1.1	3.7	0.95
	Dentists	282	1.0	3.9	0.95
Permanent Teeth					
Strehlow and Kneip, 1969	United States				
	New Jersey	190	1.08	(– 1)[b]	0.80
Wilkinson and Palmer, 1975	United States				
	Delaware	336	0.7	4.4	0.96
Al-Naimi et al, 1980	United Kingdom[c]				
	Sheffield	16	1.7	0.0	0.81
	Birmingham	26	2.0	0.0	0.83

Table 20-3 (continued)

Author (year)	Location (age)[a]	Tooth Lead Levels			
		n	*m*	*K*	*r*
Permanent Teeth (continued)					
Steenhout and Pourtois, 1981	Belgium				
	Arlon (rural)	38	0.45	6.6	0.93
	Brussels	42	0.78	3.7	0.96
	Hoboken (indust)	51	1.03	2.1	0.99
Lappalainen and Knuuttila, 1981	Finland				
	Eastern area	89	0.31	1.8	0.58
	Kuopio	50	0.58	3.7	0.74

n = number of teeth sampled.
r = correlation coefficient.
K = Tooth Pb − m(age).
m = regression slope.
[a]ages taken as midpoint of ranges given in decades.
[b]based on unweighted raw values for pooled samples; regression slope obtained by multiplying tooth ash Pb slope value by 0.695 (mean proportion of ash).
[c]circumpulpal dentine.
Data by activation analysis; all other data by atomic absorption.

listed another group of studies for which no age trend was reported; there is a third category in which tooth Pb values have been expressed only in terms of the concentration increase with time. The rate of increase in tooth Pb reported by Altshuller et al, (1962) is more than double that suggested by the data of Table 20-3. As discussed below there are variations related to region and degree of urbanization as well as those due to age. Similar positive age trends have been shown by Malik and Fremlin (1974), Stephens and Waldron (1976), Fremlin and Tanti-Wipawin (1976), Derise and Ritchey (1974), and Curzon (1977). Altshuller et al (1962) proposed that a more normal distribution of tooth Pb values resulted when these were divided by the time elapsed from tooth eruption. Their values of 2.2 ppm/yr for their control group is about one-half that found for 300 controls in the x-ray fluorescence study of Shapiro et al (1978). Blood Pb levels correlated well with free blood erythroporphyrin (FEP) values, but only with tooth Pb data when these had been age adjusted (see also Bloch et al, 1977).

Table 20-4
Summary of Reports of Tooth Lead Surveys in Which No Age Trend was Clear

Author (year)	n	Mean lead conc. in ppm (μg/g)	Remarks
Habercam et al, 1974	18	92	Atomic absorption used. Probable environmental Pb.
Kaneko et al, 1974	99	7.3	Urban and rural specimens
Chatman et al, 1975	115	5	Geographic trend observed
Attramadal and Jonsen, 1976	32 44	5.6[a] 2.4	Anodic stripping voltametry
Proud, 1976	80	29	Negative correlation to caries
Kuhnlein et al,1977	12	16.6	Pre-industrial and recent data analyzed by X-ray fluorescence

[a]Deciduous teeth.

Young deciduous and permanent teeth show similar Pb levels, but the age effect is more clearly seen in permanent teeth, for which resorption problems do not obtrude. For example, the mean Pb (ash basis) was reported to be 6.3 ppm in deciduous teeth from children aged 4 to 6 (Holtzman et al, 1969) whereas it was 23.0 to 9.2 ppm in adults aged 24 to 64. The differences were less marked in a study by Storozheva (1963);

values of 7 and 4 ppm were noted, respectively, for deciduous teeth from children aged 5 to 10 and 11 to 15, those for permanent teeth from donors aged over 40 averaging 15 ppm.

Geographic and regional variations Data from 2233 deciduous teeth analyzed by atomic absorption spectrometry (Fosse and Berg Justesen, 1978), classified by Norwegian county of origin, provide the best evidence so far available of a nationwide variation in tooth Pb. Table 20-5 shows values in terms of overall median values, excluding the highest county mean (10 ppm) which was more than three times this median (3.15 ppm). A similar range has been reported from Finland (Lappalainen and Knuuttila, 1979).

Data from enamel specimens, prepared from more than 300 premolars and analyzed by spark source mass spectrometry (Curzon, 1977), showed differences exceeding one order of magnitude between seven states of the United States. Table 20-6 (see also Table 20-1) shows values

Table 20-5
Lead in Deciduous Teeth

County	n	Pb[a]	County	n	Pb[a]
Nordland	167	0.77	Hordaland	360	1.09
Nord-Trøndelag	111	0.77	Aust-Agder	147	1.11
Sør-Trøndelag	68	0.77	Buskerud	273	1.31
Oslo	59	0.81	Finmark	228	1.42
Sogn og Fjordane	244	0.93	Møre og Tamsdal	106	1.80
Troms	175	0.93	Telemark	88	1.87
Bergen	125	1.00	Vest Agder	82	2.86

[a]Pb values expressed as multiples of a median (3.15 ppm).
Specimens from all Norwegian counties.
After Fosse and Berg Justesen, 1978.

Table 20-6
Lead in Premolar Enamel Specimens from Seven Regions of the United States

State/Region	n	Pb[a]
South Carolina	77	0.11
New England	107	0.25
New York	29	1.00
Ohio	41	1.00
Florida	14	1.29
Oregon	24	1.48
California	23	1.52
(Montana)	(7)	(2.29)

After Curzon, 1977.
[a]Expressed in terms of median concentration (24 ppm).

in terms of a median (24 ppm) based upon all but one of the areas (Montana), this being represented by only seven specimens with a high mean.

Tooth Pb levels reported by investigators in various countries were shown by Stephens and Waldron (1976), about half the means were in the region of 10 ppm (representing their own observations). These and other values are included in Table 20-7. Significant differences were found in the levels of Cd, Pb, Zn, and particularly Cu, in enamel samples from young premolars in two groups collected from New Zealand and the United States (Curzon et al, 1975).

Urban vs suburban environmental effects Several studies have concentrated on the question of differences between urban and suburban environments relative to Pb intake (Table 20-8). Values for circumpulpal dentin Pb (Shapiro et al, 1972) were reported by Needleman et al (1974) in relation to dust Pb levels. In the more deprived of the two school districts he found the mean dust Pb levels differed by a factor of five; there was a less marked difference between tooth Pb levels. No significant correlation could be demonstrated by Hunter et al (1973) between bulk tooth Pb and dust Pb from air samples. They were able to relate the distributions of the tooth Pb values to the locations of the communities from which specimens were obtained. Three times as many children aged 5 to 6, living in urban Cleveland (Ohio), had tooth Pb concentrations exceeding 80 ppm when compared with those living in the suburbs. However, according to a more recent report (Kelsall and Hunter, 1978), based upon the largest number of teeth yet analyzed from one major urban area (over 11,000), mean values representing Cleveland suburbs were of the same order as those for the inner city districts. Lockeretz (1975) showed that tooth Pb levels were twice as high in an urban district with a high Pb risk as they were in a low risk area. A somewhat greater difference emerged from the x-ray fluorescence study of Shapiro et al (1978) in Philadelphia, a city in which Needleman et al (1972) had shown a fivefold difference between tooth Pb levels in suburban and inner city sample areas.

Tooth Pb levels were twice as high in Glasgow children who had lived in old housing as in those who had always lived in new housing (Moore et al, 1980). Similar ratios were seen in urban/suburban comparisons of deciduous teeth of children in several age groups from the population of Belfast (Stewart, 1974).

In another European study (Schildknecht, 1976), only one-third as much Pb was present in samples of circumpulpal dentin and in cementum, or premolars from children living in homes isolated from major traffic than in others with a predominantly urban environment. Median values, were, for dentin, 13.3 vs 36.5 ppm and, for cementum, 46.2 vs 124.5 ppm.

The level of Pb in teeth from rural areas was expected to be 2 to 3 ppm in a recent New Zealand study by Fergusson et al (1980). However,

Table 20-7
Comparison of Tooth Pb Concentrations in Urban and Rural Areas (Mean values in μg/g or ppm)

Author (year) Country	Sample Type	Pb Concentrations in Teeth			
		Urban		*Rural*	
Stewart, 1974 Northern Ireland	Deciduous molars	Belfast	5.55[a]	Fermanagh	1.9[a]
		Belfast	13.1[b]	Fermanagh	6.3[b]
Kaneko et al, 1974 Japan	Permanent teeth			Okitsu	4.6
				Hachijo	8.6
		Tokyo	11.6	Annaka	6.3
Rytömaa and Tuompo, 1974 Finland	Deciduous teeth	Helsinki	24	Tervola	15
Shapiro et al, 1973 United States	Sectioned[c]	Philadelphia	84	Iceland	35
Shapiro et al, 1976 United States	Sectioned[c] dentin	Philadelphia	92	Chiapas (Mexico)	4.3
Shapiro et al, 1976 United States	Reamed[c] dentin	Philadelphia	188	N. Slope (Alaska)	56
Fergusson et al, 1981 New Zealand	Dentin	Christchurch	6.7	C/Church	4.8

[a]Ages 3, 4.
[b]Ages 9 and above.
[c]Circumpulpal dentin.

Table 20-8
Comparisons of Tooth Pb Concentrations in Deciduous Teeth in Relation to Urban and Suburban Populations (Mean values in ppm)

Author (year)	Location	Pb Concentrations in Teeth	
		Urban	*Suburban*
Needleman et al, 1972	Philadelphia United States	51	11
Stewart, 1974	Belfast (Northern Ireland)	5.5[a] 13.1[b]	4.0[a] 10.0[b]
Lockeretz, 1975	St. Louis United States	14.3	9.5
Shapiro et al, 1978	Philadelphia United States	14.5[c]	0.4[c]
Hunter et al, 1973	Cleveland United States	13.3[d]	4.3[d]

[a]Ages 3, 4.
[b]Ages 9 and above.
[c]ppm/yr.
[d]% of values above 80 μg/g/yr.

they observed a mean of 4.8 ppm in primary teeth of children living in post-1940 housing located in nonindustrial areas, whereas 6.7 ppm was evident when the housing was pre-1940, or donors lived in industrial areas.

Slopes of regression lines relating tooth Pb to tooth-age have been recommended (Steenhout and Pourtois, 1981) as providing a suitable index of population exposure. In two groups of teeth with mean tooth-ages of above 30 years the mean Pb level was 35.3 ppm in the case of a population living in the vicinity of a nonferrous smelter, and 21.3 ppm in a rural population of the province of Luxembourg. However, the regression line for data from the industrial district was of unit slope (ie, 1 ppm/yr) (Table 20-3), whereas that for the rural area was less than half this.

Similar tooth Pb values were reported in a Yugoslavian study (Milošević et al, 1980). The mean value for the exposed group living 1 to 5 km from a Pb smelter was 30.2 ppm (n = 111), but 18.6 ppm in a much smaller control group living 20 to 30 km distant. The difference in Cd levels was also significant at the 1% level. Much greater differences were demonstrated in hair Pb levels, which were of the order of 4:1 for males and 3:1 for females, when 400 specimens from the two sampling areas were compared.

Archeological findings Typical concentrations of Pb in the earth's crust are between 10 and 20 ppm. The analysis of tooth specimens that may have been interred for centuries in soil containing Pb at concentrations greater by an order of magnitude than those in the teeth cannot therefore seem likely to be reliable. However, mature enamel is extremely impermeable and is the tissue most likely to provide a record of human environmental experience regarding trace elements. But very low Pb levels have also been found in whole tooth specimens recovered from a var iety of burial conditions, and bone Pb levels have been found to vary independently from soil levels.

A spectrographic study by Steadman et al (1959) showed considerable variations in some trace metal ratios when enamel chips from teeth representing two Amerindian populations were compared. Very different ratios were found for Zn and Pb (7:1 and 1:5), the teeth being from a Pueblo people of about one millennium ago, and from inhabitants of a site at Indian Knoll, dated to six millennia ago (Steadman et al, 1959). Surface layers of the Pueblo enamel chips had been removed, and the Pb levels in the remainder (12 ppm) were considered to refer to the deeper layers. Several times as much Pb were found in contemporary teeth from young individuals.

Four times more Pb was reported by Kühnlein and Calloway (1977) in the teeth of contemporary Hopi Indians than in the teeth of their 17th century predecessors from the same locality. Mean Zn levels did not differ by more than one standard deviation. Both Pb and Zn were present in intermediate concentrations in teeth from present-day Californians. The most interesting feature of the analysis of the ten ancient teeth was that the Sr concentrations approached 500 ppm, a value five times greater than those found in the contemporary specimens. Similar findings for Pb have been noted for Peruvian teeth (Ericson et al, 1979).

Working with the circumpulpal dentin of contemporary and ancient teeth, Shapiro et al (1976) demonstrated, by analysis of pieces fractured from tooth sections, that the median level of Pb in such permanent tooth fragments was nearly 50 times greater in samples from urban Philadelphia than in specimens obtained similarly from Lacondon and Tzeltales Amerindians living in a remote forest area of Chiapas, Mexico. No Pb was detected in the incisors of the forest-dwellers, although Pb concentrations are usually higher in these teeth than in canines and molars.

The same workers prepared samples by reaming out circumpulpal dentin, to cause minimal damage to rare teeth, and recorded median Pb levels twice as high as those found from carefully fractured enamel of the same interfacial region, of the teeth of modern Philadelphians. A median value of 175 ppm was compared with that obtained (47 ppm) from the analysis of similar samples prepared from teeth of Alaskan (North Slope) Eskimos (range 10 to 100 ppm). However, the teeth they obtained from 12th century Peruvian Amerindians (Machu Picchu) contained much less Pb (median 3.7 ppm) in circumpulpal dentin. In another small group of specimens, from Egyptian mummies of the first and second millenia BC, they found a median value below 7 ppm. In the comparison between the latter groups it was pointed out that the Pb content was unlikely to have been altered in teeth from mummified burials.

In a further study of teeth from naturally mummified burials in ancient Nubia median tooth Pb values of 1, 2, and 5 ppm, respectively, were found for groups dated 3300 to 2900 BC, 2000 to 1600 BC, and 1650 to 1350 BC (Grandjean et al, 1978). Bone Pb levels showed a similar trend (0.6, 1, and 2 ppm).

A much larger group dated to the first eight centuries AD was characterized by intermediate values (tooth Pb 3.2, bone Pb 1.2 ppm). Specimens from a contemporary Danish population contained much more Pb in bone (5.5 ppm) and very much more in circumpulpal dentin (over 25 ppm). In the case of a group of children buried in ancient Nubia (Shapiro et al, 1980) the dentin Pb concentration was high enough to allow the use of x-ray fluorescence in analysis, the value (26.7 ppm) agreeing to within 2% of that obtained by atomic absorption spectrometry; there was also good agreement between bone and dentin values ($r = 0.65$) in the group. These authors have also reported analyses of teeth from a modern but isolated population in which one-third of the samples showed less than 1 ppm Pb in circumpulpal dentin from pieces of sectioned teeth, whereas the mean from corresponding material from an urban American population was 90 ppm.

In his book on the biochemistry of animal fossils, Wyckoff (1972) has pointed out that in several instances there have been findings of one or more unusual elements in a large percentage of the fossils from a locality. Thus, Pb and As have been found in such bones, whereas they were not present in the surrounding terrain.

Attramadal and Jonsen (1978) analyzed 26 ancient teeth for Pb, Cd, Zn, and Cu. They were from Norwegian burials of 1 to 17 centuries ago. No significant differences were apparent between the values and those they had reported for contemporary Norwegian teeth (1976). Most of the Pb levels in teeth from burials were below 10 ppm, and some low values (below 2 ppm) were found in teeth from one excavation. Values for contemporary teeth were 1 to 8 ppm, with a mean of 2.4 ppm. Analyses of these specimens were undertaken by differential pulse anodic stripping voltametry.

Analyses of bones from Roman York, in the United Kingdom, showed Pb levels to be not materially different from present-day values (Mackie et al, 1975). A trend with age was found in one group from a Romano-British cemetery in the south of England but not in another group (Waldron et al, 1979). Some rather high values were encountered in specimens from young children (cf Shapiro et al, 1980). Waldron et al (1979) showed that bone Pb levels did not seem to be affected by soil Pb differences. Teeth from the above Romano-British cemetery showed a relatively low median Pb of 6 ppm (and 0.2 ppm for Cd) when they were thoroughly cleaned (Whittaker and Stack, 1982), whereas there was a median value of 20 ppm when the cleaning was superficial. No correlation was evident between Pb levels in deciduous and permanent teeth when these were compared in the same jaw.

It may be concluded that concentrations of Pb in buried teeth are likely to be representative of those present during life, and these are, in general, lower than those in contemporary populations. In the event of encountering high Pb values in such teeth there is reason to implicate the living environment rather than the burial environment. This is of considerable interest with regard to present-day episodes of Pb poisoning, which continue to present a considerable public health problem.

Enamel solubility modification It has been shown by a number of investigators that the solubility of *powdered* enamel is reduced after exposure to Pb salts (Buonocore and Bibby, 1945; Muhler and van Huysen, 1947; Adler and Csoban, 1949; Muhler et al, 1950; and Manly and Bibby, 1949). For example, Muhler and van Huysen observed a reduction by two-thirds; values obtained by Bibby and coauthors were similar.

However, topical applications of Pb salts designed to inhibit dental caries in groups of volunteers have not proved encouraging. Klinkenberg and Bibby (1950) obtained a solubility reduction by one-third with PbF_2 solution. Neither have topical applications of Pb solutions to intact surfaces affected acid solubility unless the surfaces were etched before exposure (Brudevold and Steadman, 1956). In untreated enamel, the ratio of Pb to phosphate in leaching treatments of outer enamel at pH4 (acetate buffer) was higher than expected; this suggested that Pb might be lost from the surface more rapidly when acid conditions prevail.

The factors affecting an increase or decrease in enamel solubility include whether, or to what degree, Pb will be incorporated into hydroxyapatite; the form of the Pb will also be important.

Incorporation into hydroxyapatite It has been suggested that Pb may first adsorb to hydroxyapatite crystal surfaces and later take up positions within the structure. Some of the properties of Pb hydroxyapatite produced under laboratory conditions have been established (Posner and Perloff, 1957; Bhatnagar, 1970). No alteration of the unit cell size of the apatitic phase was discerned (Claman et al, 1975) when there was isomorphic substitution of Pb for Ca in teeth of rats receiving Pb supplementation to their diet.

In synthetic Pb hydroxyapatite, however, unit cell size was found to be somewhat greater (*a*-axis by nearly 5%, *c*-axis by nearly 8%) than that of Ca hydroxyapatite (Bhatnagar, 1970). When the two hydroxyapatites were formed together as a solid solution, by coprecipitation, followed by heating at 800°C (Verbeeck et al, 1981) a second Pb phosphate phase ap-

Table 20-9
Estimates of Pb and F* at Corresponding Points Below Surface

Cumulative Depth Score	Estimates from Regression Line *Pb (X 3)*	*F*
	values in μg/g (ppm)	
3.0	1100	1030
6.8	700	650
12.0	375	340
15.9	235	210

*Data from Brudevold et al, 1956a, 1956b.

peared if the number of Pb atoms exceeded the number of phosphate groups. Lattice parameters varied linearly with the number of Pb atoms, up to 6, these being located in the sixfold position for cations. Apatites containing 4 to 6 Pb atoms were thought to represent the minimum free energy state of solid solutions between Ca and Pb hydroxyapatites.

Wide dispersal of Pb ions is registered by electron microscopy when Ca hydroxyapatite is formed synthetically in the presence of Pb ions (Featherstone et al, 1981), whereas they concentrate into defect areas when Pb is allowed access after completion of formation.

Other trace element associations Mineralized tissues incorporate Pb and F readily, and although evidence for the interaction of these elements is not convincing they show very similar gradients in surface regions of enamel. Brudevold and colleagues (1956a, 1956b) determined Pb and F levels in enamel from extracted teeth. Their data indicate that there is a logarithmic decrease in concentrations of both elements with increasing depth scores, and also that the slopes of the logarithmic plots do not differ significantly (Table 20-9). Estimates of Pb from the regression line have been multiplied by three to facilitate comparison with similarly calculated F estimates.

However, in a recent study (Brudevold et al, 1977), in which surface enamel biopsies of children in a large city were reported, the concentrations of Pb and F were similar. Two subgroups were segregated, with above-average (0.277%) and below-average (0.107%) enamel Pb levels. Surface enamel F levels were, however, of the same order (0.236 and 0.211%), and no correlation between individual Pb and F values was evident. A higher caries incidence was noted in the group with above average Pb in the enamel surface region. They pointed out that "after eruption, enamel appears to be surprisingly inert to these chemically very active, bone-seeking elements."

A correlation coefficient of 0.25 was shown for Pb and F in a proton-induced x-ray emission (PIXE) study by Bodart et al (1981) based on 180 enamel samples. This value was similar to others they found for metal/metal correlations.

A number of investigators have chosen to determine concentrations of several trace metals in teeth at the same time, such as Cd, Pb, Zn, and Cu. Wide differences in Cu levels have been noted, probably related to the use of copper piping for domestic water supplies (see Chapter 13). Values for the other three trace metals are tabulated in Table 20-10, in ascending order of Cd concentration. There appears to be a range of Cd levels spanning two orders of magnitude, but the range of Pb values is relatively narrow, when the general range reported is considered. The Ca:Zn ratio is of the same order as the Zn:Cd ratio, with Pb showing a geometric mean position between the latter metals. Thus, the median values for Cd, Pb, and Zn are, according to this table, approximately in the ratios 1:40:1600. The presence of Pb, as well as Cd, will also affect the uptake of Zn and Cu (Wesenberg et al, 1980).

LEAD AND DISEASE

Toxemia

When toxemia has resulted from undue Pb exposure there may be problems of sampling teeth in order to obtain the maximum yield of information regarding the period of maximum Pb exposure. Knowing that dentin is usually laid down at a regular rate, and with knowledge of the probable age of tooth development, the time of formation of layers of dentin may be assessed. Analysis of these layers can show probable times of exposure during childhood. The highest concentrations of Pb are to be found at interfaces such as the dentin-enamel junction, and the circumpulpal region of dentin, as well as the enamel surface.

The first important study which showed the striking difference between concentrations of Pb in the teeth of Pb-poisoned children and those of others living in the same community was that of Altshuller et al (1962). Levels of Pb were eight times higher than normal in the teeth of survivors, and ten times higher in fatalities. There appeared to be a loss of Pb from erupted teeth, related to the time elapsed since the acute episode of Pb poisoning. The location of the Pb in such teeth has been investigated by Carroll et al (1972) and Ohm (1969).

Very much greater concentrations of Pb in circumpulpal dentin of teeth of Pb-poisoned children was observed by Shapiro et al (1973). Samples consisted of 0.3 mm wide zones fractured off the edges of 0.6 mm thick longitudinal sections; these relate to the layer analysis referred to above.

Table 20-10
Cadmium, Lead and Zinc Concentrations in Teeth (ppm)

Investigator	Location	Tooth Type	n	Cadmium	Lead	Zinc
Coles, 1979	Bristol, United Kingdom	Premolar	37	(0.01)	3.8	145
Pinchin et al, 1978	Newcastle upon Tyne, United Kingdom	Deciduous	23	0.03	4.4	–
Oehme et al, 1978	Oslo, Norway	Premolar	10	0.04	3.0	160
Kaneko et al, 1974	Okitsu (rural)	Permanent		0.07	4.65	172
	Annaka (high Cd)			0.10	6.27	186
Stone, Fletcher,[a] 1979	Bristol, United Kingdom	Perm. molar	30	0.1	3.8	140
Attramadal et al, 1978	Oslo, Norway	Deciduous	32	0.1	5.6	130
		Permanent	44	0.1	2.4	130
Losee, Cutress, and Brown, 1974	United States	Premolar enamel	28	0.22	3.6	190
Fosse and Berg Justesen, 1978	All Norway	Deciduous	2233	0.43	3.7	122
Langmyhr et al,1974	Oslo, Norway	Premolar		0.5	2.0	175
Curzon et al, 1975	United States	Premolar enamel	36	1.0	3.1	203
Oshio, 1973	Tokyo, Toyama	Deciduous	795	1.0	6.5	120
Approximate median values				0.1	4	160

[a]Personal communication, 1979.

Remarkably high levels of Pb were reported by Brudevold et al (1975, 1977) in the superficial enamel of teeth from young donors living in an urban environment such that blood Pb estimates indicated that a domiciliary follow-up should be undertaken for 10% of the children. Levels of Pb as high as 4000 ppm were typical at a mean depth of less than 1 μm, using a biopsy technique.

Albert et al (1974) found median tooth Pb levels (c 30 ppm) to be no greater in children hospitalized because of Pb poisoning, and undergoing chelation therapy, than in children with various categories of low or high blood Pb or tooth Pb. In the absence of diagnosed Pb poisoning or elevated blood Pb values (above 60 μg/dl), excess Pb exposure, as presumed by the finding of high tooth Pb values, was not associated with deleterious health effects.

Other pathologic conditions in relation to Pb in teeth have been reported. Tooth Pb levels well above those of age controls were demonstrated over 30 years ago in cases of multiple (disseminated) sclerosis (Campbell et al, 1950). Half of the 46 values for teeth from the local population in the study area in the west of England were within the range 20 to 40 ppm, but only 3 were within this range when tooth Pb values for 34 patients were established. One-third of the teeth from patients showed Pb levels as much as four times those of controls. There was a range of 100 to 140 ppm, with a trend toward higher values in the urban patients, but not with an age trend. Tooth Pb values are significantly correlated with age in the control group (Table 20-3).

The involvement of Pb in pathologic states has been referred to in this section, and its possible involvement in neurophysiologic impairment in another section. There is an intermediate situation in which Pb levels in teeth have been related to physical and mental development. Parallel analyses for Pb in blood, hair, and teeth were undertaken in a small group of black children aged 6 to 13 by Habercam et al (1974). No significant association was established between blood Pb levels and Pb associated with housing maintenance (expressed as Pb/cm^2 house paint), or with local soil. But the data suggested that in children with growth deficit nearly one-quarter of this deficit might be related to excess Pb.

Deciduous and permanent teeth of normal and mentally retarded children showed similar Pb (also Cd and Cu) levels, according to Pinchin et al (1978). Slightly higher values were found for incisors (5.4 ppm) than for the three other types of teeth analyzed (4.3 to 4.7 ppm) from young children. Rather lower values for permanent teeth (2.4 ppm) were noted.

Effects on teeth themselves are noted in a number of references to delayed eruption. A recent case (Pearl and Roland, 1980) is of interest because chelation therapy was instituted during the early months of life of an infant whose mother consumed wall plaster "for several months" during pregnancy. Dental development was normal at two years, although the lower central incisors erupted nine months later than seemed normal for the family members. There is also evidence of enamel hypoplasia that could be related to Pb poisoning (Lawson et al, 1971).

Periodontal disease

The Pb line of the gingivae in people exposed to excessive Pb intake is well known. Periodontal disease is more prevalent in the more highly exposed sections of the population. Danilevski and Gitina (1970) noted that root Pb levels were significantly higher in teeth associated with periodontal disease in Pb workers than in the affected teeth of others from the same community. Kobylańska et al (1966), however, did not find the caries incidence in Pb battery factory workers to be above expectation. However, there were pathologic changes in the gingivae of all but 8 of the 81 workers studied. One-third of them had "black-livid" rough calculus, found to contain 38 ppm Pb, a value comparable with that reported by Sonnabend et al (1976) in the calculus on teeth of men who worked in the chemical and automobile industries, with 47 and 53 ppm Pb respectively.

Sagara (1957) and Chrusciel (1975) drew attention to pathologic changes in the oral cavity of workers exposed to Pb. Increased periodontal disease and gingival ischemia were identified in groups of Pb foundry workers categorized with respect to coproporphyrinuria, basophilia, and plumbism (Cherchi et al, 1966). Despite this indication of an effect of Pb on periodontal disease, our knowledge of the subject remains sparse.

SUMMARY

The analysis of Pb in teeth provides evidence of exposure of the individual particularly during the tooth-forming years. Further information accrues by subsampling at interfacial regions, such as circumpulpal dentin, and tooth type and age need to be studied in population surveys. Tooth Pb levels tend to be higher in inner city areas and in industrial workers involved in processing of Pb. High tooth Pb has been associated with increased incidence of periodontal disease, but not related with certainty to dental caries. Confounding factors have made the association between elevated tooth Pb and a degree of neurophysiologic impairment difficult to prove in normal children.

Factors associated with the promotion or inhibition of Pb absorption have been investigated in laboratory animals, particularly rodents. Basic levels of Pb in teeth and in "clean" diets are similar. These levels are below, by one order of magnitude, those that can be estimated directly (in vivo) by an x-ray fluorescence method. Atomic absorption spectrometry and anodic stripping voltametry are techniques that have been preferred for tooth Pb analysis. They are required to operate near their detection limits when analyzing low Pb levels, for example, those from many of the teeth sampled from archeological material.

The effects of solutions of Pb salts on dental enamel, and of the partial substitution of Ca by Pb in the hydroxyapatite of tooth and bone mineral, are described. Comparable steep gradients of Pb and F in enamel are mentioned, and levels of Cd, Pb, and Zn in teeth, as determined by a number of investigators, are shown to be present in ratios which match those in the environment.

REFERENCES

Adler P, Csoban G: Protective agents against enamel dissolution. *Fogorv Szemle* 1949;42:318–324.

Albert RE, et al: Follow-up of children exposed to lead. *Environ Health Perspect* 1974;7:33–39.

Al-Naimi T, Edmonds MI, Fremlin JH: The distribution of lead in human teeth, using charged particle activation analysis. *Phys Med Biol* 1980;25:719–726.

Altshuller LF, Halak DM, Landing BH, et al: Deciduous teeth as an index of body burden of lead. *J Pediatr* 1962;60:224–229.

Anda LP: Röntgenemissions-spektralanalytische Untersuchungen über den Pb-Gehalt der Zähne von Patienten aus industriellen und industriarmen Gebieten in Beziehung zur Karies-intensitat. *Dtsch Zahnärztl Z* 1976;31:200.

Anderson RJ, Davies BE, James PMC: Dental caries prevalence in a heavy metal contaminated area of the west of England. *Br Dent J* 1976;141:311–314.

Anderson RJ, Davies BE, Nunn JH, et al: The dental health of children from five villages in north Somerset with reference to environmental cadmium and lead. *Br Dent J* 1979;147:159–161.

Anderson RJ, Davies BE: Dental caries prevalence and trace elements in soil, with special reference to lead. *J Geolog Soc Lond* 1980;137:547–558.

Aston ER: Dental study of employees of five lead plants. *Ind Med Surg* 1952; 21:17–19.

Attramadal A, Jonsen J: The content of lead, cadmium, zinc and copper in deciduous and permanent teeth. *Acta Odont Scand* 1976;34:127–131.

Attramadal A, Jonsen J: Heavy trace elements in ancient Norwegian teeth. *Acta Odont Scand* 1978;36:97–101.

Aub JC, Fairhall LT, Minot AS, et al: Lead poisoning. *Medicine* 1925;4:1–250.

Barmes DE: Caries etiology in Sepik villages—Trace element, micronutrient and macronutrient content of soil and food. *Caries Res* 1969;3:44–59.

Barry PSI: Concentration of lead in the tissues of children. *Br J Ind Med* 1981; 38:61–71.

Bhatnagar VM: The preparation, x-ray and infra-red spectra of lead apatites. *Arch Oral Biol* 1970;15:469–480.

Bloch P, Garavaglia G, Mitchell G, et al: Measurement of lead content of childrens' teeth *in situ* by x-ray fluorescence. *Phys Med Biol* 1977;22:56–63.

Bodart F, Doconninck G, Martin MT: Large scale study of enamel. *IEEE Trans Nucl Sci* 1981;28:1401–1403.

Brudevold F, Steadman LT: The distribution of lead in human enamel. *J Dent Res* 1956;35:430–437.

Brudevold F, Gardner DE, Smith FA: The distribution of fluoride in human enamel. *J Dent Res* 1956;35:420–429.

Brudevold F, Reda A, Aasenden R, et al: Determination of trace elements of human teeth by a new biopsy technique. *Arch Oral Biol* 1975;20:667–673.

Brudevold F, Aasenden R, Srinavasian BN, et al: Lead in enamel and saliva, dental caries and the use of enamel biopsies for measuring past exposure to lead. *J Dent Res* 1977;56:1165–1171.

Bryce-Smith D, Deshpande RR, Hughs J, et al: Lead and cadmium levels in stillbirths. *Lancet* 1977;1:1159.

Bryce-Smith D, Stephens R: In Lead or *Health Review for the Conservation Society,* London, 1980, p 34.

Buonocore MG, Bibby BG: The effect of various ions on enamel solubility. *J Dent Res* 1945;24:103–108.

Burchfield JL, Duffy FH, Bartels PH, et al: The combined discriminating power of quantitative electroencephalography and neuropsychologic measures in evaluating central nervous system effects of lead at low levels, in Needleman HL (ed): *Low Level Lead Exposure: The Clinical Implications of Current Research.* New York, Raven Press, 1980.

de la Burdé B, Choate McLS: Early asymptomatic lead exposure and development at school age. *J Pediatr* 1975;87:638–642.

de la Burdé B, Shapiro IM: Dental lead, blood lead and pica in urban children. *Arch Environ Health* 1975;30:281–284.

Campbell AMG, Herdan G, Tatlow WFT, et al: Lead in relation to disseminated sclerosis. *Brain* 1950;73:52–71.

Cantarow A, Trumper M: *Lead Poisoning.* Baltimore, Williams and Wilkins, 1944.

Carroll KG, Needleman H, Tuncay O, et al: The distribution of lead in human deciduous teeth. *Experientia* 1972;28:434–435.

Casey CE, Robinson MF: Cu, Mn, Zn, Ni, Cd and Pb in human foetal tissues. *Br J Nutr* 1978;39:639–646.

Chatman T, Wilson DJ: Lead levels in human deciduous teeth in Tennessee. *Environ Letters* 1975;8:173.

Cherchi P, Cortis I, Cotti S, Piredda S: Sulla comportamento dell' appartato bucco dentario in soggetti espositi al rischio saturnino. *Mondo Odontostomat* 1966;8:406–409.

Chrusciel H: The effects of exposure to toxic environmental elements on the periodontium during working hours in zinc and lead processing plants. *Czas Stomat* 1975;28:111–118.

Claman L, Foreman D, App G: A study of the presence and nature of lead deposition in calcifying tissues. *J Dent Res* 1975;54:535 (Abstract).

Coles SG: The determination of certain trace metals in teeth with special reference to lead as a pollutant that accumulates in calcified tissues. MSc Thesis, Bristol, England, 1979.

Curzon MEJ, Bibby BG: Effect of heavy metals on dental caries and tooth eruption. *J Dent Child* 1970;37:463–465.

Curzon MEJ, Losee FL, Macalister AD: Trace elements in the enamel of teeth from New Zealand and the U.S.A. *NZ Dent J* 1975;71:80–83.

Curzon MEJ: Trace element composition of human enamel and dental caries. PhD Thesis, University of London, 1977.

Curzon MEJ, Crocker DC: Relationships of trace elements in human tooth enamel to dental caries. *Arch Oral Biol* 1978;23:647–653.

Czegledi P: Natural contents of RaD (lead-210) and RaF (polonium-210) in young human teeth and bones. *Isotopenpraxis* 1977;13:124–130.

Danilevski ND, Gitina LI: (A study of lead accumulation in the hard dental tissues by means of spectral analysis.) *Stomatologiia* (Mosk) 1970;49:1-3.

Derise NL, Ritchey SJ: Mineral composition of normal human enamel and dentine and the relation of composition to dental caries. II. Micro-minerals. *J Dent Res* 1974;53:853-858.

Ericson JE, Shirahata H, Patterson CC: Skeletal concentrations of lead in ancient Peruvians. *N Engl J Med* 1979;300:946-951.

Ernhart CB, Schell NB: Subclinical levels of lead and developmental deficient — A multivariate follow-up reassessment. *Pediatrics* 1981;67:911-919.

Ewers U, Brockhaus A, Genter E, et al: Untersuchungen über den Zahnbleigehalt von Schulkindern aus zwie unterschiedlich belasteten Gebieten in Nordwestdeutschland. *Int Arch Occup Environ Health* 1979;44:65-80.

Featherstone JDB, Nelson DGA, McLean JD: An electron microscope study of modifications to defect regions in dental enamel and synthetic apatites. *Caries Res* 1981;15:278-288.

Fergusson JE, Jansen ML, Sheat AW: Lead in deciduous teeth in relation to environmental lead. *Environ Technol Letters* 1980;1:376-383.

Fosse G, Berg Justesen NP: Lead in deciduous teeth of Norwegian children. *Arch Environ Health* 1978;33:166-175.

Fremlin JH, Tanti-Wipawin W: Distribution of lead in teeth. *Proc Anal Div Chem Soc* 1976;13(7):195-197.

Grandjean P, Nielsen OV, Shapiro IM: Lead retention in ancient Nubian and contemporary populations. *J Environ Pathol Toxicol* 1978;2:781-787.

Habercam JW, Keil JE, Reigart JR, et al: Lead content of human blood, hair, and deciduous teeth: Correlation with environmental factors and growth. *J Dent Res* 1974;53:1160-1163.

Holtzman RB, Lucas HF Jr, Ilcewicz FH: *The Concentration of Lead in Human Bone.* Argonne National Laboratory of Radiology, Phys. Divn, Annual Report, 1969, ANL-7615, pp 43-49.

Hrdina K, Winneke G: Neurophysiologische Untersuchungen an Kindern mit erhöhtem Zahnbleigehalt. Contributed to meeting of Deutsch Gesellschaft für Hygiene u Mikrobiologie, 1978.

Hunter RE, Kelsall MA, Bishop WJ, et al: Lead in baby teeth and air samples in Cleveland. 4th Int. Conf. Atomic Spectroscopy, Toronto, 1973.

Jones S, Stephens R, Townshend A: Lead concentrations in deciduous teeth. Personal communication from Stephens, 1981.

Kaneko Y, Inamori I, Nishimura M: Zinc, lead, copper and cadmium in human teeth from different geographical areas in Japan. *Bull Tokyo Dent Coll* 1974;15:233-243.

Kaplan ML, Peresie HJ, Jeffcoat MK: The lead content of blood and deciduous teeth in lead-exposed beagle pups, in Needleman HL (ed): *Low Level Lead Exposure.* New York, Raven Press; 1980, pp 221-230.

Kehoe RA: The metabolism of lead in man in health and disease. *J Roy Inst Publ Health Hyg* 1941;24:81-121, 129-143, 177-203.

Kelsall MA, Hunter RE: *Tooth Lead in Children Living in Cleveland and Its Suburbs.* Environmental Protection Agency Report 600/1-78-053, 1978.

Klinkenberg E, Bibby BG: The effect of topical applications of fluorides on dental caries in young adults. *J Dent Res* 1950;29:4-8.

Kobylańska M, Rajewska D, Strżyxowska U: (Influence of lead compounds upon the dentition.) *Czas Stomatol* (Polish) 1966;19:74-80.

Kühnlein HV, Calloway DH: Minerals in human teeth: Differences between preindustrial and contemporary Hopi Indians. *Am J Clin Nutr* 1977;30: 883-886.

Langmyhr FJ, Sundli A, Jonsen J: Atomic absorption spectrophotometric determination of cadmium and lead in dental material by atomization directly from the solid state. *Anal Chim Acta* 1974;73:81–85.

Lappalainen R, Knuuttila M: The distribution and accumulation of Cd, Zn, Pb, Cu, Co, Ni, Mn and K in human teeth from five different geological areas of Finland. *Arch Oral Biol* 1979;24:363–368.

Lappalainen R, Knuuttila M: The concentrations of Pb, Cu, Co and Ni in extracted permanent teeth related to donors' age and elements in the soil. *Acta Odont Scand* 1981;39:163–167. Corrected data in press, 1982.

Lawson BF, Stout FW, Ahern DE, et al: The incidence of enamel hypoplasia associated with chronic pediatric lead poisoning. *South Carolina Dent J* 1971;29:5–10.

Lawther PJ (Chman). In *Lead and Health* (U.K. Dept. Health and Social Security, Working Party Report), 1980, Table 14.

Lazansky JD: The effect on hamster caries of toothbrushing with various chemicals. *J Dent Res* 1947;26:446, (Abstract).

Lockeretz W: Lead content of deciduous teeth of children in different environments. *Archs Environ Health* 1975;30:583–587.

Losee FL, Cutress TW, Brown R: Natural elements of the periodic table in human dental enamel. *Caries Res* 1974;8:123–134.

Mackie A, Townshend A, Waldron HA: Lead concentrations in bones from Roman York. *J Archaeol Sci* 1975;2:235–237.

Mackie AC, Stephens R, Townshend A, et al: Tooth lead levels in Birmingham children. *Archs Environ Health* 1977;32:175–183.

Malik SR, Fremlin JH: A study of lead distribution in human teeth using charged particle activation analysis. *Caries Res* 1974;8:283–292.

Manly RS, Bibby BG: Substances capable of decreasing the acid solubility of tooth enamel. *J Dent Res* 1949;28:160–171.

Maruta H: Distribution of lead in hard tissues of rats of different ages. *J Dent Health Tokyo* 1978;27:351–359, (Engl. Summ.).

Milošević M, Petrović LJ, Petrović D, et al: Epidemiological significance of determination of lead, cadmium, copper, and zinc in hair and permanent teeth in persons living in the vicinity of a lead smeltery. *Arh Hig Rada Toksikol* 1980;31:209–217.

Momčilović B, Kostial K: Kinetics of lead retention and distribution in suckling and adult rats. *Environ Res* 1974;8:214–220.

Moore MR: Diet and lead toxicity. *Proc Nutr Soc* 1979;38:243–250.

Moore MR: Prenatal exposure to lead and mental retardation, in Needleman HL (ed): *Low Level Lead Exposure*. New York, Raven Press, 1980.

Moses HA, Leno JH, Hill AV Jr, et al: Toxic metal deposition in teeth. *J Dent Res* 1976;55:Spec. Iss. Abst. #407.

Muhler JC, Boyd TM, Van Huysen G: Effect of fluorides and other compounds on the solubility of enamel, dentin, and tricalcium phosphate in dilute acids. *J Dent Res* 1950;29:182–193.

Muhler JC, Van Huysen G: Solubility of enamel protected by sodium fluoride and other compounds. *J Dent Res* 1947;26:119–127.

Needleman HL, Davidson I, Sewell EM, et al: Subclinical lead exposure in Philadelphia schoolchildren. Identification by lead analysis. *N Engl J Med* 1974;290:245–248.

Needleman HL, Shapiro IM: Dentine lead in asymptomatic Philadelphia schoolchildren: Subclinical exposure in high and low risk groups. *Environ Health Perspect* 1974;7:27–31.

Needleman HL, Gunnoe C, Leviton A, et al: Deficits in psychologic and classroom performance of children with elevated dentine lead levels. *N Engl J Med* 1979;300:689–695.

Needleman HL, Tuncay OC, Shapiro IM: Lead levels in deciduous teeth of urban and suburban American children. *Nature* 1972;235:111–112.

Nishimura M: Trace metals in the environment: The distribution of lead in human teeth. *Shikwa Gakuho* 1978;78:1119–1126, (Review in Japanese).

Niwa M: Effects of protein nutrition on Pb, Ca and P contents in hard tissues of lead-administered rats. *J Dent Health Tokyo* 1980;30:17–22.

Oehme M, Lund W, Jonsen J: The determination of copper, lead, cadmium and zinc in human teeth by anodic stripping voltametry. *Anal Chim Acta* 1978;100:389–398.

Ohm HJ: Die Lokalisation des in menschlichen Zähnen abgeagerten Bleis. *Dtsch. zahnärztl Z* 1969;24:202–209.

Oshio H: The cadmium, zinc and lead content of deciduous teeth from two different geographical areas of Japan. *J Dent Health Tokyo* 1973;23:208–222.

Pearl M, Roland NM: Delayed primary dentition in a case of congenital lead poisoning. *J Dent Child* 1980;47:269–271.

Pinchin MJ, Newham J, Thompson RPJ: Lead, copper and cadmium in teeth of normal and mentally retarded children. *Clin Chim Acta* 1978;85:89–94.

Posner AS, Perloff A: Apatites deficient in divalent cations. *J Res Natl Bur Stand* 1957;58:279–286.

Proud M: A study of the lead levels in deciduous teeth in Leicester children. *J Dent Res* 1976;55:D109, (Abs).

Rytömaa I, Tuompo H: Lead levels in deciduous teeth. *Naturwissenschaften* 1974;61:363.

Sagara Y: Studies of lead poisoning—I. (Clinical study of oral changes on workers exposed to lead.) *Igaku Kenkyu* 1957;6:1360–1380.

Schroeder HA, Tipton IH: The human body burden of lead. *Arch Environ Health* 1968;17:965–973.

Schildknecht E: Der Bleigehalt von Dentin und Zement an jugenlichen Praemolaren. Dr. Med. Dent. Thesis, Zürich, 1976.

Shah BG, Momcilovic B, McLaughlan HH: Increased retention of lead in young rats fed suboptimal protein and minerals. *Nutr Repts Int* 1980;21:1–9.

Shapiro IM, Dobkin B, Tuncay OC, Needleman HL: Lead levels in dentine and circumpulpal dentine of deciduous teeth of normal and lead poisoned children. *Clin Chim Acta* 1973;46:119–123.

Shapiro IM, Needleman HL, Tuncay OC: The lead content of human deciduous and permanent teeth. *Environ Res* 1972;5:467–470.

Shapiro IM, Mitchell G, Davidson I, Katz SH: Lead content of teeth: Evidence establishing new minimal levels of exposure in a living preindustrialized population. *Arch Environ Health* 1975;30:483–486.

Shapiro IM, Burke A, Mitchell G, Bloch P: X-ray fluorescence analysis of lead in teeth of urban children in situ: Correlation between the tooth lead level and the concentration of blood lead and free erythroporphyrin. *Environ Res* 1978;17:46–52.

Shapiro IM, Grandjean P, Nielsen OV: Lead levels in bones and teeth of children in ancient Nubia: Evidence of both *minimal* lead exposure and lead poisoning, in Needleman HL (ed): *Low Level Lead Exposure.* New York, Raven Press, 1980, pp 35–42.

Sonnabend E, Sansoni B, Herzog J, Kracke W: Untersuchungen über den Blei- und Cadmiumgehalt im Zahnstein bei verschiedenen Berufsgruppen. *Dtsch Zähnarztl Z* 1976;31:189–191.

Steadman LT, Brudevold F, Smith FA, et al: Trace elements in ancient Indian teeth. *J Dent Res* 1959;38:285–292.

Steenhoùt A, Pourtois M: Lead accumulation in teeth as a function of age with different exposures. *Br J Ind Med* 1981;38:297–303.

Stephens R, Waldron HA: Body burdens of lead in Birmingham. *Roy Soc Health J* 1976;96:176–180.

Stewart DJ: Teeth as indicators of exposure of children to lead. *Arch Dis Child* 1974;49:895–897.

Storozheva NN: (Lead and tin content of human teeth in health and dental caries.) *Stomatologiia* (Mosk) 1963:42:44–48.

Strehlow CD: The use of deciduous teeth as indicators of lead exposure. *J Dent Res* 1974;53:1078, (Abstract) (See also Thesis, New York, 1972).

Strehlow CD, Kneip TJ: The distribution of lead and zinc in the human skeleton. *Am Ind Hyg Assoc* 1969;30:372–378.

Venugopal B, Luckey TD, in: *Metal Toxicity in Mammals,* vol 2, New York, Plenum Press, 1978, p 185.

Verbeeck RMH, Lassuyt CJ, Heijligers HJM, et al: Lattice parameters and cation distribution of solid solutions of calcium and lead hydroxyapatite. *Calcif Tissue Int* 1981;33:243–247.

Waldron HA, Khera A, Walker G, et al: Lead concentrations in bones and soil. *J Archaeol* 1979;6:295–298.

Wesenberg GBR, Fosse G, Berg Justesen N-P, Rasmussen P: Lead and cadmium in teeth, bone and kidneys of rats with a standard Pb-Cd supply. *Int J Environ Stud* 1979;13:441–446.

Wesenberg GBR, Fosse G, Rasmussen P, et al: The effect of Pb and Cd uptake on Zn and Cu levels in hard and soft tissues of rats. *Int J Environ Stud* 1980;14:191–196.

Whittaker DK, Stack MV: The lead, cadmium and zinc content of some Romano-British teeth (in preparation, 1982).

Wilkinson DR, Palmer W: Lead concentrations in U.S.A. permanent human teeth. *International Laboratory* May/June 1975;41–46.

Wisotzky J, Hein JW: The effect of cations above hydrogen in the electromotive series on experime ntal caries in the Syrian hamster. *JADA* 1958;57:796–800.

Wyckoff RWG: *The Biochemistry of Animal Fossils.* Bristol, Scientechnica, 1972, p 46.

Zakson ML: (Trace elements in teeth of elderly and senile persons.) *Dopov Akad Nauk Ukr RSR Ser. B* 1968;3C(9):813, (Chem. Abstr. 69:104401).

$_{48}$ **Cd**

112.40

18-18-2

21 Cadmium

M. V. Stack

Cadmium (Cd) is a transition metal, related to Zn and Hg (group IIb). It is important in the metal-plating industry since it is rustproof, and there are several low-melting alloys of value. Complexes are formed with ammines, sulfur complexes, and chelates, and natural isotopes are numerous. Typical concentrations in the earth's crust fall between 0.1 and 1 gram per ton, where it is usually associated with Zn and Pb ores and is recovered from oxide fumes during smelting.

The element Cd is toxic to virtually every system in the animal body, whether ingested, injected, or inhaled (Underwood, 1977) and as such has attracted some attention for its role in environmental pollution. Its major interest medically, besides its implication in "itai-itai (ouch-ouch) disease," has been its association with systolic hypertension (Schroeder and Buckman, 1967).

Metabolism

For man the weekly tolerable intake proposed by WHO is approximately 1 μg/kg body weight or 55 to 70 μg Cd/day. Diets might vary in their Cd composition depending on the type of food used. Generally, however, food sources are low except for some grains and particular types of shellfish, such as oysters (Table 21-1). Small amounts of Cd are found in most water supplies when they have been standing in galvanized iron pipes, and the concentrations will be found to be higher in areas of soft water, while in fresh running water Cd is usually undetected. The pH of the water supplies may have a bearing on the amount of Cd in drinking water since hard, alkaline, waters tend to deposit insoluble salts in pipes and protect them from corrosion. Acute Cd poisoning has, however, been reported by the contact of acid (citrus) fluids with Cd-coated ice trays. Waters which are naturally soft, such as those occurring in New England, may pick up Cd from galvanized pipes where Cd is a natural contaminant of Zn. Higher levels of Cd have been noted in teeth from the New England areas.

A further source of intake of Cd of some importance in humans is cigarettes. The amount of Cd taken in by inhalation will obviously vary according to smoking habits but could well be compounded in an industrially polluted environment.

The main accumulation of Cd in the body takes place in the liver and kidney. This is explained by the formation of a metalloprotein with the sulfur-containing thionein in the renal cortex (Margoshes and Vallee, 1957). There is little evidence that normal environmental levels of Cd are deleterious. Minimal kidney damage is considered to be unlikely unless current levels increase by a factor of between 4 and 6 (Fleischer et al, 1974). Chronic exposure leads to a characteristic emphysema, and exces-

Table 21-1
Cadmium Concentrations in Various Foods

Food	Item	Cadmium μg/g
Organ meats	Beef liver	0.2
Meat	Beef	0.27
Grains	Wheat, winter	2.63
Milk	Whole	0.15
Vegetables	Kale	1.25
	Beans, lima	0
Sea foods	Oysters	6.05
	Clams	2.35
	Tuna	0.7

Data from Schroeder and Buchman, 1967.

sive Cd oxide fume inhalation can cause pneumonitis after a few days in industrial workers. The relationship between Cd intake and hypertension or cardiovascular disease, although demonstrated in rats (Schroeder, 1964), has not yet been substantiated in man. Mild hypochromic anemia has been observed following Cd exposure, and this has been reproduced in rabbits. Testicular atrophy can be induced by orally administered $CdCl_2$, but the damage can be avoided by preliminary treatment with very low doses, which presumably induce formation of metallothionein. It has been found that Zn affords dramatic protection against exposure to Cd (Pǎrizek, 1957).

CARIES

Man

From a study of the literature it becomes clear that there has been little or no interest in studying the effects of Cd on caries in man. This is so even though Cd toxicity has been known for some time, and investigations on its effects on the health of rats were published 40 years ago.

Five cases of chronic Cd poisoning in industrial workers were reported by Hardy and Skinner (1947), and the bad state of the men's teeth was associated with the time when these men were exposed. Kobylańska et al (1968) reported poor oral hygiene and active caries in 40% of the 80 workers in a battery factory where Cd was processed. The same proportion complained of a metallic taste, and 70% had dry mouths. Half of them had light yellow to orange brown discoloration of their teeth, and 20% showed a more characteristic golden yellow ring of stain at the cervical margin. This has been associated with excessive intake of Cd (Østergaard, 1974; Tarasenko and Vorob'eva, 1973).

High Cd soils were identified, during a geochemical survey, in a hill village in Somerset, a county in the southwest of England. Studies on the dental health of 12- and 15-year-old children there showed no statistical difference in caries prevalence when compared with data for neighboring villages (Anderson et al, 1979). Significantly higher tooth Cd values were apparent ($p < 0.01$) in teeth from young residents than in teeth from residents of Bristol, the nearest city (Stone and Fletcher, personal communication, 1979); however, intermediate tooth Cd levels (0.12 ppm) were noted in further analyses of teeth from children living in Lewis, the largest island northwest of the Scottish mainland, presumed to be relatively free from environmental Cd.

Determinations of Cd concentrations in more than 300 enamel samples by spark source mass spectrometry (Curzon et al, 1977) showed Cd levels in enamel from young premolars of individuals with high caries living in high caries areas to be more than 20 times those in corresponding samples from subjects with low caries in low caries areas. In both areas, those with high caries had tooth Cd levels about three times those with low caries. In multiple regression analyses (Curzon and Crocker, 1978), there were significant interactions between Cu and Cd, also between Mn and Cd. Moses et al (1976), however, found no correlation between Cd levels in teeth and the presence or severity of caries in an urban population.

Animals

A banding type of pigmentation has been seen in the incisor teeth of rats at the same Cd level in drinking water (5 ppm) as has been found effective for fluoride when used in protection against dental caries (Castillo-Mercado and Bibby, 1973). Several other metals produced similar effects, but at concentrations ten times greater. Early observations on the potentiation of caries in rats by Cd supplementation (Leicester, 1946; Ginn and Volker, 1944) were found to apply to the Syrian hamster (Wisotzky and Hein, 1958). More recently, subcutaneous injection of solutions of Cd salts before weaning resulted in a considerable increase in caries in rats (Tamura and Moriya, 1973). The results of these experiments are summarized in Table 21-2, together with recent observations showing the effect of Cd on caries and on the cariostatic properties of fluoride in rats (Shearer et al, 1980a).

Shearer and colleagues stress that the absolute concentration of Cd in teeth should not be taken as predictive of the effect of Cd on caries. The importance of uptake during early postnatal life was also manifested. In rats receiving 0.25, 0.5 and 0.75 ppm up to the 19th day of life the increase in Cd level in the enamel was threefold (in the dentin, fourfold). Buccal carious lesions in molars were increased threefold in female rats receiving the highest Cd dose. In another experiment it was observed that the normally beneficial effect of fluoride was partly negated by Cd. When Cd was supplied later to weanling rats the effect on caries was not appreciable. But Cd uptake by teeth was "remarkable"—150 to 300 times those of controls.

Similar values to those found in the teeth of control rats (c 1 ppm) have been reported in bovine and human enamel surface samples. There was little evidence of a gradient, but there was a significantly higher level in the surface at a depth of less than 0.1 mm (Shearer et al, 1980b).

MECHANISM OF ACTION

Cadmium in Human Tissues

Hair A high correlation has been shown between levels of Cd and Pb in human hair (Petering et al, 1973), an important fact when considering the use of biologic indicators of exposure to toxic metals. Hair

Table 21-2
Effect of Cadmium on Dental Caries in Rodents

Author (year)	Cadmium Used		Period of Administration	Effect of Caries
	(Salt)	*Conc.*		
Ginn and Volker, 1944	$CdCl_2$	0.005%	150 days PW	increase
Leicester, 1946	$CdCl_2$	a) 0.002%	100 days PW	no effect
		b) 0.004%	100 days PW	no effect
		c) 0.004%	during calcification	increases caries progress
		d) 0.004%	calcific + 100 days	
Wisotzky and Hein, 1958[a]	$CdSO_4$	½–¼ mEq	84/98 days PW	increase
Pappalardo and Gulisano, 1968		0.5 mg/kg body weight	NG	halved (molars)
Tamura and Moriya, 1973	$CdCl_2$	0.5, 1.0 mg	subcutaneous inj. preweaning	considerable
Shearer et al, 1980	$CdCl_2$	0.5, 0.75, 0.05 ppm	3–19	trebled buccal caries
		with 15 ppm F′		negated effect of F′

[a]used hamsters; all others used rats.
PW = postweaning.
NG = Not given.

samples have also been used to differentiate among workers who have been exposed occupationally or residentially (Oleru, 1975; Milosevic et al, 1980). Levels of Cd and Pb in teeth were higher by factors of nearly 2 in teeth of an exposed adult population sampled (within 5 km of a lead smelter), compared with a smaller number of controls (living 20 km or more distant). Concentrations of Cd in hair were in the ratio 5:3 in the two groups, but hair Pb levels were much more dissimilar. Milosevic et al (1980) reported significant differences in the content of the metals in the two groups of hair samples, representing a total of 400 persons. No trend in hair Cd levels was seen by these workers, unlike Petering et al (1973) and Oleru (1975). A typical hair value of 2 ppm, and some absorption of trace elements, including Cd, are considered to occur directly from the environment; this tends to show near the tip of the hair (Valković, 1977).

Bone The Cd content of bone is relatively low compared with that in kidney. Rib bone levels showed a mean Cd level below 0.1 ppm. The increase observed in the case of a patient who died from Cd-induced disease ("itai-itai byo") was very much less than that evident in liver and kidney (Nogawa et al, 1975). Similar findings were reported for Cd-exposed rats, in which bone Cd levels were increased from about 0.1 ppm to above 1.5 ppm after exposure, when liver and kidney levels were 300 to 400 times those of controls. Doubling of rat bone Cd levels (1.3–1.8 to 2.2–3.9 ppm) was noted by Itokawa et al (1973) in response to dietary Cd supplementation by about 0.2 g/kg body weight. In a later study (Itokawa et al, 1974), Cd intakes were doubled over a period of four months instead of one month, and an upper value of nearly 20 ppm was attained in femur samples. Much lower normal values for Cd in rat femurs (0.03–0.08 ppm) have been given (Okano et al, 1978). Accumulation of Cd by hard tissues has been reported by several other investigators (eg, Kimura et al, 1974; Nordberg, 1974; Cousins et al, 1973). The effect of Cd on Ca accretion in rats has also been studied (Ando et al, 1978).

The concentrations of Cd reported in bones in cases of stillbirth (Bryce-Smith et al, 1977) seem rather high (vertebra 2.2 ppm, rib 2.2 ppm); similar Pb values were reported for these samples. McKenzie (1974) observed a male/female difference in Cd levels in fat-free rib from adult autopsy specimens. Values more within the expected range were reported recently (Casey and Robinson, 1978), the range being 0.02 to 0.23 ppm, and the mean 0.08 ppm (SD 0.04). Corresponding values for Pb were 0.4 to 4.3 ppm, with the mean 2.2 ppm (SD 0.9) on the same dry weight basis.

Cadmium in Dental Tissues

It is uncertain at present whether Cd, at concentrations not greatly above normal, has any effect on the structure of teeth during development. Mineralization can be affected and apatite crystal formation modified, as Bird and Thomas (1979) have shown, while morphologic changes in crystal growth of dicalcium phosphate dihydrate (brushite) have been introduced in the presence of Cd and other ions (LeGeros et al, 1978). The solubility of enamel was shown to be slightly reduced in response to treatment with Cd solutions (Manly and Bibby, 1949). Superficial uptake of Cd on rat teeth has also been demonstrated (Leicester et al, 1953) following oral or intraperitoneal administration.

Most of the experimental studies of the effects of subacute and acute doses of Cd have been concerned with rats, but the demonstration that the Cd content of dental tissues was dependent on the level of Cd supplementation arose from feeding experiments with swine (Cousins et al, 1973) where tooth levels were 0.2 ± 0.05 times dietary levels in animals receiving 50 to 1350 ppm in diets. Pindborg (1950) described changes in rat enamel organ cells following Cd administration and noted defective formation of incisor dentin, and may also be related to mottled enamel (DeEds, 1941). New hypercalcified tissue became apparent between the dentin margin and the odontoblast layer in subacute dosage (1.7 mg/kg body weight) in a study (Furuta, 1978) in which acute doses were also given. It was considered that the changes were not primarily due to the induced calcium deficiency (Bawden and Hammarström, 1975). Deficient and deformed odontoblasts were also noted by Furuta, and there was accumulation of radioactive Cd in both ameloblasts and odontoblasts. In the study by Bawden and Hammarström the distribution of Cd in the enamel organ matched those of alkaline phosphatase activity and of mitochondria.

Concentrations of Cd and Pb have been determined in teeth, bones, and kidney cortex of rats receiving drinking water containing 25 ppm $PbCl_2$ and 5 ppm $CdCl_2$; the effect of supplementation was tested during pregnancy, lactation, and the weeks following (Wesenberg et al, 1979). The food pellets were found to have a Pb:Cd ratio of at least 5:1 (1.5 to 2.0 mg/kg and 0.3 mg/kg, respectively). In the two control groups and in two groups receiving supplementation only during weeks 5 to 9 the geometric mean Pb:Cd ratio was 17.5 for both incisors and molars. The geometric mean of the five samples with the lowest Pb levels was 1.2 ppm (Cd 0.1 ppm). Mean Pb:Cd ratios were 6.6 and 9.6 for epiphyseal and diaphyseal samples, excluding exceptionally high values for one of the groups receiving supplementation. During weeks 5 to 9 the groups so treated also showed the lowest Pb:Cd ratios (1.5) in kidney cortex. Levels of Cd in molars and kidneys were comparable in control groups, but the Pb levels were 3 to 4 times higher in molars than in kidney cortex. Levels of both metals in kidney could be assessed with reasonable confidence from levels in molars.

In electron probe microanalyses of molar enamel from mice that had received 25 ppm $CdCl_2$ throughout life (Fosse and Berg Justesen,

Table 21-3
Reported Concentrations of Cadmium in Human Teeth

Author (Year)	n	Method	Mean Conc. ($\mu g/g \pm$ SE)	Range
Oshio, 1973	795	Atomic absorption	0.61 ± 0.82	NG
Langmyhr et al, 1974	14	Atomic absorption	0.41 ± 0.13	0.07–2.2
Kaneko et al, 1974	97	Atomic absorption	0.08 ± 0.015[b]	0.01–0.32
Losee et al, 1974a	28[a]	Spark mass spectr.	0.51 ± 0.12	0.03–2.4
Losee et al, 1974b	56[a]	Spark mass spectr.	0.99 ± 0.15	0.03–6.7
Fosse and Berg Justesen, 1977	2169	Atomic absorption	0.43 ± 0.02	NG
Attramadal and Jonsen, 1976	35[c]	Anodic voltametry	0.1	0.03–0.24
Attramadal and Jonsen, 1978	26[d]	Anodic voltametry	0.2	0.02–0.63
Curzon et al, 1977	334[(a)]	Spark mass spectr.	1.86 ± 0.16	NG
Oehme et al, 1978	10	Anodic voltametry	< 0.05	< 0.04–16.2

[a]Data for whole enamel.
[b]Data from Tokyo, Okitsu, Hachijo Isl.
[c]Same mean for deciduous and permanent teeth.
[d]Archeological specimens.
NG = not given in paper.

1979) the Ca:P ratio was 1.91 (SD 0.05), significantly below that found for control mice, for which the corresponding ratio was 2.35 (SD 0.18). Itokawa et al (1974) did not observe change in Ca:P ratios in femur samples from rats treated with Cd.

A number of studies have reported on the concentrations of Cd in human tooth enamel (Table 21-3). The element seems to be readily detected and is present in the majority of samples. There is some variation according to method of analysis but generally Cd concentrations found by different authors are similar.

A comprehensive study of Cd concentrations in teeth (also of Pb, Zn, and Cu) involved the analysis of over 2000 specimens which were collected throughout Norway. The corrected mean value was about 0.43 ppm. Concentrations did not seem to reflect the degree of urbanization or industrialization. One-third of all values were below 0.1 ppm, and only 21 teeth showed values above 3.4 ppm. The concentration of Cd in the enamel (one-fifth of the tooth substance) was stated to be 14.4 times that in the dentin (Fosse and Berg Justesen, 1977).

Earlier observations (Kaneko, et al 1974) also led to the finding of a mean value of 0.43 ppm for Cd in teeth from a rural community where fishing was the main occupation. However, much lower levels were found for groups from Tokyo (0.08 ppm), the rural area of Okitsu (0.07 ppm), and even Annaka (0.10 ppm), where the community is known to have a Cd contamination problem. Higher values in teeth from inhabitants of a Cd-contaminated region in Toyama Prefecture (0.25 ppm in enamel and 0.40 ppm in dentin) have been reported in another investigation (Iwakura, 1972), in which much lower values were observed in specimens from a region thought to be free of such contamination (0.135 ppm Cd in enamel, 0.10 ppm in dentin). However, other values (Oshio, 1973) from both Toyama City and Tokyo were about 1 ppm (795 teeth).

Referring again to Norwegian surveys, a range of 0.07 to 2.2 ppm Cd was reported for permanent teeth, only two values falling below 0.1 ppm (Langmyhr et al, 1974). Similar levels were reported both in deciduous (0.04 to 0.24) and permanent teeth (0.03 to 0.51) in another survey (Attramadal and Jonsen, 1976). Here, the means—both 0.1 ppm—were at ten times the detectable limit. Further analyses, again using differential pulse anodic stripping voltametry, of 26 ancient Scandinavian teeth, showed a mean value below 0.2 ppm, apart from those from a single community (mean 0.4 ppm) (Attramadal and Jonsen, 1978). Considerable tooth-to-tooth variation was evident in other Cd analyses of Scandinavian teeth (Oehme et al, 1978), but for the majority of teeth the concentration did not exceed that of Cd in the reagents used (0.04 ppm).

In a study of over 450 enamel specimens from premolars collected from various American states (Curzon, 1977) it was noted that about

40% of the samples contained insufficient Cd to permit determination by spark source mass spectrometry. The overall mean was estimated to be 1.86 ppm, approximately one-tenth as much as was found for Pb in the same study. There was a wide range in the mean values for groups of specimens from various states—0.80 ppm for South Carolina, 7.33 ppm for Montana (Curzon et al, 1977).

High Cd levels were seen in the incisal tips of 1 mm sections of a young child's incisors (Pinchin et al, 1978), five to seven times those found for the whole tooth material; the distinction was not seen in the cusp of a deciduous molar from the same child (whole tooth Cd 0.115 ppm). High tip levels for Pb and Cu were not seen in the incisor specimens. Whole deciduous teeth from the northeast of England (Newcastle upon Tyne) were also analyzed in this survey, and a mean value of 0.03 ppm was found for 12 specimens, using anodic stripping voltametry; no Cd was seen in 3 of the 18 specimens, and high values (0.14 to 0.28 ppm) were found for the remaining 3 specimens.

A possible role of Cd in dental caries still remains to be adequately investigated. Knowing that Cd has a definitely toxic role in animal metabolism, it may be that this element enhances the susceptibility to dental decay. It is also possible that the action is complex, with Cd interacting with other elements. Interactions of Cd with other elements such as Cu, and reversal mechanisms whereby Cd replaces Cu, Zn, and Se in enzyme systems have been suggested (Pǎrizek et al, 1974). In studies on trace elements in whole human enamel Curzon and Crocker (1978) found indication of Cd interactions with Cu, Mn, Pb, and Se with varying relationships to dental caries.

More work is needed to clarify the role of Cd in dental disease. Much work has been carried out in the medical field on this element because of its increasing presence as an environmental pollutant. There should now be comprehensive studies on the action of Cd on oral bacteria, in surface enamel, and on the caries process. In addition to these studies further consideration needs to be given to the probable complex interactions of Cd with other elements.

SUMMARY

At typical environmental levels Cd is incorporated less into mineralized tissues than into kidney and liver. At a level between 0.1 and 1 ppm it probably plays a minor role, if any, in promoting dental caries. However, high caries subjects in high caries areas have been found to show very much higher enamel Cd levels than those with few caries living in low caries areas. At the high environmental levels of industrial Cd processing the clinical consequences have been noted as tooth discolorations, severe dental caries, and periodontal disease. Tooth pigmentation and increased caries susceptibility have also been noted in rodents receiving Cd supplementation to their diets, and the beneficial effect of fluoride supplementation has been partly counteracted. The possibility of interactions between Cd and other trace elements with varying relationships to dental caries remains to be evaluated.

REFERENCES

Anderson RJ, Davies BE, Nunn JH, et al: The dental health of children from five villages in North Somerset with reference to environmental cadmium and lead. *Br Dent J* 1979;147:159–161.

Ando M, et al: Studies on the disposition of cadmium in bones of rats after continuous oral administration. *Toxicol Appl Pharmacol* 1978;46:625–632.

Attramadal A, Jonsen J: The content of lead, cadmium, zinc and copper in deciduous and permanent human teeth. *Acta Odont Scand* 1976;34:127–131.

Attramadal A, Jonsen J: Heavy trace elements in ancient Norwegian teeth. *Acta Odont Scand* 1978;36:97–101.

Bawden JW, Hammarström LE: Distribution of cadmium in developing teeth and bone in young rats. *Scand J Dent Res* 1975;83:179–186.

Bird ED, Thomas WC: Inhibition of mineralization and apatite crystal formation by various ions. *Proc Soc Exp Biol Med* 1979;112:640–643.

Bryce-Smith D, Deshpande RR, Hughes J, Waldron HA: Lead and cadmium levels in still-births. *Lancet* 1977;1:1159.

Casey CE, Robinson MF: Copper, manganese, zinc, nickel, cadmium and lead in human foetal tissues. *Br J Nutr* 1978;39:639–646.

Castillo-Mercado R, Bibby BG: Trace element effects on enamel pigmentation, incisor growth, and molar morphology in rats. *Arch Oral Biol* 1973;18: 629–635.

Coles SG: *The Determination of Certain Trace Metals in Teeth with Special Reference to Lead as a Pollutant that Accumulates in Calcified Tissues.* MSc Thesis, Bristol, England, 1979.

Cousins RJ, Barber AK, Trout JR: Cadmium toxicity in growing swine. *J Nutr* 1973;103:964–972.

Curzon MEJ: *Trace Element Composition of Human Enamel and Dental Caries.* PhD Thesis, London, 1977.

Curzon MEJ, Spector PC, Losee FL, et al: Dental caries related to Cd and Pb in whole human dental enamel. *Trace Substances in Environmental Health* – XI:1977, pp 23–28.

Curzon MEJ, Crocker DC: Relationships of trace elements in human tooth enamel to dental caries. *Arch Oral Biol* 1978;23:647–653.

DeEds F: Factors in the etiology of mottled enamel. *JADA* 1941;28:1804–1814.

Fleischer M, et al: Environmental impact of cadmium: A review by the panel on hazardous trace substances. *Environ Health Perspect* 1974;13:253–323.

Fosse G, Berg Justesen N-P: Cadmium in deciduous teeth of Norwegian children. *Int J Environ Stud* 1977;11:17–27.

Fosse G, Berg Justesen N-P: Teeth as indicators of pediatric cadmium exposure, in Nriagu JO (ed): *Biogeochemistry of Cadmium.* New York, Elsevier, vol 2, 1979, pp 46–50.

Furuta H: Cadmium effects on bone and dental tissues of rats in acute and subacute poisoning. *Experientia* 1978;34:1317–1325.

Ginn JT, Volker JF: Effect of cadmium and fluoride on the rat dentition. *Proc Soc Exp Biol Med* 1944;57:59–61.

Hardy HL, Skinner JB: The possibility of chronic cadmium poisoning. *J Ind Hyg Toxicol* 1947;29:321–324.

Itokawa Y, Abe T, Tanaka S: Bone changes in experimental cadmium poisoning. *Arch Environ Health* 1973;26:241–244.

Itokawa Y, Abe T, Tabei R, et al: Bone changes in experimental chronic cadmium poisoning. *Arch Environ Health* 1974;28:149–154.

Iwakura M: (Quantitative analysis of cadmium in teeth of the inhabitants of a cadmium-contaminated region) *Koku Eisei Gakkai Zasshi* 1972;22:1–9 (Jap.).

Kaneko Y, Inamori I, Nishimura M: Zinc, lead, copper and cadmium in human teeth from different geographical areas of Japan. *Bull Tokyo Dent Coll* 1974;15:233–234.

Kimura M, Otaki N, Yoshiki S, et al: The isolation of metallothionein and its protective role in cadmium poisoning. *Toxicol Appl Pharmacol* 1974; 46:625–632.

Kobylańska M, Rajewska D, Strzyxowska U: (Clinical studies on the effects of cadmium on the teeth and oral mucosa) *Czas Stomatol (Polish)* 1968;21:913–918.

Langmyhr FJ, Sundli A, Jonsen J: Atomic absorption spectrophotometric determination of cadmium and lead in dental material by atomization directly from the solid state. *Anal Chim Acta* 1974;73:81–85.

LeGoros R, Quirolgico G, Go P: $CaHPO_4 \bullet 2H_2O$ (DCPD): Effect of some "trace elements" on its crystal growth. *J Dent Res* 1978;57:A 89, (Abst).

Leicester HM: The effect of cadmium on the production of caries in the rat. *J Dent Res* 1946;25:337–340.

Leicester HM, Thomassen PR, Denzler GJ: Tracer studies on the uptake of cadmium in the rat molar. *J Dent Res* 1953;32:73, (Abst).

Losee FL, Cutress TG, Brown R: Trace elements in human dental enamel. *Trace Sub Environ Health* 1974a;7:19–24.

Losee FL, Cutress TG, Brown R: Natural elements of the periodic table in human dental enamel. *Caries Res* 1974b;8:123–124.

McKenzie JM: Tissue concentrations of cadmium, zinc and copper from autopsy samples. *NZ Med J* 1974;79:1016–1019.

Manley RS, Bibby BG: Substances capable of decreasing the acid solubility of tooth enamel. *J Dent Res* 1949;28:160–171.

Margoshes M, Vallee BL: A cadmium protein from equine kidney cortex. *J Am Chem Soc* 1957;79:4813–48148.

Milošević M, Petrovic LJ, Petrovic D, et al: Epidemiological significance of determination of lead, cadmium, copper and zinc in hair and permanent teeth, in persons living in the vicinity of a lead smeltery. *Arh Hig Rad Toksikol* 1980;31:209–217.

Moses HA, Leno JH, Hill AV Jr, et al: Toxic metal deposition in teeth. *J Dent Res* 1976;55:407 (Abst).

Nogawa K, Ishizawa A, Fukushima M: Studies on the women with acquired Fanconi syndrome observed in the Ichi River basin polluted by cadmium. *Environ Res* 1975;10:280–307.

Nordberg G: Health hazard of environmental cadmium pollution. *AMBIO* 1974; 3:55–66.

Oehme M, Lund W, Jonsen J: The determination of copper, lead, cadmium and zinc in human teeth by anodic stripping voltametry. *Anal Chim Acta* 1978;100:388-398.

Okano T, Ikebe K, Ichikawa T, et al: (Determination of trace amounts of cadmium in bone) *Eisei Kagazu* 1978;24(4):159-162;24(5):231-234, (Jap).

Oleru UG: Epidemiological implications of environmental cadmium: I. The probable utility of human hair for occupational trace metal (cadmium) screening. *Am Med Hyg Assoc* 1975;36:229-233.

Oshio H: The cadmium, zinc, and lead content of deciduous teeth from two different geographic areas in Japan. *J Dent Health Tokyo* 1973;23:208-222.

Østergaard K: Cadmium, an environmental poison. *Ugeskr Laeg* 1974;136: 858-862.

Pappalardo G, Gulisano S: Cadmio e carie sperimentale nel ratto. *Min Stomatol* 1968;17:59-62.

Pǎrizek J: The destructive effect of cadmium ion on testicular tissue and its prevention by zinc. *J Endocrinol* 1975;15:656-663.

Pǎrizek J, Kalovsková J, Babicky J, et al: Interaction of selenium with mercury, cadmium and other toxic metals, in Hoekstra WG, et al (eds): *Trace Element Metabolism in Animals—2.* Baltimore, University Park Press, 1974.

Petering HG, Yaeger DW, Witherup SO: Trace metal content of hair. II. Cd and Pb of human hair in relation to age and sex. *Arch Environ Health* 1973;27:327-330.

Pinchin MJ, Newham J, Thompson RPJ: Lead, copper and cadmium in teeth of normal and mentally retarded children. *Clin Chim Acta* 1978;85:89-94.

Pindborg JJ: (The effect of chronic poisoning with fluoride and cadmium upon incisors of the white rat with special reference to the enamel organ) *Tandlaegebladet* (suppl.) 1950;1-136, (Danish).

Schroeder HA, Balassa JJ, Hogencamp JC: Abnormal trace elements in man: Cadmium. *J Chron Dis* 1961;14:236-238.

Schroeder HA, Buckman J: Cadmium hypertension. *Arch Environ Health* 1967;14:693-697.

Shearer TR, Britton JL, DeSart DJ, et al: Influence of cadmium on caries and the cariostatic properties of fluoride in rats. *Arch Environ Health* 1980a; 35:176-180.

Shearer TR, Johnson JR, DeSart DJ: Cadmium gradient in human and bovine enamel. *J Dent Res* 1980b;59:1072.

Tamura T, Moriya Yu: Studies on the toxicity of drugs. XI. Toxicity of cadmium. *Jap J Pharmacol* 1973;23:129, (Abst).

Tarasenko NY, Vorob'eva RS: Hygienic problems in the use of cadmium. *Vestn Akad Nauk SSSR* 1973;28(10):37-43 (Russ).

Underwood EJ: *Trace Elements in Human and Animal Nutrition,* ed 4. London, Academic Press, 1977.

Valković V: in *Trace Elements in Human Hair.* New York, Garland STPM Press, 1977, p 72.

Wesenberg GBR, Fosse G, Berg Justesen N-P, Rasmussen P: Lead and cadmium in teeth, bone and kidneys of rats with a standard Pb-Cd supply. *Int J Environ Stud* 1979;13:438-444.

Wisotzky J, Hein JW: The effect of cations above hydrogen in the electromotive series on experimental caries in the Syrian hamster. *JADA* 1958;57:796-800.

INDEX